Model Predictive Control

Model Predictive Control

Baocang Ding and Yuanqing Yang
Chongqing University of Posts & Telecommunications

This edition first published 2024.

Registered Offices
John Wiley & Sons, Inc., 111 River Street, Hoboken, NJ 07030, USA
John Wiley & Sons Ltd, The Atrium, Southern Gate, Chichester, West Sussex, PO19 8SQ, UK

For details of our global editorial offices, customer services, and more information about Wiley products visit us at www.wiley.com.

Wiley also publishes its books in a variety of electronic formats and by print-on-demand. Some content that appears in standard print versions of this book may not be available in other formats.

Library of Congress Cataloging-in-Publication Data:

Names: Ding, Baocang, author. | Yang, Yuanqing (College teacher), author.
Title: Model predictive control / Baocang
Ding and Yuanqing Yang.
Description: Hoboken, NJ : Wiley-IEEE Press, 2024. | Includes
bibliographical references and index.
Identifiers: LCCN 2023049301 (print) | LCCN 2023049302 (ebook) | ISBN
9781119471394 (cloth) | ISBN 9781119471424 (adobe pdf) | ISBN
9781119471318 (epub)
Subjects: LCSH: Predictive control.
Classification: LCC TJ217.6 .D55 2024 (print) | LCC TJ217.6 (ebook) | DDC
629.8–dc23/eng/20240206
LC record available at https://lccn.loc.gov/2023049301
LC ebook record available at https://lccn.loc.gov/2023049302

Cover Design: Wiley
Cover Image: © Abstract Aerial Art/Getty Images

Set in 9.5/12.5pt STIXTwoText by Straive, Chennai, India
Printed and bound by CPI Group (UK) Ltd, Croydon, CR0 4YY

C9781119471394_130324

Contents

About the Authors

Baocang Ding, PhD, teaches model predictive control (MPC) to both undergraduate and graduate students in the School of Automation, Chongqing University of Posts and Telecommunications, China. His research interests include MPC, control of power networks, process control, and control software development.

Yuanqing Yang, PhD, teaches MPC to both undergraduate and graduate students in the School of Automation, Chongqing University of Posts and Telecommunications, China. His research interests include MPC, fuzzy control, networked control, and distributed control systems.

Preface

As a class of model-based control algorithms, model predictive control (MPC) has been extensively researched and applied to numerous real processes/equipment. Among various MPC approaches applied in practice, dynamic matrix control (DMC) is perhaps the best representative. DMC has been developed from the single-layered to the two-layered. The steady-state targets (generally called setpoints in control) are tracked in the single-layered MPC. The single-layered MPC is the lower layer of the two-layered MPC. In the upper layer of the two-layered MPC, the steady-state targets are calculated. Among various studies on MPC theory, the robust MPC for models with uncertainties is one of the best representatives. In the community of robust MPC, the approaches for linear parameter-varying (LPV) model have their own deep impacts.

We began our research on MPC early in 1997. We not only applied MPC in petrochemical/chemical processes but also published over 200 MPC papers. In this book, we talk about MPC but concentrate on our contributions/ideas. Chapter 1 tells how MPC is developed, beginning as a substitution/upper level of proportional integral derivative (PID), from the single-layered to the two-layered, and as a lower level of real-time optimization (RTO). In general, the two-layered DMC has three modules: open-loop prediction, steady-state target calculation (SSTC), and dynamic control (DC). Chapter 2 concerns the identification of the basic model for MPC but emphasizes finite step response (FSR) and, being based on FSR model, the open-loop prediction. Chapter 3 explains the general steps of SSTC. Chapters 4 and 5 concern the two-layered DMC when there are, respectively, only stable controlled variables (CVs) and both stable and integral CVs. Chapter 6 extends the ideas in Chapters 4 and 5 to the state-space model, also named two-layered DMC. The open-loop prediction in Chapter 2 can be seen as a bridge between DMC for FSR and DMC for state-space, where Kalman filter (KF) is the link. Chapter 7 discusses some important issues in two-layered MPC: offset-free property, static nonlinearity, and variable structure. Hammerstein nonlinearity is an important type of static nonlinearity, and it incurs necessary conditions in order to retrieve closed-loop stability. In Chapter 8, the two-step MPCs for Hammerstein model, including the two-step MPC with the state feedback, the two-step MPC with the dynamic output feedback, the two-step generalized predictive control (GPC), are given with stability analyses. In the first step of two-step MPC, the control law or the control move, being based on the linear model, is given. In the second step of two-step MPC, Hammerstein nonlinearity is handled by the inaccurate inversion. Chapter 9 gives two heuristic approaches of

MPC for LPV model, with state feedback and dynamic output feedback, respectively. The controllers cannot guarantee closed-loop stability due to their "heuristic" nature, like DMC. Chapter 10 retrieves closed-loop stability (as compared with Chapter 9) with state-feedback for LPV model. This was a hot topic in MPC. Chapter 11 retrieves closed-loop stability (as compared with Chapter 9) with the dynamic output feedback for LPV model. Chapter 11 shows our unique contribution to robust MPC.

This book may have missed citing some important materials, and we sincerely apologize for that.

Baocang Ding
College of Automation
Chongqing University of Posts and Telecommunications
Chongqing, P. R. China

Acronyms

CARIMA	controlled auto-regressive integral moving average
CARMA	controlled auto-regressive moving average
CCA	cone-complementary approach
CLTVQR	constrained linear time-varying quadratic regulation
CRHPC	constrained receding horizon predictive control
CSTR	continuous stirred tank reactor
CV	controlled variable
CVET	external target of controlled variable
CVss	ss of controlled variable
DbCC	double convex combination
DC	dynamic control, dynamic control module, dynamic calculation
DepV	dependent variable
DMC	dynamic matrix control
DV	disturbance variable
ET	external target
FIR	finite impulse response
FSR	finite step response
GNB	generalized binary noise
GPC	generalized predictive control
HHL	high high limit, upper engineering limit
HL	high limit, upper operating limit
ICCA	iterative cone-complementary approach
iCV	first-order integral controlled variable
iDepV	first-order integral dependent variable
IndepV	independent variable
IRV	ideal resting value
KBM	Kothare–Balakrishnan–Morari
KF	Kalman filter
LA	Lu–Arkun
LGPC	linear generalized predictive control
LHS	left-hand side
LL	low limit, lower operating limit

LLL	low low limit, lower engineering limit
LMI	linear matrix inequality
LP	linear programming
LPV	linear parameter-varying
LQ	linear quadratic
LQR	linear quadratic regulator
LS	least square
LTI	linear time-invariant
MIMO	multiple-input multiple-output
MISO	multiple-input single-output
MPC	model predictive control, multivariable predictive controller
MV	manipulated variable
MVET	external target of manipulated variable
MVss	ss of manipulated variable
NLGPC	nonlinear generalized predictive control
NSGPC	nonlinear separation generalized predictive control
OFRMPC	output feedback robust MPC
ol	open-loop
PID	proportional integral derivative
PRBS	pseudo-random binary sequence
QB	quadratic boundedness
QP	quadratic programming
RHC	receding horizon control
RHS	right-hand side
RoA	region of attraction
RTO	real-time optimization
sCV	stable controlled variable
sDepV	stable dependent variable
SIORHC	stabilizing input/output receding horizon control
SISO	single-input single-output
sp	setpoint
ss	steady-state target, steady-state targets
ssKF	steady-state Kalman filter
SSTC	steady-state target calculation
SVD	singular value decomposition
TSGPC	two-step generalized predictive control
TSMPC	two-step MPC, two-step state feedback MPC
TSOFMPC	two-step output feedback MPC
TTSS	time to steady-state

Introduction

This book discusses the two-layered dynamic matrix control (DMC) for finite step response (FSR) and state-space models, the two-step model predictive control (MPC) for the state-space and the input–output models with Hammerstein static nonlinearity, and the robust MPC for linear parameter-varying (LPV) model with/without bounded disturbance. The topics are linked by Kalman filter and the state-space equivalence. It covers both real applied algorithms and theoretical results. This book represents the author's main contributions to MPC, with appropriate extensions.

1

Concepts

When we talk about model predictive control (MPC), we should know that MPC has other names, e.g., receding horizon control (RHC). What are the differences between the two names? It is usually called MPC in the industrial circle. When we apply the state space paradigm to study MPC with stability guarantee, it is sometimes called RHC for emphasizing the feature of receding-horizon optimization. On the application aspects, besides receding-horizon optimization, MPC has other features, such as model-based prediction and feedback correction. If MPC has no feedback correction, and its prediction is naturally obtained from the model, it is named RHC. Thus, RHC is often used in academic/theoretical research studies.

1.1 PID and Model Predictive Control

It is said that in industry, more than 80% of automatic control loops are utilizing proportional integral derivative (PID). Someone says that this number should be 85% or even 90%. The percentage cannot be very authoritative. PID control strategy is widely used not only in civil industry but also in aerospace, military, and electronic mechanical devices. The use of PID is shown in Figure 1.1. Figure 1.1 has PID, the actuator, the controlled process (controlled device) {plant}, the controlled output $\{y\}$, sp (the setpoint) of controlled output $\{y_{sp}\}$, a plus sign and a minus sign. The measured output feedback. A measurement (meter) block can be added. However, for theoretical research studies, it assumes including the measurement (meter) in the plant.

For many factories, it is optimistic to apply PID for above 80%, since many actuators are manually operated where PID is not operable. The manual operation of actuators is shown in Figure 1.2. Since there are many PIDs, s is added, and both y_{sp} and y are vectors.

What is the situation for MPC? According to the statistics, as compared with PID, MPC occupies about 10–15% of automatic control loops in the process control. We should not count based on the upper bounds (PID 85%, MPC 15%), since other control strategies (different from PID and MPC) misleadingly seem useless. In many factories, 80–90% of actuators are manually operated, i.e., with neither PID nor MPC. Let us concern a modern

Model Predictive Control, First Edition. Baocang Ding and Yuanqing Yang.

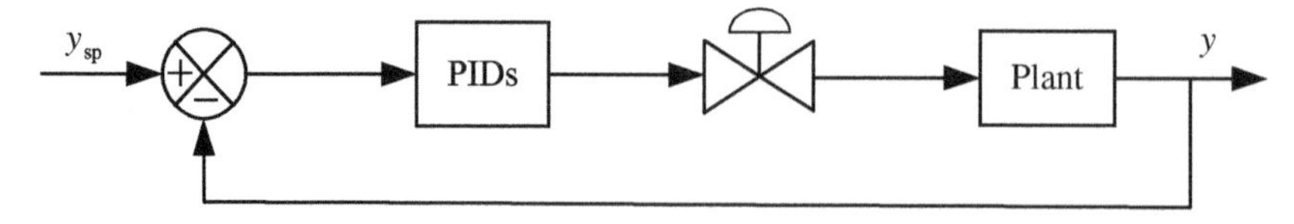

Figure 1.1 Control system using PID.

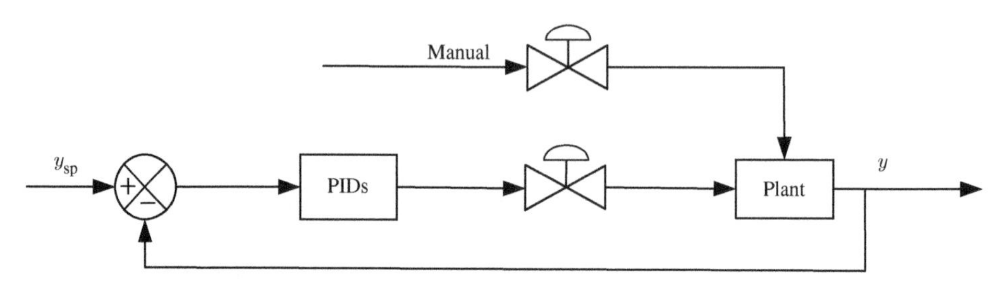

Figure 1.2 Control system using PID+manual.

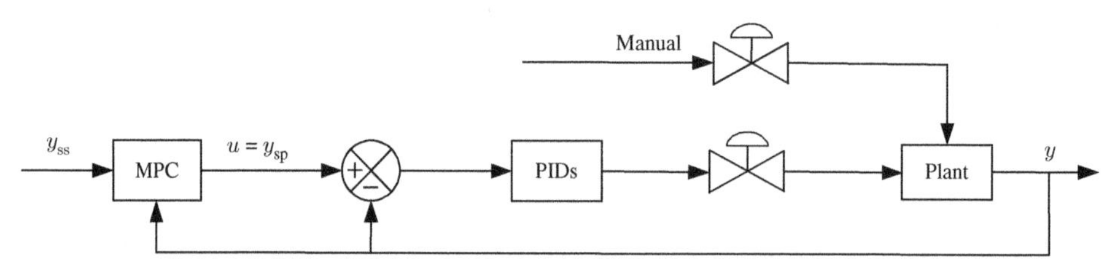

Figure 1.3 Control system using MPC based on PID.

factory with high-level automation and immense courage to accept advanced control strategies like MPC; according to the statistics in these factories, PID occupies approximately 80% and MPC approximately 10%, while other control strategies are definitely non-mainstream.

How does MPC play its role? Sometimes there is misunderstanding. Based on Figure 1.1, MPC acts as in Figure 1.3. In MPC, $u = y_{sp}$. MPC lies before PID. y_{ss} is in front of MPC, which is called ss (the steady-state target) of y, i.e., sp of MPC. The measurement of y is sent to MPC.

In the process control, the controllable input u is called manipulated variable (MV); the controlled output y is called controlled variable (CV); the measurable disturbance f is called disturbance variable (DV), sometimes called the feedforward variable. These names are conventional in the industrial MPC.

Can the "manual" of Figure 1.2 become automatic? If MPC is well-applied and "manual" is also given by MPC, then u of MPC includes both $u^1 = y_{sp}$ and $u^2 =$ manual, as shown in 1.4. Before applying MPC, "manual" implies that the operator directly operates the valves; PID implies using PID algorithm to manipulate the valve, and the operator has to operate PID sp. By applying MPC, MPC manipulates both valve and PID sp. MPC primarily manipulates PID sp, and secondly directly manipulates some valves.

Figure 1.4 is not a general situation. In practice, some projects cannot utilize MPC on all PID sps (setpoints), i.e., some PID sps are still manually adjusted, as shown in Figure 1.5.

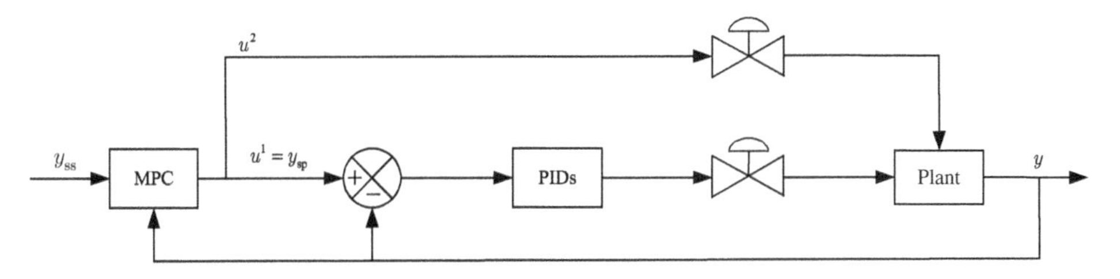

Figure 1.4 Control system manipulating "manual" by MPC.

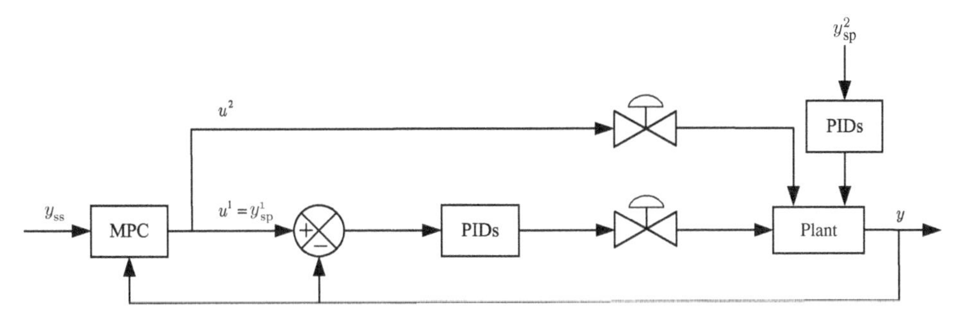

Figure 1.5 Control system using MPC+PID.

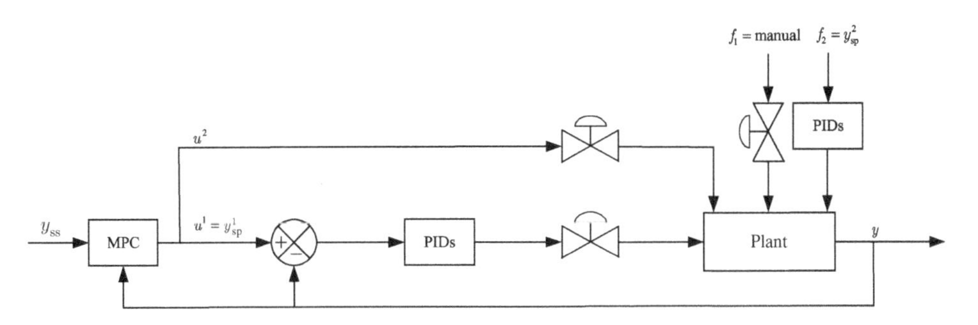

Figure 1.6 Control system using MPC+PID+manual.

Some PID sps are manipulated by MPC, denoted as y^1_{sp}; the other PID sps, denoted as y^2_{sp}, are not manipulated by MPC.

Figure 1.5 is still not a general situation. Some "manuals" may not be manipulated by MPC, as shown in Figure 1.6. Applying MPC for all "manuals" represents a high level of control, but all projects are not achievable.

In industry, all u of MPC are not actuator positions; some are the actuator positions, and the others are PID sps. There were misunderstandings about this fact.

What does the controlled object of MPC become? It is shown in the dashed box in Figure 1.7. Hence, applying MPC to a real system requires establishing a mathematical model of the object in the dashed box rather than merely building the "plant" model. The model ready for MPC must take PID into account, i.e., the model includes the role of PID. The "manual" in Figure 1.7 becomes DV of MPC, and so is y^2_{sp}.

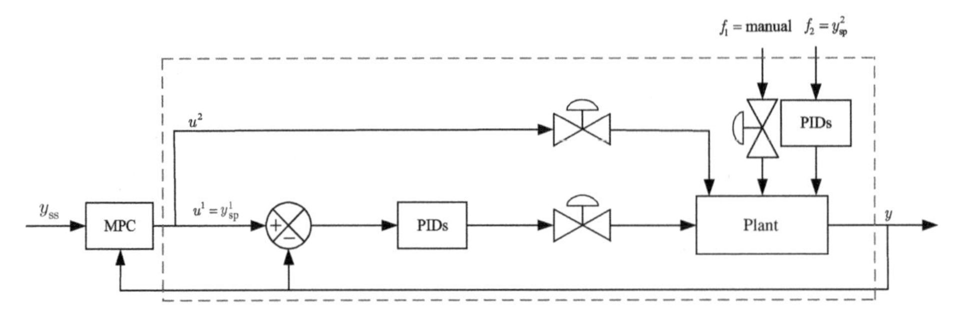

Figure 1.7 In the control system using MPC+PID+manual, a controlled object of MPC is in dashed box.

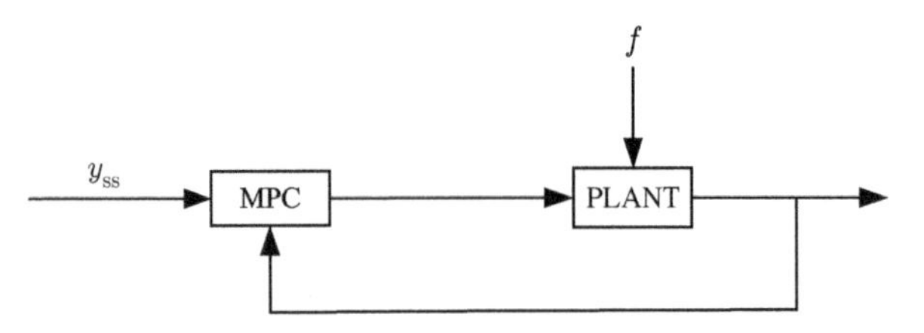

Figure 1.8 Control system using MPC.

Suppose the model in the dashed box is obtained, including the portions for both DV-to-CV and MV-to-CV. In the literature, most researchers are concerned, given y_{ss}, studying

(1) how to optimize $\{u(k), u(k+1|k), \ldots, u(k+M-1|k)\}$, i.e., the algorithms, which were dominating in the 1970s and 1980s;
(2) whether or not the sequence $\{y(k)\}$ (k from 0 to ∞) converges, i.e., stability, which is dominating in the academic theory;
(3) whether or not $y(\infty) = y_{ss}$, i.e., the offset-free, which is not mainstream but has some papers.

In the 1990s, there were many mature results in stability. The offset-free is only valid after assuming stability. It seems that the research studies on offset-free has less patterns than on stability.

Let us abbreviate the controlled object of MPC (often referred to as a generalized object in process control), in the dashed box in Figure 1.7, as PLANT wich is capitalized. Then, Figure 1.7 reduces to Figure 1.8. Thus, all three types of studies, mentioned earlier, are for Figure 1.8.

1.2 Two-Layered Model Predictive Control

Are the aforementioned three types of studies closely consistent with state-of-the- art industrial applications? The answer is negative. There are more issues to tackle. When MPC is applied, as in Figure 1.8, it is only called dynamic control (DC) or dynamic tracking, or often, dynamic move calculation in industrial software.

The biggest issue is where y_{ss} comes from. Before using MPC, both valve and PID sp are manually operated. By applying MPC, if y_{ss} is again manually operated, can we have high enough knowledge for well-operating? If we cannot well-operate PID sps, we might not gain big benefits by operating y_{ss} in Figure 1.8. If obtaining y_{ss} is not automatic, it requires a lot of operation experiences and a high-level engineer in order to enhance MPC efficiency. Hence, automating the calculation of y_{ss} is a key to simplifying MPC operation.

Let us take an example, where "PLANT" takes a simple form, i.e., the transfer function model $y(s) = G(s)u(s)$. According to the final-value theorem, we obtain $y_{ss} = Gu_{ss}$, where G is the steady-state gain matrix. In applying PID, for every y being controlled, there must be a sp. MPC manipulates not only PID sps, but also some valves. Imagine that the numbers of MVs and CVs may be unequal. When they are unequal, for any y_{ss}, is there u_{ss} satisfying $y_{ss} = Gu_{ss}$? By setting a y_{ss} arbitrarily, does MPC necessarily drive CV to y_{ss}? Obviously not. Looking at the equation $y_{ss} = Gu_{ss}$ with u_{ss} as the unknown, it unnecessarily has a solution for any y_{ss}. Uniqueness of the solution for this equation is rare. In most cases, either there is no solution, or there are infinitely many solutions. For the industrial applications, MPC should automatically calculate not only y_{ss}, but also u_{ss}. The term compatibility refers to, with the given y_{ss}, whether or not there is a solution u_{ss}. The term uniqueness refers to, with the given y_{ss}, whether or not there is a unique solution u_{ss}. In the case of multiple solutions, it should be given the principle to choose u_{ss}.

In industrial operations, it is evident that both y_{ss} and u_{ss} may be related to economy. Any variable related to economy could be involved in an optimization. In MPC, economy is a broad concept; it may include, e.g., increasing money, reducing energy consumption, reducing exhaust gas emission, and reducing pollutants.

Figure 1.8 can be modified. Besides y_{ss}, there is u_{ss}, satisfying $y_{ss} = Gu_{ss}$. For any y_{ss}, it may fail to find u_{ss}. For any u_{ss}, there is a unique y_{ss} as long as G is a real matrix. Since PLANT is $y(s) = G(s)u(s)$, a failure to satisfy $y_{ss} = Gu_{ss}$ brings troubles. The small trouble could be steady-state error (non-offset-free), and the big could be dynamic instability. $y_{ss} \neq Gu_{ss}$ can easily cause dynamic instability.

In summary, it is important to automatically calculate y_{ss}, u_{ss}. MPC in Figure 1.8 is renamed as DC. In order to take out a set of $\{y_{ss}, u_{ss}\}$ for DC, it needs a so-called steady-state target calculation (SSTC) in front of DC, as shown in Figure 1.9. The term SSTC is somewhat academic, and its alias in industry is steady-state optimization.

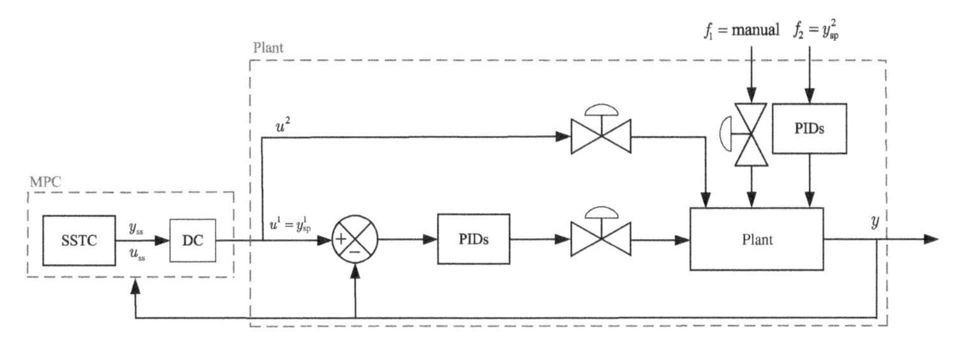

Figure 1.9 Control system using MPC with SSTC.

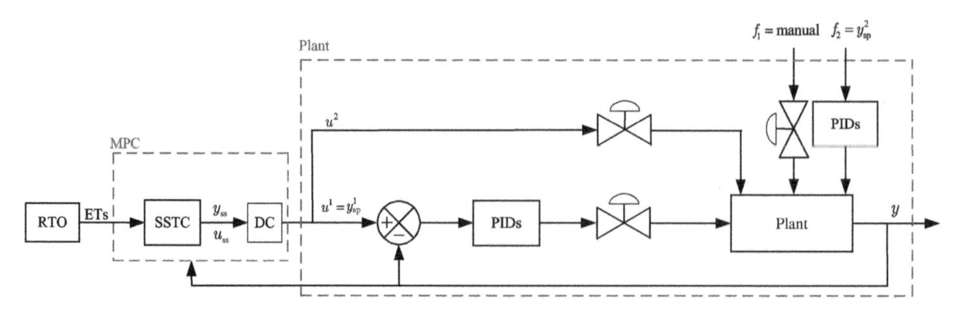

Figure 1.10 Control system using RTO+MPC including SSTC.

Recall that there is an expected value y_{sp} for PID and an expected value y_{ss} for DC. Is there an expected value for SSTC? Yes, there are $\{y_t^{sm}, u_t^{sm}\}$ before SSTC (some y have y_t, i.e., y_t^{sm}; some u have u_t, i.e., u_t^{sm}), as shown in Figure 1.10. $\{y_t^{sm}, u_t^{sm}\}$ are called the ideal values or external targets (ETs). SSTC calculates $\{y_{ss}, u_{ss}\}$ based on these ideal values and some other factors. u_{ss} is ss of the actuator position or PID sp. Therefore, ss is calculated, which is the steady-state PID sp or actuator position. Then, DC calculates the dynamic PID sp and actuator position, which moves the physical devices.

Recall that manually giving $\{y_{ss}, u_{ss}\}$ might be difficult. By introducing $\{y_t^{sm}, u_t^{sm}\}$, will the difficulty turn to manually operating $\{y_t^{sm}, u_t^{sm}\}$? It depends on the key difference between $\{y_t^{sm}, u_t^{sm}\}$ and $\{y_{ss}, u_{ss}\}$. $\{y_{ss}, u_{ss}\}$ (satisfying $y_{ss} = Gu_{ss}$) is related to control, which is lower-layered. In a factory, calculating sp of the controller is low-layered work. $\{y_t^{sm}, u_t^{sm}\}$, on the other hand, is related to the optimization, or related to economy. The relationship that $\{y_t^{sm}, u_t^{sm}\}$ satisfies is written as

$$F(y_t^{sm}, u_t^{sm}, \cdot) = 0, \tag{1.1}$$

where $\cdot$ denotes some other variables. Then, $\{y_t^{sm}, u_t^{sm}\}$ is obtained by solving

$$\min J(\text{of economic, etc.}) \text{ s.t. } (1.1), \tag{1.2}$$

where $J(\text{of economic, etc.}) = L(y_t^{sm}, u_t^{sm})$ is the performance index related to economy. Equation (1.2) is usually called real-time optimization (RTO) in the industry. Here, "real-time" mainly reflects that some physical parameters in F, relating to the real system, are updated in real-time. There are a lot of research studies on RTO in the industrial and academic circles. RTO technology could be more difficult than MPC.

Remark 1.1 The nature of (1.1) is steady-state. What is the difference between (1.1) and $y_{ss} = Gu_{ss}$? In general, they are nonequivalent. Some research works linearize $F(y_t, u_t, \cdot) = 0$ to yield $y_t = Gu_t$. Academically, it should take such a linearization; otherwise, linearity and nonlinearity can be inconsistent. In theory, if linearizing $F(y_t, u_t, \cdot) = 0$ does not give $y_t = Gu_t$, it implies that the model is not well built. However, in the industrial practices, we might not worry about this inconsistency. While RTO is developed by some technicians, MPC can be developed by different groups. The people for RTO are engaged in system dynamics, energy technology or technological processes, while the group for MPC is in system control technique. The two groups of persons are unnecessarily consistent, although cooperation is encouraged. The two groups should maintain some independence.

In other words, $\{y_t^{\rm sm}, u_t^{\rm sm}\}$ has its advanced source, coming from thoughtful people. A thoughtful module RTO provides $\{y_t^{\rm sm}, u_t^{\rm sm}\}$, which may be better than from control technical circles.

Let us make a visual analogy. The Ministry of Education shows some assessment indices to a university; the university partitions these indices into the colleges, and each college then assigns tasks to the departments. The college corresponds to MPC level, and the department is at PID level. If each PID corresponds to a supervisor or a team, then each department has several PIDs. The target of PID level, or the college level, as compared with that of a higher level (university, Ministry of Education) planning, will have some deviation. Indeed, what PIDs and colleges should do is to carry out higher-level optimums as closely as possible. $\{y_{\rm ss}, u_{\rm ss}\}$ is driven by the higher-leveled $\{y_t^{\rm sm}, u_t^{\rm sm}\}$.

1.3 Hierarchical Model Predictive Control

Let us enclose MPC with SSTC in a box, which is referred to as the two-layered MPC in academia. If RTO is included, then it is called the hierarchical MPC. f appears in Figure 1.11, including the previous f_1 and f_2; in addition, some DVs denoted as f_3 are not mentioned above, which are neither f_1 =manual nor $f_2 = y_{\rm sp}^2$. f in Figure 1.11 should contain $\{f_1, f_2, f_3\}$, mainly f_3. Why is it called hierarchy? Because this is a high-to-low layered framework.

Actually, the two-layered MPC does not have a hierarchical framework. Two-layered, being also named dual-moduled, implies that at each control interval, first calculate ss, and then track ss; see Figure 1.12.

In summary, for industrial MPC, actuator positioning is not a complete understanding. We need to change our minds in two manners.

(1) MPC mainly controls PID sp, i.e., $u^1 = y_{\rm sp}^1$;
(2) $\{y_{\rm ss}, u_{\rm ss}\}$ is also automatically calculated.

Then, we can elevate from the theoretical MPC (as seen in a large number of literatures) to the industrial MPC.

In order to highlight hierarchy, reshape Figure 1.11 as Figure 1.13. Within the dashed box, it is MPC, and below and above, the real industrial process and RTO, respectively. In fact, RTO may do more than giving $\{y_t^{\rm sm}, u_t^{\rm sm}\}$, but we only care about its role on MPC. Other $y_{\rm ss}$ without y_t, and $u_{\rm ss}$ without u_t, are handled by SSTC. In order to calculate ss, it is necessary to set a standard, which will be detailed later.

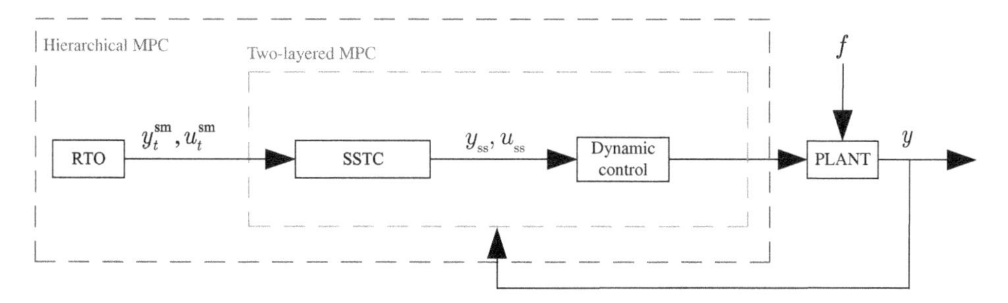

Figure 1.11 The contents of hierarchical MPC and two-layered MPC.

In "model predictive control", what does "model predictive" represent? MPC depends on a model, which implies building a model for the part below the dashed box in Figure 1.13. MPC can be renamed prediction-based control. How is the prediction given? Not only the basic KF (Kalman filter), but also the extended KF, the unscented KF, the information fusion KF, and the particle filter, can be used. As long as there is a causality model, we can make the prediction. Since the dynamic u is developed by one module, and ss is calculated by another, it is better to prepare a separate module for prediction, as shown in Figure 1.14. Figure 1.13 sends $\{f, y\}$ to DC. In Figure 1.14, since $\{f, y\}$ is sent to the prediction module, DC does not have to receive $\{f, y\}$.

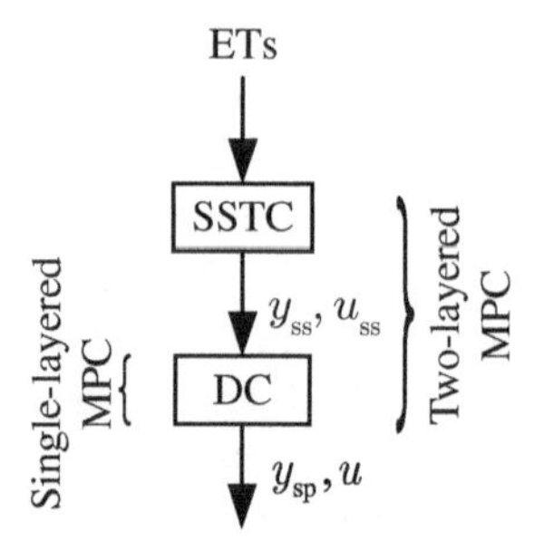

Figure 1.12 Two-layered structure composed of SSTC and DC.

The predictions have two types. One is called the dynamic prediction, which refers to predicting over a period of time, i.e.,

$$y^{ol}(k+1|k), y^{ol}(k+2|k), \ldots, y^{ol}(k+p|k), \tag{1.3}$$

where the superscript ol represents open-loop; when the control move is unchanged, it is ol. DC does not send back information to the prediction module. The prediction module unidirectionally sends information to DC. The prediction module is unaware of the immediate result of DC. A prediction without feedback is called ol. By applying as much disturbance information as it knows, while assuming $u(k), u(k+1|k), u(k+2|k), \ldots = u(k-1)$ (i.e., all the future control moves to be equivalent to $u(k-1)$, i.e., the control moves to remain unchanged), the prediction is called ol. If we know the future f series for a period of time, then apply the series for prediction. If we only know the current f value, then only apply this value for prediction.

The second type of prediction is called steady-state prediction. The steady-state prediction is also ol, denoted as y_{ss}^{ol}. SSTC uses ol prediction, and then y_{ss} from SSTC is not ol. Does the

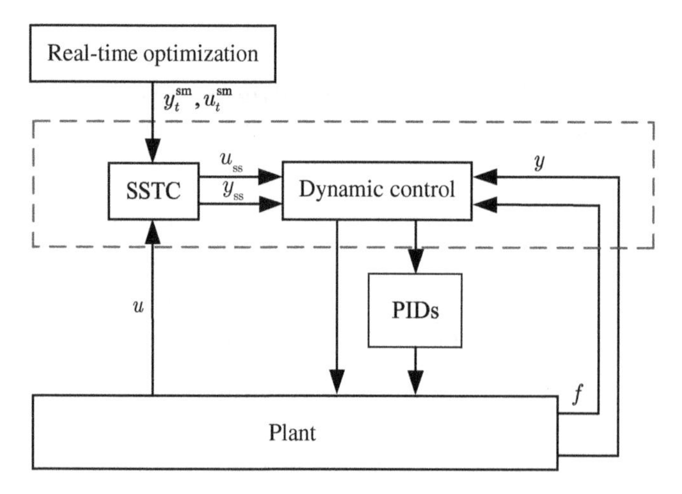

Figure 1.13 Transformation of system structure of RTO+two-layered MPC.

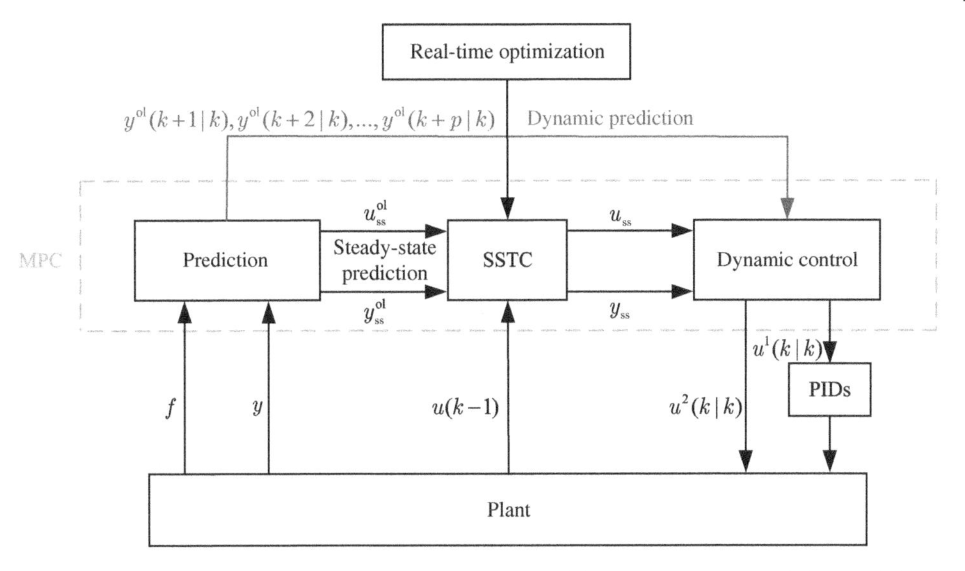

Figure 1.14 The system structure of RTO+two-layered MPC with open-loop prediction module.

prediction module give u_{ss}^{ol}? We should consider it. Usually, $u_{ss}^{ol} = u(k-1)$, but sometimes $u_{ss}^{ol} \neq u(k-1)$.

There are various methods for the prediction module, as long as it can give the dynamic and steady-state ol predictions. Since the prediction is a completely independent module, for both academic research and industrial application, it is not restricted to the traditional prediction methods.

Why is $u(k-1)$ drawn in Figure 1.14? For some theoretical research, it may be unnecessary to inform the controller of the value $u(k-1)$, as the controller should know $u(k-1)$ as that having been sent for implementation at $k-1$. However, for the real control, $u(k-1|k-1)$ given at $k-1$ may unnecessarily be equal to $u(k-1)$. For example, suppose the control period is 1 minute. At 11 : 59 it gives a PID sp. This sp may not be realized until 12 : 00. In the real industry, a PID sp is limited to be "slowly varying". If the PID sp given at 11 : 59 varies significantly (as compared with the instantaneous value of PID sp at 11 : 59), it may be clipped (cannot be realized at 12 : 00). We can consider this limitation in MPC, i.e., not allowing a significant variation; this still may not avoid $u(k-1|k-1) \neq u(k-1)$. Let us take another example, i.e., $u^2(k|k)$ directly applies to the actuator. It cannot guarantee that the actuator position sp given by 11 : 59 is achieved at 12 : 00. In this situation, at 12 : 00, it is necessary to read the actuator position again, in order to get a true position value. In both theory and simulation, since u is sent in the previous control interval, and the computer program runs accordingly, there is no need to doubt whether or not it is achieved in the next control interval. The computers cannot make mistakes, but the physical plants may fail to react as quick.

The dashed box in Figure 1.14 is called MPC, which is in line with the literature. It should be called the two-layered MPC, which is often referred to as constrained multivariable control in industry. Why is there such a big difference between the two names? In industry, PID accounts for around 80% of automatic control loops, while MPC accounts for around 10%, and there are rarely other control algorithms. Thus, the constrained multivariable control is generally referred to as MPC. So far, MPC is the only method that can systematically handle constraints and be applied in industry. MPC has an almost equivalent name, the constrained multivariable control.

This book uses f to denote DV (other literature may use d). In mathematics, $y = f(x)$ is generally applied as a function, while this book is a little different: f denoting DV, u denoting MV, y denoting CV, and x denoting the state when the state space model is applied.

This book includes three types of two-layered MPC. The ordinary two-layered dynamic matrix control (DMC) is given in Chapter 4, which is the usual DMC for open-loop stable system. Chapter 5 has special complexity for the two-layered DMC with integral controlled variable (iCV). Chapter 6 adopts the state space model for the two-layered DMC, which also has its complexity. Relatively, the basic two-layered DMC in Chapter 4 is easier, at least as compared with those for iCV and state space model.

2

Parameter Estimation and Output Prediction

2.1 Test Signal for Model Identification

The industrial process test produces the steady-state and the dynamic information by exciting the process with the test signals. For the model-based control, the richer the information being contained in the model, the greater the help to the controller designer. Since MPC is in the middle (coordination and optimization) layer of the hierarchical control, it only needs to contain accurate low-frequency information, the high-frequency dynamics can be controlled by PID, as shown in Figure 2.1. Hence, in the model identification for MPC, the data produced in the test stage should accurately reflect the steady-state gain and the slow dynamics of the process.

2.1.1 Step Test

When the industrial process stays at a steady state, a step change is applied to MV, and then the process gradually arrives at a new steady state. Such a perturbation of MV is called a step test. After applying a step change to MV, the change of curve for CV stands for the step response, as shown in Figure 2.2. The step response contains rich steady-state gain and slow dynamic information, but only a little fast dynamic information, i.e., the energy level of the high-frequency signal is far lower than that of the low-frequency signal. When using the step signal, it is necessary to consider its amplitude and duration. Some industrial process takes too long to reach the steady state; if we use the step test, the data may be polluted by the disturbances, which leads to serious degradation of the identified model.

2.1.2 White Noise

In theory, the best test signal may be the white noise. The white noise is the simplest stochastic process with zero mean and nonzero spectral density. The power of white noise is equally distributed over the entire frequency band, from $-\infty$ to $+\infty$. The mathematical definition of white noise is as follows. Consider the stochastic process $w(t)$. Suppose its autocorrelation function is

$$r_w(t) = \sigma^2 \delta(t), \tag{2.1}$$

Model Predictive Control, First Edition. Baocang Ding and Yuanqing Yang.

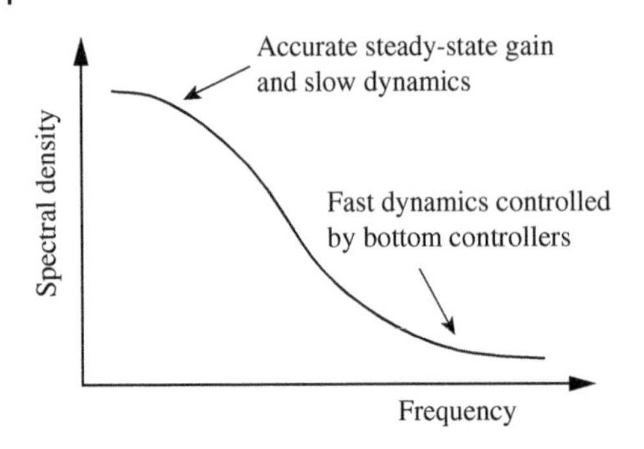

Figure 2.1 Expected coverage by MPC model.

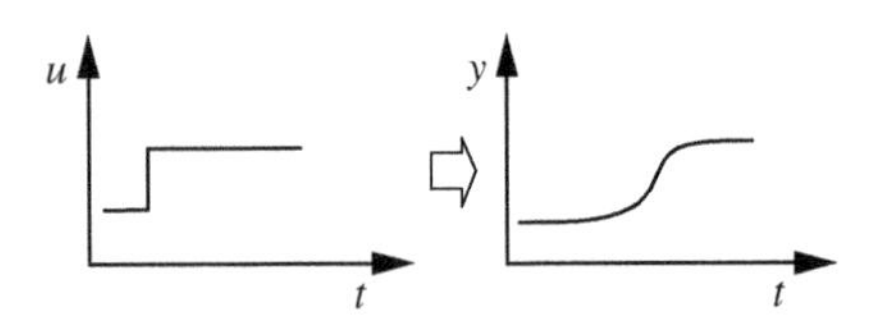

Figure 2.2 Step test.

where σ is a constant and $\delta(t)$ is the Dirac function or the impulse function, i.e.,

$$\delta(t) = \begin{cases} \infty, & t = 0 \\ 0, & t \neq 0 \end{cases}, \tag{2.2}$$

and

$$\int_{-\infty}^{\infty} \delta(t)dt = 1. \tag{2.3}$$

Then, $w(t)$ is the white noise process. The spectral density of white noise $w(t)$ is the constant σ^2.

The discretized white noise sequence, as shown in Figure 2.3, can be used as the test signal. The white noise sequence can be produced by transformation of a random number with a certain distribution. In theory, as long as there is a pseudo-random number with a continuous distribution, any other pseudo-random number with an arbitrary distribution can be obtained through an appropriate function transformation. The uniformly distributed pseudo-random number over (0, 1) is a basic and simple pseudo-random number, which is produced by the multiplication congruence method, as in the following two steps:

(1) Apply the multiplication congruence method to produce a sequence of positive integers $\{x_i\}$. Taking $x_i = (Ax_{i-1})\text{mod}M$. The initial integer x_0 is called the seed, which is generally a positive odd number. Usually, one can take $x_0 = 1, A = 5^{13}$, and $M = 10^{36}$.
(2) Let $c_i = x_i/M$ $(i = 1, 2 \ldots)$. $\{c_i\}$ is the uniformly distributed pseudo-random sequence over (0, 1). The white noise with amplitude 0.5 can be obtained by subtracting 0.5 from each $\{c_i\}$.

However, the white noise sequence is not favored for realization in engineering, because the movements as the white noise can lead to serious wear of physical equipment (e.g., valves). In fact, pseudo-random binary sequence (PRBS) and GBN (generalized binary noise) are the typical industrial process test signals.

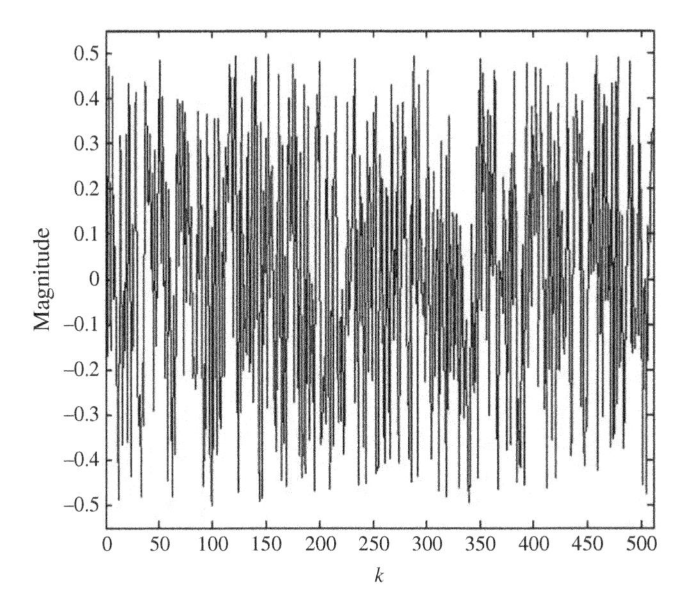

Figure 2.3 A white noise.

2.1.3 Pseudo-Random Binary Sequence

PRBS is a long-cycle signal whose element is either 0 or 1. Within one cycle, PRBS is a random binary signal; outside of each cycle, the signal repeats the previous cycle. PRBS is produced by

$$\begin{cases} s_1(l+1) = b_1 s_1(l) \oplus b_2 s_2(l) \oplus \cdots \oplus b_n s_n(l) \\ s_j(l+1) = s_{j-1}(l), \quad 2 \leq j \leq n \\ u(l) = s_n(l) \end{cases}, \tag{2.4}$$

where $l \geq 0$ (the unit of l being the clock period); $\oplus$ is the logical XOR operator; $b_j \in \{0,1\}$ $(1 \leq j \leq n-1)$ is the feedback coefficient; $b_n = 1$; all the initial states $s_j(0) \in \{0,1\}$ $(1 \leq j \leq n)$ are the nonzero random binary signals. Choosing the appropriate feedback coefficients b_j yields the longest period $M = 2^n - 1$ (the unit being the clock period), and then the corresponding PRBS is called the M sequence.

In real applications, in order to tune the amplitude, we can choose $u(l) = a(1 - 2s_n(l))$ so that the test signal $u(l)$ is either a or $-a$. In computing $u(l)$, $s_n(l)$ changes from a logical value to a real.

In the longest period M, the number $u(l) = -a$ appears 1 less time than the number $u(l) = a$. Hence, the average of $u(k)$ is (the unit of k being the sampling period)

$$|\mathrm{E}[u(k)]| = \frac{a}{M}, \tag{2.5}$$

which indicates that the M sequence contains a "direct current" component. In order to remove direct current component, let PBRS generated by M-sequence $s_n(l)$ XOR with $\{0,1,0,1,0,1,\ldots\}$ (i.e., $\{0,1\}$ square wave with cycle 2) element-wise, which yields an inverse M sequence $s'_n(l)$ (with a long cycle $2M$). In order to tune the amplitude of $s'_n(l)$, take $u'(l) = a(1 - 2s'_n(l))$.

When M is large enough, the autocorrelation function of PRBS is close to that of the pulse sequence. Hence, PRBS can replace the white noise to test the industrial processes.

2.1.4 Generalized Binary Noise

The element of GBN is either $-a$ or a. The shortest time interval, within which GBN element remains unchanged, is $T_{\min}$. At each time l, GBN element switches with probability, i.e.,

$$\begin{cases} \mathrm{P}\{u(l) = -u(l-1)\} = p_{\mathrm{sw}} \\ \mathrm{P}\{u(l) = u(l-1)\} = 1 - p_{\mathrm{sw}} \end{cases}, \tag{2.6}$$

where $l \geq 0$ (the unit of l being $T_{\min}$); p_{sw} is the switching probability. Obviously, the mean value of GBN is 0. $T_{\min}$ is called the minimum switching time, which can be taken as the control interval or multiple control intervals. Define the switching time T_{sw} as the interval between two switchings, then the average switching time of GBN is

$$\mathrm{E}[T_{\mathrm{sw}}] = \frac{T_{\min}}{p_{\mathrm{sw}}}. \tag{2.7}$$

When $T_{\min} = 1$, $a = 0.5$ and $p_{\mathrm{sw}} = 1/2$ (e.g., using each element of $[0, 1]$ uniformly distributed pseudo-random sequence, switch if and only if the element is greater than 0.5), use the MATLAB function `spa` to obtain the spectral density of the white noise GBN, as shown in Figure 2.4. However, the white noise signal is not the best test signal. By reducing the switching probability, i.e., increasing the average switching time, a low-pass GBN signal is obtained. When $p_{\mathrm{sw}} = 1/8$, the spectral density of GBN is shown in Figure 2.5. From Figure 2.5, it is shown that GBN with a larger average switch time is a low-pass signal. Moreover, the length of GBN can be set arbitrarily, and two different GBN signals are completely uncorrelated. Therefore, GBN can be conveniently applied to the multivariable model identification.

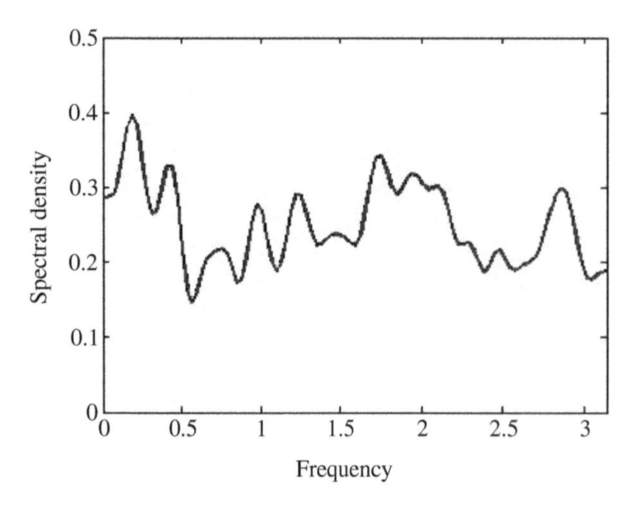

Figure 2.4 Spectral density of GBN with $p_{\mathrm{sw}} = 1/2$.

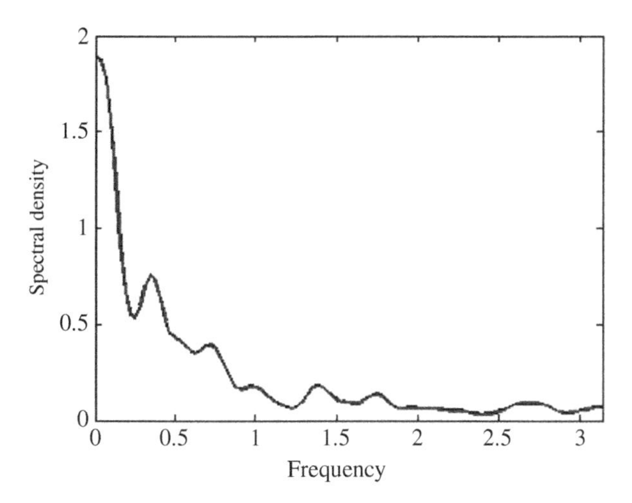

Figure 2.5 Spectral density of GBN with $p_{sw} = 1/8$.

2.2 Step Response Model Identification

Notations of this section: $y \in \mathbb{R}^{n_y}$ is DepV (dependent variable), $y^s \in \mathbb{R}^{n_y^s}$ is sDepV (stable DepV), $y^r \in \mathbb{R}^{n_y^r}$ is iDepV (the first-order integral DepV), $w = \begin{bmatrix} u \\ f \end{bmatrix} \in \mathbb{R}^{n_w}$ is IndepV (independent variable), $n_w = n_u + n_f$.

2.2.1 Model

DepV y is divided into two types: sDepV and iDepV, i.e., $y = \begin{bmatrix} y^s \\ y^r \end{bmatrix}$, where the superscripts s and r represent stable and integral (ramp), respectively. Without loss of generality, suppose $n_y^s + n_y^r = n_y$. Suppose the equilibrium point of the system is

$$\{y_{\text{eq}}, w_{\text{eq}}\} = \left\{ \begin{bmatrix} y^s_{\text{eq}} \\ y^r_{\text{eq}} \end{bmatrix}, \begin{bmatrix} u_{\text{eq}} \\ f_{\text{eq}} \end{bmatrix} \right\}.$$

The purpose of identification is to obtain an approximate linear model of the system near the equilibrium point. In the identification, we should treat the sDepV and iDepV distinctively. For sDepV, apply the following finite impulse response (FIR) model:

$$\nabla y^s(k) = \sum_{l=1}^{N} H_l^s \nabla w(k-l), \tag{2.8}$$

where H_l^s is the impulse response coefficient matrix, $\nabla y^s(k) = y^s(k) - y^s_{\text{eq}}$, $\nabla w(k-l) = w(k-l) - w_{\text{eq}}$. The necessary condition for using (2.8) to denote the linear ol (open-loop) stable system is that $H_l^s \approx 0$ for all $l > N$, where N is the model horizon, i.e., (2.8) is a valid approximation to $\nabla y^s(k) = \sum_{l=1}^{\infty} H_l^s \nabla w(k-l)$.

For iDepV, apply the FIR model in its incremental form, i.e.,

$$\nabla(\Delta y^r(k)) = \sum_{l=1}^{N} \Delta H_l^r \nabla w(k-l), \tag{2.9}$$

where $\Delta H_l^r = H_l^r - H_{l-1}^r$, $\nabla(\Delta y^r(k)) = \Delta(\nabla y^r(k)) = \Delta y^r(k)$, and $\Delta y^r(k) = y^r(k) - y^r(k-1)$, $\nabla y^r(k) = y^r(k) - y_{\text{eq}}^r$. The necessary condition for using (2.9) to denote the linear ol first-order integral system is that $\Delta H_l^r \approx 0$ for all $l > N$, i.e., (2.9) is a valid approximation for $\nabla(\Delta y^r(k)) = \sum_{l=1}^{\infty} \Delta H_l^r \nabla w(k-l)$.

For applying (2.8) and (2.9), we need to know y_{eq}^s and w_{eq}, which are difficult to achieve in most cases. If we use the incremental data, it is unnecessary to know the values y_{eq}^s and w_{eq}. For this reason, multiply by Δ on both sides of (2.8) and (2.9). Noting that $\Delta(\nabla y^s(k)) = \Delta y^s(k)$, $\Delta(\nabla w(k-l)) = \Delta w(k-l)$, and $\Delta(\nabla(\Delta y(k))) = \Delta^2 y(k)$, we obtain

$$\Delta y^s(k) = \sum_{l=1}^{N} H_l^s \Delta w(k-l), \tag{2.10}$$

$$\Delta^2 y^r(k) = \sum_{l=1}^{N} \Delta H_l^r \Delta w(k-l), \tag{2.11}$$

where $\Delta^2 y^r(k) = y^r(k) - 2y^r(k-1) + y^r(k-2)$. Combing (2.10) and (2.11) yields

$$\begin{bmatrix} \Delta y^s(k) \\ \Delta^2 y^r(k) \end{bmatrix} = \sum_{l=1}^{N} \begin{bmatrix} H_l^s \\ \Delta H_l^r \end{bmatrix} \Delta w(k-l).$$

The least square method in this section utilizes the data set of $\left\{ \begin{bmatrix} \Delta y^s \\ \Delta^2 y^r \end{bmatrix}, \Delta w \right\}$ to regress the coefficients $\begin{bmatrix} H_l^s \\ \Delta H_l^r \end{bmatrix}$ for all $l \in \{1, 2, \dots, N\}$.

Each IndepV does not necessarily have relationship with each DepV. Denote H_l as

$$H_l = \begin{bmatrix} h_{11,l} & h_{12,l} & \cdots & h_{1,n_w,l} \\ h_{21,l} & h_{22,l} & \cdots & h_{2,n_w,l} \\ \vdots & \vdots & & \vdots \\ h_{n_y,1,l} & h_{n_y,2,l} & \cdots & h_{n_y,n_w,l} \end{bmatrix}.$$

If there exist $i \in \{1, 2, \dots, n_w\}$ and $j \in \{1, 2, \dots, n_y\}$ so that $h_{ji,l} = 0$ for all $l \in \{1, 2, \dots, N\}$, then $h_{ji,l}$ can disappear in identified parameters. When a zero-in-truth parameter is identified, the identification result may be nonzero, which may have a side affect for identification. In order to avoid identifying zero-in-truth parameters, FIR identification executes each least square (LS) for a multiple input single output (MISO) model. For the convenience of MISO execution, (2.10) and (2.11) can be denoted as

$$\Delta y_j^s(k) = \sum_{i \in \pi_j} \sum_{l=1}^{N} h_{ji,l}^s \Delta w_i(k-l) = \sum_{i \in \pi_j} \vec{h}_{ji}^s \Delta \vec{w}_i(k-1),$$
$$j \in \{1, 2, \dots, n_y^s\}, \tag{2.12}$$

$$\Delta^2 y_j^r(k) = \sum_{i\in\pi_j}\sum_{l=1}^{N}\Delta h_{ji,l}^r \Delta w_i(k-l) = \sum_{i\in\pi_j}\Delta\vec{h}_{ji}^r \Delta\vec{w}_i(k-1),$$
$$j \in \{n_y^s+1, n_y^s+2, \dots, n_y\}, \tag{2.13}$$

where π_j is the index set of IndepVs, which have a causal relationship with the jth DepV. $\Delta\vec{w}_i(k-1) = [\Delta w_i(k-1), \Delta w_i(k-2), \dots, \Delta w_i(k-N)]^T$, $\vec{h}_{ji}^s = [h_{ji,1}^s, h_{ji,2}^s, \dots, h_{ji,N}^s]$, $\Delta\vec{h}_{ji}^r = [\Delta h_{ji,1}^r, \Delta h_{ji,2}^r, \dots, \Delta h_{ji,N}^r]$.

Remark 2.1 If the time delay τ_{ji} for the ith IndepV and the jth DepV is known, then we can remove the first τ_{ji} items in $\{\vec{h}_{ji}^s, \Delta\vec{h}_{ji}^r, \Delta\vec{w}_i(k-1)\}$. The details are omitted.

2.2.2 Data Processing

By the process test, one obtains the test data set $\{y(k), w(k)|k = 0, 1, 2, \dots, L\}$ which is written as a matrix (matrix of sampled data), i.e.,

$$\mathcal{M}_{\rm sd} = \begin{bmatrix} y_1(0) & y_1(1) & \cdots & y_1(L) \\ y_2(0) & y_2(1) & \cdots & y_2(L) \\ \vdots & \vdots & & \vdots \\ y_{n_y}(0) & y_{n_y}(1) & \cdots & y_{n_y}(L) \\ w_1(0) & w_1(1) & \cdots & w_1(L) \\ w_2(0) & w_2(1) & \cdots & w_2(L) \\ \vdots & \vdots & & \vdots \\ w_{n_w}(0) & w_{n_w}(1) & \cdots & w_{n_w}(L) \end{bmatrix}, \tag{2.14}$$

where each row is the data for one variable.

2.2.2.1 Marking or Interpolation of Bad Data

Some data may be "bad" (e.g., jump, out of range, non-value), which cannot be utilized for identification. We need to exclude "bad" or replace it with interpolation. For a long queue of "bad," as well as "bad" at the beginning/end of the sequence, we will exclude it by revising it as b. For a short queue of "bad," we will replace it with interpolation. The simplest linear interpolation is

$$x(l) = x(t_1-1) + \frac{l-t_1+1}{t_2-t_1+2}[x(t_2+1) - x(t_1-1)],$$
$$l \in \{t_1, t_1+1, \dots, t_2\}, \tag{2.15}$$

where $x \in \{y_1, y_2, \dots, y_{n_y}, w_1, w_2, \dots, w_{n_w}\}$; t_1 and t_2 are the beginning and end time of the "bad," respectively.

By handling the "bad," (2.14) becomes

$$\mathcal{M}'_{\rm sd} = \begin{bmatrix} y'_1(0) & y'_1(1) & \cdots & y'_1(L) \\ y'_2(0) & y'_2(1) & \cdots & y'_2(L) \\ \vdots & \vdots & & \vdots \\ y'_{n_y}(0) & y'_{n_y}(1) & \cdots & y'_{n_y}(L) \\ w'_1(0) & w'_1(1) & \cdots & w'_1(L) \\ w'_2(0) & w'_2(1) & \cdots & w'_2(L) \\ \vdots & \vdots & & \vdots \\ w'_{n_w}(0) & w'_{n_w}(1) & \cdots & w'_{n_w}(L) \end{bmatrix}, \tag{2.16}$$

where some elements are b.

2.2.2.2 Smoothing Data

Due to the inevitable presence of noise in the data, smoothing is necessary. Here, we take the first-order inertial filter as an example. In (2.16), non-b data is divided into multiple segments, and we smooth each segment separately. The first-order inertial filter is

$$y''_j(l) = \begin{cases} y'_j(l), & y'_j(l-1) \text{ is b or } l = 0 \\ \alpha_{y,j} y'_j(l) + (1-\alpha_{y,j}) y''_j(l-1), & \text{others} \end{cases}, \quad j \in \{1, 2, \ldots, n_y\}, \tag{2.17}$$

$$u''_j(l) = \begin{cases} u'_j(l), & u'_j(l-1) \text{ is b or } l = 0 \\ \alpha_{u,j} u'_j(l) + (1-\alpha_{u,j}) u''_j(l-1), & \text{others} \end{cases}, \quad j \in \{1, 2, \ldots, n_u\}, \tag{2.18}$$

$$f''_j(l) = \begin{cases} f'_j(l), & f'_j(l-1) \text{ is b or } l = 0 \\ \alpha_{f,j} f'_j(l) + (1-\alpha_{f,j}) f''_j(l-1), & \text{others} \end{cases}, \quad j \in \{1, 2, \ldots, n_f\}, \tag{2.19}$$

where $\{\alpha_{y,j}, \alpha_{u,j}, \alpha_{f,j}\}$ are smoother coefficients. For practical applications, use smoother coefficients, at least being different between CV, MV, and DV. If MV is equal to the test signal, then do not smooth it. We can directly set $\alpha \in (0, 1)$, or calculate by $\alpha = 1 - e^{-\frac{T_s}{T}}$ where T_s and T are, respectively, the sampling interval and the filter time constant. After smoothing, (2.16) becomes

$$\mathcal{M}''_{\rm sd} = \begin{bmatrix} y''_1(0) & y''_1(1) & \cdots & y''_1(L) \\ y''_2(0) & y''_2(1) & \cdots & y''_2(L) \\ \vdots & \vdots & & \vdots \\ y''_{n_y}(0) & y''_{n_y}(1) & \cdots & y''_{n_y}(L) \\ w''_1(0) & w''_1(1) & \cdots & w''_1(L) \\ w''_2(0) & w''_2(1) & \cdots & w''_2(L) \\ \vdots & \vdots & & \vdots \\ w''_{n_w}(0) & w''_{n_w}(1) & \cdots & w''_{n_w}(L) \end{bmatrix}, \tag{2.20}$$

where some elements are b.

2.2.3 Model Identification

The multivariable model reflects the interactive relationship between multiple IndepVs and multiple DepVs. A long production process usually has upstream and downstream. Usually, a downstream IndepV does not affect an upstream DepV. If IndepV and DepV data from upstream and downstream are utilized all at once, an erroneous conclusion can be drawn that some downstream IndepV influences some upstream DepV.

Moreover, all upstream IndepVs do not necessarily affect each downstream DepV.

If all IndepVs affect each DepV, in principle, we can identify MIMO models once for all. However, considering the general case, we divide the whole MIMO into a number of cases, and then compute once per case.

2.2.3.1 Case Grouping

If there is a complete causal relationship between $y^{(j)}$ (a part of DepVs) and $w^{(j)}$ (a part of IndepVs), then the identification of the dynamic model between $y^{(j)}$ and $w^{(j)}$ is regarded as a case, being called case j. This always includes the situation when it is unsure whether or not a variable in $y^{(j)}$ is truly affected by a variable in $w^{(j)}$.

The case j may include both sDepV and iDepV. After identifying case j, if we find that there is no causal relationship between a DepV in $y^{(j)}$ and an IndepV in $w^{(j)}$, then this case j will be partitioned into more than one case. Hence, case grouping may be adjusted several times in the whole flow of identification.

2.2.3.2 Cased Data Preparation for Stable Dependent Variables

Suppose y_j^s belongs to the case ℓ, in identifying the dynamics between y_j^s and $w^{(\ell)}$, first, $\mathcal{M}_j$ is constructed. Each row of $\mathcal{M}_j$ is formulated as

$$\mathcal{L}_l = \left[\Delta y_j^{\prime\prime s}(N+l)\ \Delta w_i^{\prime\prime}(N+l-1)^T\ \Delta w_i^{\prime\prime}(N+l-2)^T\ \cdots\ \Delta w_i^{\prime\prime}(l)^T | i \in \pi_j\right],$$

where $l \in \{1, 2, \dots, L-N\}$; the order of the subscript i in $\mathcal{L}_l$ is always increasing. If $\mathcal{L}_l$ includes b, then it is not put into $\mathcal{M}_j$, otherwise then put it below $\mathcal{M}_j$. Each time $l \in \{1, 2, \dots, L-N\}$ increases by 1, $\mathcal{M}_j$ either gets one more line or remains unchanged. Finally, $\mathcal{M}_j$ is formulated as

$$\mathcal{M}_j = \left[\,Y_j\ \Phi_j\,\right], \tag{2.21}$$

where Y_j and Φ_j are the data matrices for Δy_j^s and $\Delta w^{(\ell)}$, respectively.

Rewrite the part of (2.12) for the jth DepV as

$$\Delta y_j^s(k) = \sum_{i\in\pi_j} \left[\,\Delta w_i(k-1)\ \Delta w_i(k-2)\ \cdots\ \Delta w_i(k-N)\,\right] (\vec{h}_{ji}^s)^T. \tag{2.22}$$

If the model is precise, the data has no b, and there is no noise, then

$$\begin{bmatrix} \Delta y_j^s(N+1) \\ \Delta y_j^s(N+2) \\ \vdots \\ \Delta y_j^s(L) \end{bmatrix} = \sum_{i\in\pi_j} \begin{bmatrix} \Delta w_i(N) & \Delta w_i(N-1) & \cdots & \Delta w_i(1) \\ \Delta w_i(N+1) & \Delta w_i(N) & \cdots & \Delta w_i(2) \\ \vdots & \vdots & & \vdots \\ \Delta w_i(L-1) & \Delta w_i(L-2) & \cdots & \Delta w_i(L-N) \end{bmatrix} (\vec{h}_{ji}^s)^T. \tag{2.23}$$

Equation (2.23) is abbreviated as

$$\Delta y_j^s(N+1 : L) = \sum_{i\in\pi_j} \times$$

$$\left[\Delta w_i(N : L-1)\ \Delta w_i(N-1 : L-2)\ \cdots\ \Delta w_i(1 : L-N)\right] (\vec{h}_{ji}^s)^T. \tag{2.24}$$

After excluding b, interpolation, and smoothing, instead of utilizing (2.23) and (2.24), the following is applied:

$$\Delta y_j^{\prime\prime s}\begin{pmatrix}\tau_0+N+2:\tau_1-1\\ \hline \tau_2+N+2:\tau_3-1\\ \hline \vdots\\ \hline \tau_{2\text{nsr}}+N+2:\tau_{2\text{nsr}+1}-1\end{pmatrix}$$

$$=\sum_{i\in\pi_j}\left[\Delta w_i^{\prime\prime}\begin{pmatrix}\tau_0+N+1:\tau_1-2\\ \hline \tau_2+N+1:\tau_3-2\\ \hline \vdots\\ \hline \tau_{2\text{nsr}}+N+1:\tau_{2\text{nsr}+1}-2\end{pmatrix}\Delta w_i^{\prime\prime}\begin{pmatrix}\tau_0+N:\tau_1-3\\ \hline \tau_2+N:\tau_3-3\\ \hline \vdots\\ \hline \tau_{2\text{nsr}}+N:\tau_{2\text{nsr}+1}-3\end{pmatrix}\cdots\right.$$

$$\left.\Delta w_i^{\prime\prime}\begin{pmatrix}\tau_0+2:\tau_1-N-1\\ \hline \tau_2+2:\tau_3-N-1\\ \hline \vdots\\ \hline \tau_{2\text{nsr}}+2:\tau_{2\text{nsr}+1}-N-1\end{pmatrix}\right](\vec{h}_{ji}^{s})^T, \tag{2.25}$$

where τ_i $(i\in\{0,1,\ldots,2\text{nsr}+1\})$ is the integer; nsr is the abbreviation of "number of sections removed." Here, τ is different from t in (2.15).

Example 2.1 Suppose $\tau_0=-1$, $\tau_1=L_1+2$, $\tau_2=L_2$, and $\tau_3=L+1$, then (2.25) becomes

$$\Delta y_j^{\prime\prime s}\begin{pmatrix}N+1:L_1+1\\ \hline L_2+N+2:L\end{pmatrix}$$

$$=\sum_{i\in\pi_j}\left[\Delta w_i^{\prime\prime}\begin{pmatrix}N:L_1\\ \hline L_2+N+1:L-1\end{pmatrix}\Delta w_i^{\prime\prime}\begin{pmatrix}N-1:L_1-1\\ \hline L_2+N:L-2\end{pmatrix}\right.$$

$$\left.\cdots\Delta w_i^{\prime\prime}\begin{pmatrix}1:L_1-N+1\\ \hline L_2+2:L-N\end{pmatrix}\right](\vec{h}_{ji}^{s})^T, \tag{2.26}$$

which is the abbreviation of

$$\begin{bmatrix}\Delta y_j^{\prime\prime s}(N+1)\\ \Delta y_j^{\prime\prime s}(N+2)\\ \vdots\\ \Delta y_j^{\prime\prime s}(L_1+1)\\ \hline \Delta y_j^{\prime\prime s}(L_2+N+2)\\ \Delta y_j^{\prime\prime s}(L_2+N+3)\\ \vdots\\ \Delta y_j^{\prime\prime s}(L)\end{bmatrix}=\sum_{i\in\pi_j}\times$$

$$\left[\begin{array}{cccc} \Delta w_i''(N) & \Delta w_i''(N-1) & \cdots & \Delta w_i''(1) \\ \Delta w_i''(N+1) & \Delta w_i''(N) & \cdots & \Delta w_i''(2) \\ \vdots & \vdots & & \vdots \\ \Delta w_i''(L_1) & \Delta w_i''(L_1-1) & \cdots & \Delta w_i''(L_1-N+1) \\ \hline \Delta w_i''(L_2+N+1) & \Delta w_i''(L_2+N) & \cdots & \Delta w_i''(L_2+2) \\ \Delta w_i''(L_2+N+2) & \Delta w_i''(L_2+N+1) & \cdots & \Delta w_i''(L_2+3) \\ \vdots & \vdots & & \vdots \\ \Delta w_i''(L-1) & \Delta w_i''(L-2) & \cdots & \Delta w_i''(L-N) \end{array}\right] (\vec{h}_{ji}^{s})^T.$$

2.2.3.3 Cased Data Preparation for Integral Dependent Variables

Suppose y_j^r belongs to the case ℓ, in identifying the dynamics between y_j^r and $w^{(\ell)}$, first $\mathcal{M}_j$ is constructed. Each row of $\mathcal{M}_j$ is formulated as

$$\mathcal{L}_l = \left[\Delta^2 y_j''^r(N+l)\ \Delta w_i''(N+l-1)^T\ \Delta w_i''(N+l-2)^T \cdots \Delta w_i''(l)^T \,|\, i \in \pi_j\right],$$

where $l \in \{1, 2, \dots, L-N\}$; the order of the subscript i in $\mathcal{L}_l$ is always increasing. If $\mathcal{L}_l$ includes b, then it is not put into $\mathcal{M}_j$, otherwise then put it below $\mathcal{M}_j$. Each time $l \in \{1, 2, \dots, L-N\}$ increases by 1, $\mathcal{M}_j$ either gets one more line or remains unchanged. Finally, $\mathcal{M}_j$ is formulated as

$$\mathcal{M}_j = \left[\, Y_j\ \Phi_j \,\right], \tag{2.27}$$

where Y_j and Φ_j are the data matrices for $\Delta^2 y_j^r$ and $\Delta w^{(\ell)}$, respectively.

Rewrite the part of (2.13) for the jth DepV, as

$$\Delta^2 y_j^r(k) = \sum_{i \in \pi_j} \left[\, \Delta w_i(k-1)\ \Delta w_i(k-2) \cdots \Delta w_i(k-N) \,\right] (\Delta \vec{h}_{ji}^{r})^T. \tag{2.28}$$

If the model is precise, the data has no b, and there is no noise, then

$$\begin{bmatrix} \Delta^2 y_j^r(N+1) \\ \Delta^2 y_j^r(N+2) \\ \vdots \\ \Delta^2 y_j^r(L) \end{bmatrix} = \sum_{i \in \pi_j} \begin{bmatrix} \Delta w_i(N) & \Delta w_i(N-1) & \cdots & \Delta w_i(1) \\ \Delta w_i(N+1) & \Delta w_i(N) & \cdots & \Delta w_i(2) \\ \vdots & \vdots & & \vdots \\ \Delta w_i(L-1) & \Delta w_i(L-2) & \cdots & \Delta w_i(L-N) \end{bmatrix} (\Delta \vec{h}_{ji}^{r})^T. \tag{2.29}$$

Equation (2.29) is abbreviated as

$$\Delta^2 y_j^r(N+1:L) = \sum_{i \in \pi_j} \times$$

$$\left[\Delta w_i(N:L-1)\ \Delta w_i(N-1:L-2) \cdots \Delta w_i(1:L-N)\right] (\Delta \vec{h}_{ji}^{r})^T. \tag{2.30}$$

After excluding b, interpolation, and smoothing, instead of utilizing (2.29) and (2.30), the following is applied:

$$\Delta^2 y_j''^r\begin{pmatrix} \tau_0+N+3:\tau_1-1 \\ \hline \tau_2+N+3:\tau_3-1 \\ \hline \vdots \\ \hline \tau_{2nsr}+N+3:\tau_{2nsr+1}-1 \end{pmatrix}$$

$$=\sum_{i\in\pi_j}\left[\Delta w_i''\begin{pmatrix} \tau_0+N+2:\tau_1-2 \\ \hline \tau_2+N+2:\tau_3-2 \\ \hline \vdots \\ \hline \tau_{2nsr}+N+2:\tau_{2nsr+1}-2 \end{pmatrix}\Delta w_i''\begin{pmatrix} \tau_0+N+1:\tau_1-3 \\ \hline \tau_2+N+1:\tau_3-3 \\ \hline \vdots \\ \hline \tau_{2nsr}+N+1:\tau_{2nsr+1}-3 \end{pmatrix}\cdots\right.$$

$$\left.\Delta w_i''\begin{pmatrix} \tau_0+3:\tau_1-N-1 \\ \hline \tau_2+3:\tau_3-N-1 \\ \hline \vdots \\ \hline \tau_{2nsr}+3:\tau_{2nsr+1}-N-1 \end{pmatrix}\right](\Delta\vec{h}_{ji}^r)^T. \tag{2.31}$$

Here, τ is different from t in (2.15). The length of each data segment in (2.31) is 1 less than that in (2.25) because the second-order difference of iDepV may reduce the data length by 1 (this reduction does not happen, e.g., when b is for w_i).

2.2.3.4 Least Square Solution to Parameter Regression

Abbreviate (2.25) or (2.31), as

$$Y_j = \Phi_j\theta_j, \tag{2.32}$$

where θ_j is composed of the parameters to be identified, i.e., $\theta_j^T = [\vec{h}_{ji}^s | i \in \pi_j]^T$ or $\theta_j^T = [\Delta\vec{h}_{ji}^r | i \in \pi_j]^T$. In θ_j, the order of the subscript i is always increasing, i.e., $\theta_j = \begin{bmatrix} \vec{h}_{j,i_1}^s \\ \vec{h}_{j,i_2}^s \\ \vdots \end{bmatrix}$ or $\theta_j = \begin{bmatrix} \Delta\vec{h}_{j,i_1}^r \\ \Delta\vec{h}_{j,i_2}^r \\ \vdots \end{bmatrix}$, $i_1 < i_2 < \cdots$ and $i_1, i_2, \cdots \in \pi_j$.

Due to the presence of, e.g., noise (whose side-effect cannot be eliminated by smoothing), truncation (N being finite), and inevitable nonlinearity and time-varying features of the system, it cannot satisfy (2.32) strictly. Introducing the residual error, (2.32) is modified as

$$Y_j = \Phi_j\hat{\theta}_j + \varepsilon_j, \tag{2.33}$$

where $\hat{\theta}_j$ denotes the estimate of θ_j, and ε_j is the residual error sequence.

The criterion for the parameter estimation is minimizing

$$J = \varepsilon_j^T\varepsilon_j + \mu\hat{\theta}_j^T\Pi\hat{\theta}_j, \tag{2.34}$$

where $\mu \geq 0$ is the smoother factor. There are multiple choices for $\Pi \geq 0$. When we expect to penalize the rate of parameters, and to minimize the rate close to the steady state, then we can take (see [Dayal and MacGregor, 1996])

$$\Pi = \Gamma^T \Lambda \Gamma, \;\; \Gamma = \text{diag}\{\Gamma_i | i \in \pi_j\}, \;\; \Lambda = \text{diag}\{\Lambda_i | i \in \pi_j\},$$

$$\Gamma_i = \begin{bmatrix} 1 & 0 & 0 & \dots & 0 \\ -1 & 1 & 0 & \ddots & \vdots \\ 0 & -1 & 1 & \ddots & 0 \\ \vdots & \ddots & \ddots & \ddots & 0 \\ 0 & \dots & 0 & -1 & 1 \end{bmatrix}, \;\; \Lambda_i = \begin{bmatrix} 1 & 0 & \dots & 0 \\ 0 & 2 & \ddots & \vdots \\ \vdots & \ddots & \ddots & 0 \\ 0 & \dots & 0 & N \end{bmatrix}.$$

Minimizing J is equivalent to finding LS solution to the linear equation

$$\begin{bmatrix} \Phi_j \\ \mu^{1/2}\Pi^{1/2} \end{bmatrix} \hat{\theta}_j = \begin{bmatrix} Y_j \\ 0 \end{bmatrix} \tag{2.35}$$

of $\hat{\theta}_j$. This LS solution is

$$\hat{\theta}_j = \left(\Phi_j^T \Phi_j + \mu \Pi\right)^{-1} \Phi_j^T Y_j. \tag{2.36}$$

If $y_{j'}$ and y_j are in the same case, then

$$\hat{\theta}_{j'} = \left(\Phi_j^T \Phi_j + \mu \Pi\right)^{-1} \Phi_j^T Y_{j'}. \tag{2.37}$$

Both (2.36) and (2.37) adopt the same $(\Phi_j^T \Phi_j + \mu \Pi)^{-1} \Phi_j^T$, so the calculation is simplified. For the same case, all $\hat{\theta}_j$ can be calculated by one computation, but this does not reduce the complexity; hence, the identification of FIR here is essentially MISO.

Remark 2.2 For (2.34), the standard QP objective function is

$$\hat{J} = \frac{1}{2} \hat{\theta}_j^T \left(\Phi_j^T \Phi_j + \mu \Pi\right) \hat{\theta}_j - Y_j^T \Phi_j \hat{\theta}_j.$$

In the following two cases, some elements of $\hat{\theta}_j$ can be determined in advance.

- The time delays between some IndepVs and DepVs are known, so the corresponding elements of $\hat{\theta}_j$ are zeros.
- Some IndepVs are sps of PID with offset-free. Hence, the steady-state gains (Nth coefficient matrix of FSR) of some sing-input single-output (SISO) models are 1, and of some other, are 0.

Suppose it is known that $\Phi_j^0 \hat{\theta}_j = \hat{\theta}_j^0$, where Φ_j^0 is the full rank matrix. Then, the optimization problem for the identification becomes

$$\min_{\hat{\theta}_j} \hat{J}, \;\; \text{s.t.} \; \Phi_j^0 \hat{\theta}_j = \hat{\theta}_j^0.$$

This is QP with the equality constraints, whose solution is

$$\begin{aligned} \hat{\theta}_j &= \left[\Omega_j^{-1} - \Omega_j^{-1} \Phi_j^{0T} (\Phi_j^0 \Omega_j^{-1} \Phi_j^{0T})^{-1} \Phi_j^0 \Omega_j^{-1}\right] \Phi_j^T Y_j + \Omega_j^{-1} \Phi_j^{0T} (\Phi_j^0 \Omega_j^{-1} \Phi_j^{0T})^{-1} \hat{\theta}_j^0 \\ &= \Omega_j^{-1} \Phi_j^T Y_j + \Omega_j^{-1} \Phi_j^{0T} (\Phi_j^0 \Omega_j^{-1} \Phi_j^{0T})^{-1} \left(\hat{\theta}_j^0 - \Phi_j^0 \Omega_j^{-1} \Phi_j^T Y_j\right), \end{aligned}$$

where $\Omega_j = \Phi_j^T \Phi_j + \mu \Pi$.

2.2.3.5 Least Square Solution by SVD Decomposition

Generally, decomposing $\begin{bmatrix} \Phi_j \\ \mu^{1/2}\Pi^{1/2} \end{bmatrix}$ by singular value decomposition (SVD) yields $\begin{bmatrix} \Phi_j \\ \mu^{1/2}\Pi^{1/2} \end{bmatrix} = Q_j \begin{bmatrix} V_{j,1} \\ 0 \end{bmatrix} R_j$, where $V_{j,1}$ is the nonsingular diagonal square matrix. Left multiplying $Q_j^{-1} = Q_j^T$ on both sides of (2.35), yields

$$\begin{bmatrix} V_{j,1} \\ 0 \end{bmatrix} R_j\hat{\theta}_j = \begin{bmatrix} Y_{j,1} \\ Y_{j,2} \end{bmatrix}, \tag{2.38}$$

where $Q_j^T \begin{bmatrix} Y_j \\ 0 \end{bmatrix} = \begin{bmatrix} Y_{j,1} \\ Y_{j,2} \end{bmatrix}$ does not implies that Y_j has the same dimension with $Y_{j,1}$. In terms of minimizing (2.34), we have LS solution of (2.38), i.e.,

$$\hat{\theta}_j = R_j^T V_{j,1}^{-1} Y_{j,1}. \tag{2.39}$$

Remark 2.3 If $\mu = 0$, and the data quality is low, then decomposing Φ_j by SVD may yield $\Phi_j = Q_j \begin{bmatrix} V_{j,1} & 0_1 \\ 0 & 0 \end{bmatrix} R_j$, where $V_{j,1}$ is the nonsingular diagonal square matrix and 0_1 is the zero matrix. In this situation, LS solution is

$$\hat{\theta}_j = R_j^T \begin{bmatrix} V_{j,1}^{-1} & 0_1 \\ 0_1^T & 0 \end{bmatrix} Y_{j,1}. \tag{2.40}$$

If the dimension of 0_1 is nonzero, then the software should warn "data incompatibility."

2.2.3.6 Filtering Pulse Response Coefficients

For iDepV, $\hat{h}_{ji,l} = \sum_{l'=1}^{l} \Delta\hat{h}_{ji,l'}$. For all DepVs, $\hat{h}_{ji,l}$ will be filtered. Taking the first-order inertial filter as an example, the step response coefficient for the controller is

$$h_{ji,l} = \begin{cases} \hat{h}_{ji,l}, & l = 1 \\ \alpha\hat{h}_{ji,l} + (1-\alpha)h_{ji,l-1}, & l > 1 \end{cases}, \quad s_{ji,l} = \sum_{l'=1}^{l} h_{ji,l'}. \tag{2.41}$$

In the existing identification software, iDepV needs to be manually specified, and in the model file, the parameters for iDepVs are marked by flag.

Example 2.2 Consider FIR model of SISO system, i.e.,

$$y(k) = \sum_{l=1}^{N} h(l)u(k-l) + e(k), \tag{2.42}$$

where $e(k)$ is the modeling error. Expanding (2.42), we have

$$\begin{aligned} y(k) &= h(1)u(k-1) + h(2)u(k-2) + \cdots + h(N)u(k-N) + e(k) \\ &= \varphi(k)\theta + e(k), \end{aligned} \tag{2.43}$$

where $\theta = \begin{bmatrix} h(1) \\ h(2) \\ \vdots \\ h(N) \end{bmatrix}$, $\varphi(k) = \left[u(k-1), u(k-2), \dots, u(k-N)\right]$.

Suppose the data sequences obtained via the test are $\{u(1), u(2), \dots, u(L)\}$ and $\{y(1), y(2), \dots, y(L)\}$. Substituting the data into (2.43) and writing in matrix form yields

$$\mathbf{y} = \Phi\theta + \mathbf{e}, \tag{2.44}$$

where $\mathbf{y} = \begin{bmatrix} y(N+1) \\ y(N+2) \\ \vdots \\ y(L) \end{bmatrix}$, $\mathbf{e} = \begin{bmatrix} e(N+1) \\ e(N+2) \\ \vdots \\ e(L) \end{bmatrix}$, $\Phi = \begin{bmatrix} u(N) & u(N-1) & \cdots & u(1) \\ u(N+1) & u(N) & \cdots & u(2) \\ \vdots & \vdots & & \vdots \\ u(L-1) & u(L-2) & \cdots & u(L-N) \end{bmatrix}$.

Minimizing the norm of the residual error, i.e.,

$$\min J = \min \boldsymbol{\varepsilon}^{\mathrm{T}}\boldsymbol{\varepsilon} = \min\, (\mathbf{y} - \Phi\hat{\theta})^T(\mathbf{y} - \Phi\hat{\theta}), \tag{2.45}$$

yields

$$\hat{\theta} = (\Phi^T\Phi)^{-1}\Phi^T\mathbf{y}, \tag{2.46}$$

where $\hat{\theta}$ is the estimate of θ.

Example 2.3 Apply the result in Example 2.2 to illustrate the effect of FIR identifications via LS method. Establish the test platform on the Simulink of Matlab, as shown in Figure 2.6. Directly using the random number as IndepV and adding a random noise to DepV; the mean value for both random signals is 0, and the variance is 0.1; the sampling interval of the platform is 1. IndepV and DepV are shown in Figure 2.7.

In order to demonstrate the effects of {data length, model horizon} on the FIR identification, three cases are taken for test, i.e., {data length, model horizon}={1200, 480}, {1200, 600}, and {1800, 600}, respectively. The identified FIRs/FSRs are shown in Figures 2.8–2.10, for the three cases, respectively; FIR diagram also contains the true FIR.

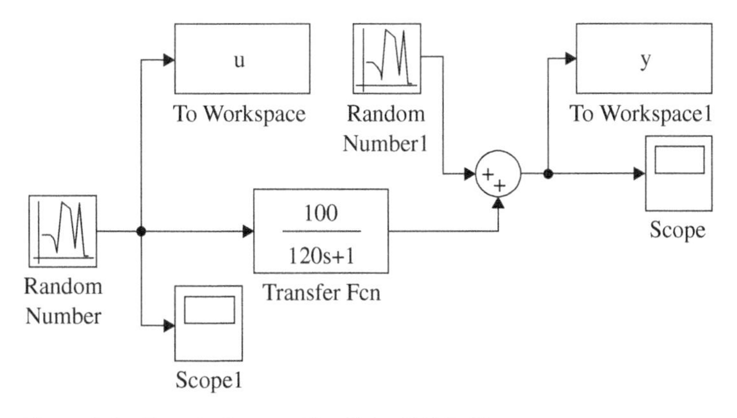

Figure 2.6 Test platform on Simulink of MATLAB.

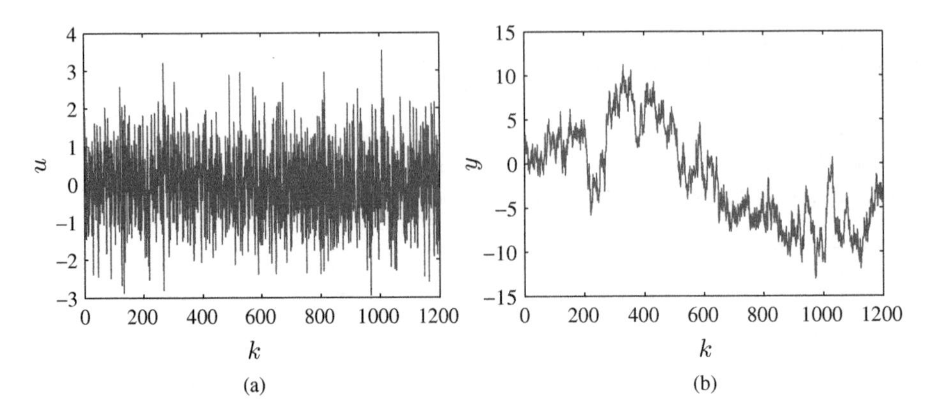

Figure 2.7 Sampled data for IndepV (a) and DepV (b).

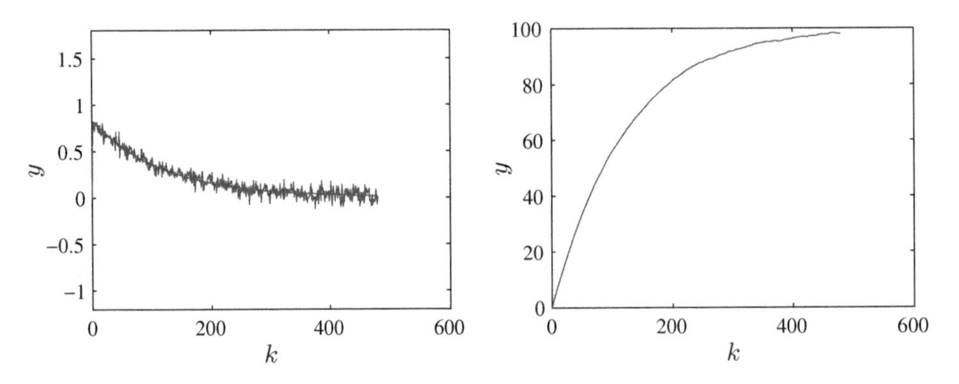

Figure 2.8 FIR and FSR obtained when data length is 1200 and model horizon is 480.

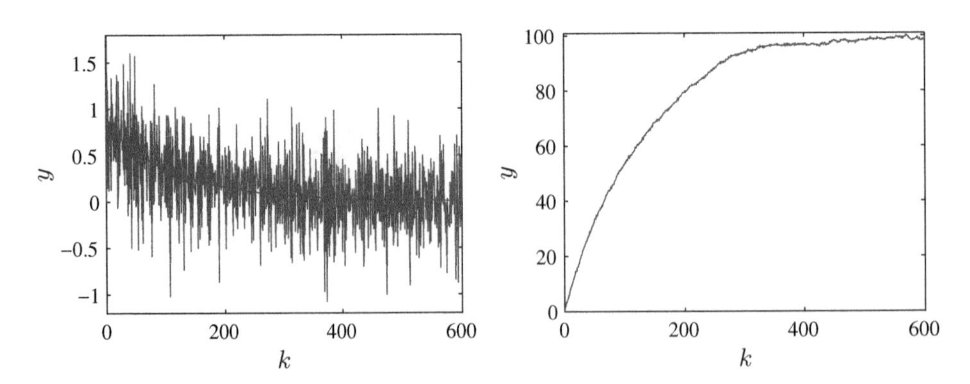

Figure 2.9 FIR and FSR obtained when data length is 1200 and model horizon is 600.

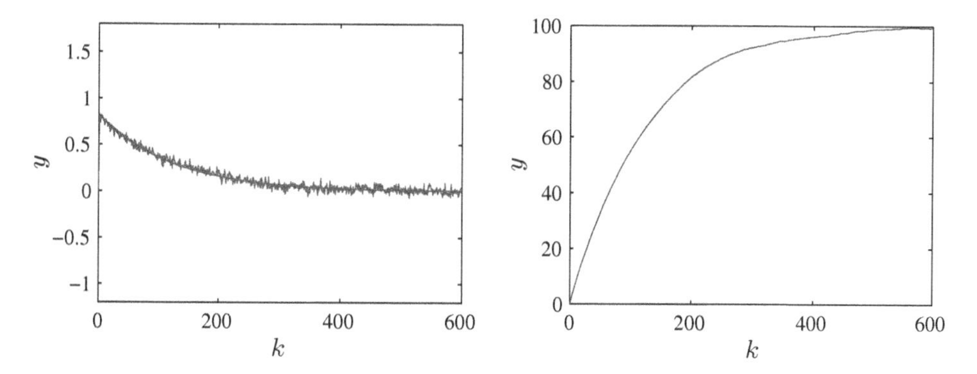

Figure 2.10 FIR and FSR obtained when data length is 1800 and model horizon is 600.

We can have the following simple conclusions: (i) the model horizon is an important parameter for FIR identification, and inappropriate choice of model horizon may lead to remarkable errors; (ii) when there are many parameters to be identified, the data length should be large enough in order to reduce error.

In this example, the purpose of choosing a large model horizon is to illustrate whether or not the identification is effective. When MPC is applied in the process industry, the sampling interval is usually in minutes; the model horizon is chosen as the longest time to steady state (TTSS) among all those for the pairs of {IndepV, DepV} in MIMO process; typically, the model horizon equals to 45 ∼ 120 sampling intervals.

2.2.4 Numerical Example

Consider the continuous-time transfer function of MV channel, $G^u(s) = \begin{bmatrix} G_a^u(s) & G_b^u(s) \\ 0 & G_c^u(s) \end{bmatrix}$, and that of DV channel, $G^f(s) = [G_a^f(s), G_b^f(s)]^T$, where

$$G_a^u(s) = \begin{bmatrix} \frac{4.05}{50s+1} & \frac{5.39}{50s+1} & \frac{6.88}{50s+1} & \frac{5.82}{50s+1} \\ \frac{1.77e^{-27s}}{60s+1} & \frac{2.49}{30s+1} & \frac{2.88e^{-18s}}{40s+1} & \frac{5.88}{50s+1} \\ 0 & 0 & 0 & 0 \\ \frac{3.57}{s(20s+1)} & \frac{6.91}{s(40s+1)} & \frac{2.09}{s(55s+1)} & \frac{4.23}{s(22s+1)} \\ 0 & 0 & 0 & 0 \\ \frac{5.88}{s(50s+1)} & \frac{5.72}{s(60s+1)} & \frac{2.54}{s(27s+1)} & \frac{2.38}{s(19s+1)} \end{bmatrix},$$

$$G_b^u(s) = \begin{bmatrix} \frac{5.88}{30s+1} & \frac{7.82}{60s+1} & \frac{5.88}{50s+1} & \frac{4.88}{20s+1} \\ \frac{4.88}{40s+1} & \frac{4.83}{50s+1} & \frac{5.38}{40s+1} & \frac{6.29}{15s+1} \\ \frac{7.91}{19s+1} & \frac{5.59}{44s+1} & \frac{3.58}{55s+1} & \frac{4.28}{35s+1} \\ \frac{8.53}{s(50s+1)} & \frac{2.38e^{-15s}}{s(19s+1)} & \frac{4.26}{s(22s+1)} & \frac{4.85}{s(30s+1)} \\ \frac{4.23}{15s+1} & \frac{7.62}{60s+1} & \frac{5.88}{25s+1} & \frac{7.53}{60s+1} \\ \frac{8.26}{s(60s+1)} & \frac{2.54}{s(27s+1)} & \frac{4.56}{s(20s+1)} & \frac{8.36}{s(22s+1)} \end{bmatrix},$$

$$G_c^u(s) = \begin{bmatrix} 0 & 0 & 0 & \frac{7.18}{55s+1} \\ 0 & 0 & 0 & \frac{7.15}{40s+1} \\ \frac{4.42}{15s+1} & \frac{4.23}{19s+1} & \frac{6.24}{34s+1} & \frac{4.26}{34s+1} \\ 0 & 0 & 0 & \frac{1}{s(55s+1)} \end{bmatrix}, G_a^f(s) = \begin{bmatrix} \frac{3.18}{32s+1} & \frac{7.35}{26s+1} & \frac{4.28}{34s+1} & \frac{8.53}{s(50s+1)} & \frac{4.65}{40s+1} & \frac{2.65}{s(30s+1)} \\ \frac{5.28}{28s+1} & \frac{7.35}{19s+1} & \frac{1.35}{20s+1} & \frac{4.59}{s(26s+1)} & \frac{7.26}{50s+1} & \frac{7.53e^{-22s}}{s(50s+1)} \end{bmatrix},$$

$$G_b^f(s) = \begin{bmatrix} \frac{6.54}{40s+1} & \frac{4.28}{26s+1} & \frac{9.26}{50s+1} & \frac{5.84}{s(26s+1)} \\ \frac{5.36}{20s+1} & \frac{7.53e^{-23s}}{50s+1} & \frac{9.26}{26s+1} & \frac{4.26}{s(60s+1)} \end{bmatrix}.$$

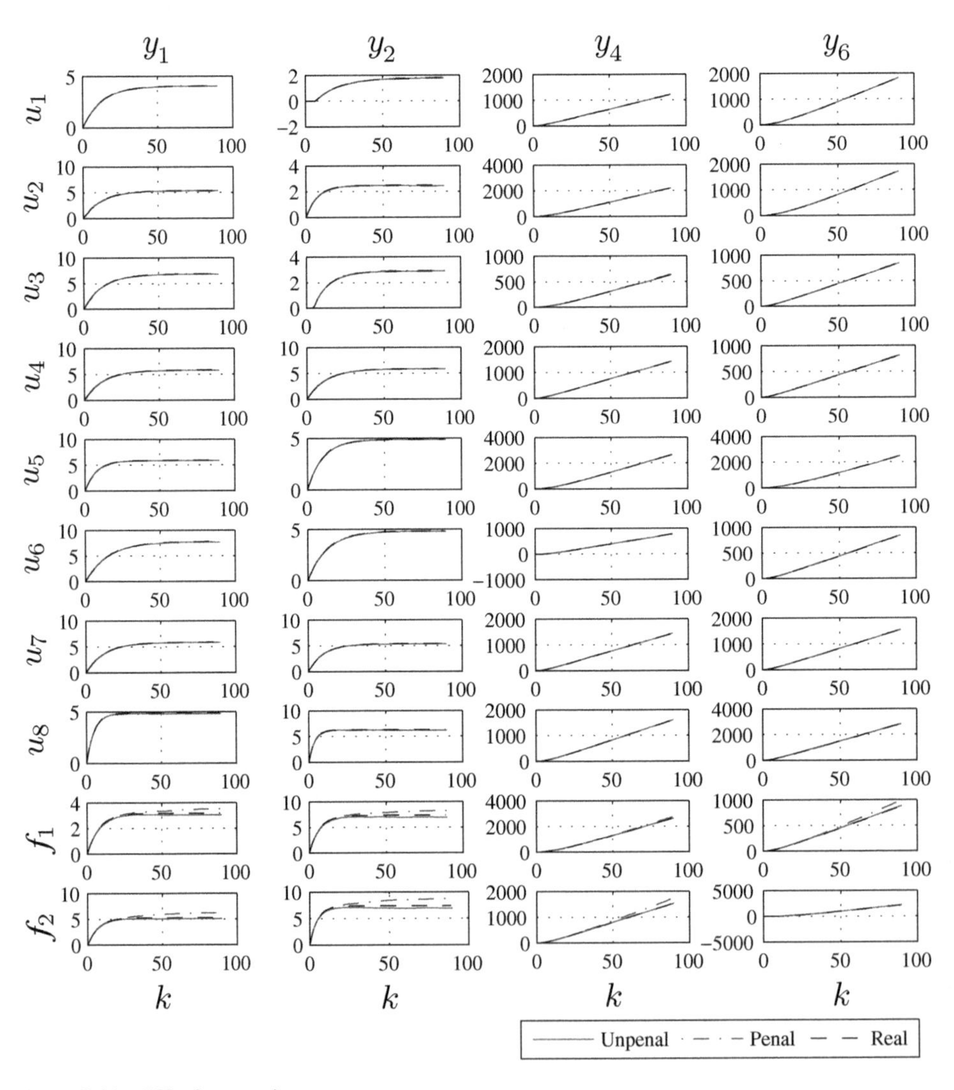

Figure 2.11 FSRs for case 1.

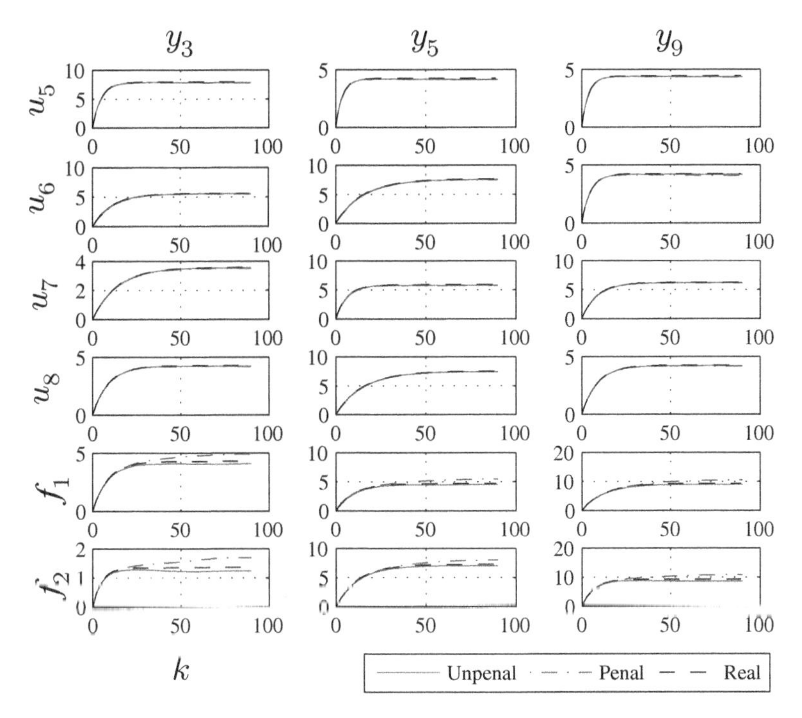

Figure 2.12 FSRs for case 2.

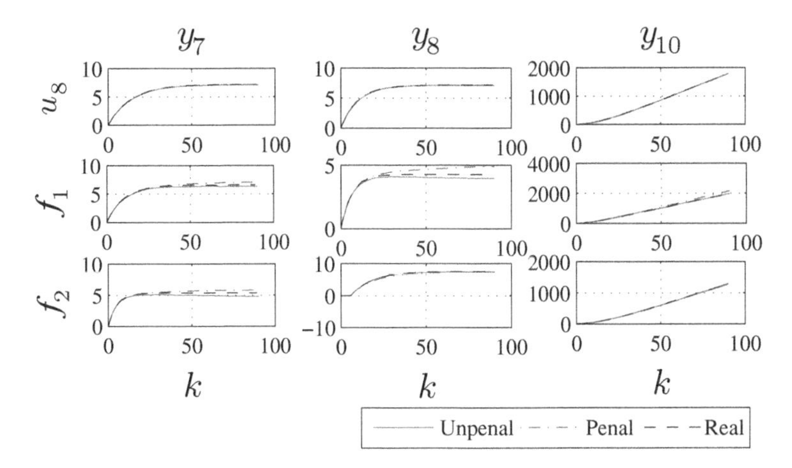

Figure 2.13 FSRs for case 3.

There are causal relationships between the {1st, 2nd, 4th, 6th} CVs and the {1st–8th} MVs, between the {3rd, 5th, 9th} CVs and the {5th–8th} MV, and between the {7th, 8th, 10th} CVs and the 8th MV. There is the causal relationship between all CVs and all DVs. We take three cases for the whole identification. The {4th, 6th, 10th} CVs are integral.

The MV test signal is a low-pass GBN; each MV has a different GBN, and 8 MVs are tested simultaneously. The switch probability of GBN is $\frac{1}{10}$, and the amplitude is 1.

$T_{\min} = T_s = 4$. A total of eight independent GBN are produced. The data length is $L = 3000$. The disturbance is taken as Gaussian white noise sequence with a variance of 0.25. Build the simulation model through the MATLAB Simulink, and then run it to obtain CV data.

Take the smoother coefficients $\alpha_{y,j} = \alpha_{u,j} = \alpha_{f,j} = 1$, the smooth factor $\mu = 5$, and the model horizon $N = 90$. Utilize LS method to obtain FSR. FSRs for the three cases are shown in Figures 2.11–2.13, respectively; in the figures, "unpenal," "penal," and "real" denote the identified FSR with $\mu = 0$, the identified FSR with $\mu = 5$, and the true FSR, respectively.

2.3 Prediction Based on Step Response Model and Kalman Filter

Consider a dynamic system described by the following state space model:

$$x(k+1) = \Phi x(k) + Bu(k) + \Gamma\eta(k), \tag{2.47}$$

$$y(k) = Hx(k) + \xi(k), \tag{2.48}$$

where $x(k) \in \mathbb{R}^{n_x}$ is the system state, $u(k) \in \mathbb{R}^{n_u}$ is the system input, $y(k) \in \mathbb{R}^{n_y}$ is the measured output, $\eta(k) \in \mathbb{R}^{n_\eta}$ is the input noise, and $\xi(k) \in \mathbb{R}^{n_y}$ is the measurement noise. Equation (2.47) is called the state equation, and (2.48) is the measurement equation. $\Phi, \Gamma, B,$ and H are known matrices. Φ is the state transition matrix. H is the observation matrix.

Assumption 2.1 $\eta(k)$ and $\xi(k)$ are uncorrelated white noises, whose variance matrices are Q and R, respectively. Moreover,

$$\mathrm{E}[\eta(k)] = 0, \ \ \mathrm{E}[\xi(k)] = 0, \tag{2.49}$$

$$\mathrm{E}[\eta(k)\eta^T(j)] = Q\kappa_{kj}, \ \ \mathrm{E}[\xi(k)\xi^T(j)] = R\kappa_{kj}, \tag{2.50}$$

$$\mathrm{E}[\eta(k)\xi^T(j)] = 0, \ \ \forall k, j, \tag{2.51}$$

where $\kappa_{kk} = 1$ for all k, and $\kappa_{kj} = 0$ for $k \neq j$.

Assumption 2.2 $x(0)$ is uncorrelated with both $\eta(k)$ and $\xi(k)$. Moreover,

$$\mathrm{E}[x(0)] = \mu_0, \ \ \mathrm{E}[(x(0) - \mu_0)(x(0) - \mu_0)^T] = P_0. \tag{2.52}$$

Assumption 2.3 $u(k) \in \mathbb{R}^{n_u}$ is either known deterministic (non-stochastic), or a linear function of $\{y(k), y(k-1), \ldots; u(k-1), u(k-2), \ldots\}$.

When H is an invertible square matrix, we can take $\mu_0 = H^{-1}y(0)$. In general, H is non-invertible; the basic Kalman filter (KF) gives the linear minimum-variance estimate of the state $x(j)$, say $\hat{x}(j|k)$, based on the measurements $\{y(1), \ldots, y(k)\}$. KF minimizes the performance index

$$J = \mathrm{E}[(x(j) - \hat{x}(j|k))^T(x(j) - \hat{x}(j|k))]. \tag{2.53}$$

For $j > k$ and $j < k$, KF is called Kalman predictor and Kalman smoother, respectively.

2.3.1 Steady-State Kalman Filter and Predictor

If the system (2.47) and (2.48) is observable and controllable, then ssKF (the steady-state KF) exists, which is asymptotically stable. The observability condition of (2.47) and (2.48) is

$$\text{rank}\begin{bmatrix} H \\ H\Phi \\ \vdots \\ H\Phi^{n_x-1} \end{bmatrix} = n_x. \tag{2.54}$$

The controllability condition of (2.47) and (2.48) is

$$\text{rank}[\Gamma L, \Phi\Gamma L, \dots, \Phi^{n_x-1}\Gamma L] = n_x, \tag{2.55}$$

where L satisfies $Q = LL^T$. For the controller designer, it should be noted that the controllability here is for input noise. Furthermore, observability can be weakened as detectability.

Theorem 2.1 Consider the system (2.47) and (2.48) with Assumptions 2.1–2.3. Suppose the system is observable and controllable. Then, the steady-state Riccati equation

$$\Sigma = \Phi[\Sigma - \Sigma H^T(H\Sigma H^T + R)^{-1}H\Sigma]\Phi^T + \Gamma Q\Gamma^T \tag{2.56}$$

has a unique positive-definite solution Σ, which satisfies

$$\Sigma = \Phi P\Phi^T + \Gamma Q\Gamma^T \tag{2.57}$$

$$K = \Sigma H^T\left(H\Sigma H^T + R\right)^{-1} \tag{2.58}$$

$$P = (I - KH)\Sigma(I - KH)^T + KRK^T. \tag{2.59}$$

The ssKF is

$$\hat{x}(k|k) = \Psi_f\hat{x}(k-1|k-1) + \Upsilon_f u(k-1) + Ky(k), \tag{2.60}$$

$$\Psi_f = (I - KH)\,\Phi,\ \ \Upsilon_f = (I - KH)\,B, \tag{2.61}$$

where K is called ssKF gain. The steady-state Kalman predictor is

$$\hat{x}(k+1|k) = \Phi\hat{x}(k|k-1) + Bu(k) + K_p\varepsilon(k), \tag{2.62}$$

$$\varepsilon(k) = y(k) - H\hat{x}(k|k-1), \tag{2.63}$$

$$K_p = \Phi K, \tag{2.64}$$

where K_p is called the steady-state Kalman predictor gain. The steady-state Kalman predictor is also denoted as

$$\hat{x}(k+1|k) = \Psi_p\hat{x}(k|k-1) + Bu(k) + K_p y(k), \tag{2.65}$$

$$\Psi_p = \Phi - K_pH = \Phi(I - KH). \tag{2.66}$$

The above transition matrices Ψ_f and Ψ_p are stable matrices with the same eigenvalues. Thus, the steady-state KF and Kalman predictor are asymptotically stable.

Theorem 2.2 Consider the system (2.47) and (2.48) with Assumptions 2.1–2.3. Suppose the system is observable and controllable. Then, the m-step ahead steady-state Kalman predictor is

$$\hat{x}(k+m|k) = \Phi^{m-1}\hat{x}(k+1|k) + \sum_{j=2}^{m}\Phi^{m-j}Bu(k+j-1),\ m > 1, \tag{2.67}$$

where $\hat{x}(k+1|k)$ is the same as in the steady-state Kalman predictor.

2.3.2 Steady-State Kalman Filter and Predictor Based on Step Response Model

For sCV (the stable CV), consider the following FIR model:

$$y^s(k) = \sum_{i=1}^{N'}(H_i^{u,s}u(k-i) + H_i^{f,s}f(k-i)), \tag{2.68}$$

where the superscript "s" stands for stable; $H_i^{u,s}$ and $H_i^{f,s}$ are FIR coefficient matrices for u and f, respectively. Suppose $H_i^{u,s} = 0$, $H_i^{f,s} = 0$ for all $i > N$, where N is the model horizon, and $N' \geq N$. Applying (2.68) yields

$$\begin{aligned} y^s(k) = {} & \sum_{i=1}^{N'}\left[H_i^{u,s}\left(\sum_{j=1}^{N'-i}\Delta u(k-N'+j) + u(k-N')\right)\right. \\ & \left. + H_i^{f,s}\left(\sum_{j=1}^{N'-i}\Delta f(k-N'+j) + f(k-N')\right)\right] \\ = {} & \sum_{i=1}^{N'-1}\left(\sum_{j=1}^{i}H_j^{u,s}\right)\Delta u(k-i) + \sum_{i=1}^{N'-1}\left(\sum_{j=1}^{i}H_j^{f,s}\right)\Delta f(k-i) \\ & + \left(\sum_{j=1}^{N'}H_j^{u,s}\right)u(k-N') + \left(\sum_{j=1}^{N'}H_j^{f,s}\right)f(k-N'). \end{aligned} \tag{2.69}$$

Using (2.69), we obtain the following FSR model (equivalent for all $N' \geq N$):

$$\begin{aligned} y^s(k) = {} & \sum_{i=1}^{N'-1} S_i^{u,s}\Delta u(k-i) + S_{N'}^{u,s}u(k-N') \\ & + \sum_{i=1}^{N'-1} S_i^{f,s}\Delta f(k-i) + S_{N'}^{f,s}f(k-N'), \end{aligned} \tag{2.70}$$

where $S_i^{u,s}$ and $S_i^{f,s}$ are FSR coefficient matrices for u and f, respectively, satisfying $S_{N+i}^{u,s} = S_N^{u,s}$ and $S_{N+i}^{f,s} = S_N^{f,s}$ for all $i \geq 0$.

Suppose, at time k, before obtaining $\Delta u(k)$, $\Delta f(k)$ has been obtained. Let $y^{s,\text{fr}}(k+p|k)$ be the prediction of $y^s(k+p|k)$ under $\Delta u(k+i-1|k) = 0$, $1 \leq i \leq p$ and $\Delta f(k+i) = 0$. $y^{s,\text{fr}}(k+p|k)$ is called the free prediction. The superscript "fr" stands for "free." The free predictions satisfy $y^{s,\text{fr}}(k+i|k) = y^{s,\text{fr}}(k+N|k)$ for all $i \geq N$. Making the free predictions for (2.70), yields

$$y^{s,\text{fr}}(k+p|k) - y^{s,\text{fr}}(k+p|k-1) = S_{p+1}^{u,s}\Delta u(k-1) + S_p^{f,s}\Delta f(k),\ p \geq 0, \tag{2.71}$$

which satisfies $y^{s,\text{fr}}(k|k) = y^s(k)$.

For iCV (the first-order integral CV), we use the increments of {CV, FIR coefficient, FSR coefficient}, i.e., (2.68) and (2.70) are replaced by

$$\begin{aligned}\Delta y^r(k) &= \sum_{i=1}^{N'}(\Delta H_i^{u,r}u(k-i) + \Delta H_i^{f,r}f(k-i)) \\ &= \sum_{i=1}^{N'-1}\Delta S_i^{u,r}\Delta u(k-i) + \Delta S_{N'}^{u,r}u(k-N') \\ &+ \sum_{i=1}^{N'-1}\Delta S_i^{f,r}\Delta f(k-i) + \Delta S_{N'}^{f,r}f(k-N'),\end{aligned} \tag{2.72}$$

where the superscript "r" stands for "ramp" (ramp is another name for first-order integral); $H_i^{u,r}$ and $H_i^{f,r}$ are FIR coefficient matrices for u and f, respectively. Suppose $H_{i+1}^{u,r} = H_i^{u,r}$ and $H_{i+1}^{f,r} = H_i^{f,r}$ for all $i > N$, so $S_{N+i+1}^{f,r} - S_{N+i}^{f,r} = S_{N+1}^{f,r} - S_N^{f,r}$ and $S_{N+i+1}^{u,r} - S_{N+i}^{u,r} = S_{N+1}^{u,r} - S_N^{u,r}$ for all $i \geq 0$. Δ cannot be removed in both sides of (2.72), since iCV cannot be represented as the weighted sum of a finite number of MVs and DVs. Accordingly, (2.71) should be replaced by

$$\Delta y^{r,\mathrm{fr}}(k+p|k) - \Delta y^{r,\mathrm{fr}}(k+p|k-1) = \Delta S_{p+1}^{u,r}\Delta u(k-1) + \Delta S_p^{f,r}\Delta f(k), \tag{2.73}$$

which satisfies $y^{r,\mathrm{fr}}(k|k) = y^r(k)$. Equation (2.73) is equivalent to

$$y^{r,\mathrm{fr}}(k+p|k) - y^{r,\mathrm{fr}}(k+p|k-1) = S_{p+1}^{u,r}\Delta u(k-1) + S_p^{f,r}\Delta f(k). \tag{2.74}$$

Thus, both (2.74) and (2.71) have the same form. The free prediction of iCV satisfies $y^{r,\mathrm{fr}}(k+i+1|k) - y^{r,\mathrm{fr}}(k+i|k) = y^{r,\mathrm{fr}}(k+N+1|k) - y^{r,\mathrm{fr}}(k+N|k)$ for all $i \geq N$.

2.3.2.1 Open-Loop Prediction of Stable CV

Denote

$$\tilde{Y}_{N+1}^{s,\mathrm{fr}}(k) \triangleq \begin{bmatrix} y^{s,\mathrm{fr}}(k|k) \\ y^{s,\mathrm{fr}}(k+1|k) \\ \vdots \\ y^{s,\mathrm{fr}}(k+N|k) \end{bmatrix} \triangleq \begin{bmatrix} y^{s,\mathrm{fr}}(k|k) \\ Y_N^{s,\mathrm{fr}}(k) \end{bmatrix} = \begin{bmatrix} \tilde{Y}_N^{s,\mathrm{fr}}(k) \\ y^{s,\mathrm{fr}}(k+N|k) \end{bmatrix}.$$

Applying (2.71) yields

$$\tilde{Y}_{N+1}^{s,\mathrm{fr}}(k+1) = M^s\left\{\tilde{Y}_{N+1}^{s,\mathrm{fr}}(k) + \begin{bmatrix} 0 \\ S_1^{u,s} \\ \vdots \\ S_N^{u,s} \end{bmatrix}\Delta u(k)\right\} + \begin{bmatrix} 0 \\ S_1^{f,s} \\ \vdots \\ S_N^{f,s} \end{bmatrix}\Delta f(k+1) \tag{2.75}$$

$$y^s(k) = \tilde{H}\tilde{Y}_{N+1}^{s,\mathrm{fr}}(k), \tag{2.76}$$

where $M^s = \begin{bmatrix} 0 & I \\ 0 & \begin{bmatrix} 0 & \cdots & 0 & I \end{bmatrix} \end{bmatrix}$, $\tilde{H} = \begin{bmatrix} I & 0 & \cdots & 0 \end{bmatrix}$. In order to cater for (2.47) and (2.48), artificially add $\Gamma\eta(k)$ and $\xi(k)$ in (2.75) and (2.76), then we obtain

$$\tilde{Y}_{N+1}^{s,\mathrm{fr}}(k+1) = M^s \tilde{Y}_{N+1}^{s,\mathrm{fr}}(k) + M^s \begin{bmatrix} 0 \\ S_1^{u,s} \\ \vdots \\ S_N^{u,s} \end{bmatrix} \Delta u(k) + \begin{bmatrix} 0 \\ S_1^{f,s} \\ \vdots \\ S_N^{f,s} \end{bmatrix} \Delta f(k+1) + \Gamma\eta(k), \tag{2.77}$$

$$y^s(k) = \tilde{H}\tilde{Y}_{N+1}^{s,\mathrm{fr}}(k) + \xi(k). \tag{2.78}$$

Analogy of (2.77) and (2.78) to (2.47) and (2.48) is shown in Table 2.1.

For (2.77) and (2.78), apply ssKF in Theorem 2.1. Denote the state estimation as

$$\tilde{Y}_{N+1}^{s,\mathrm{ol}}(k|k) \triangleq \begin{bmatrix} y^{s,\mathrm{ol}}((k|k)|k) \\ y^{s,\mathrm{ol}}((k+1|k)|k) \\ \vdots \\ y^{s,\mathrm{ol}}((k+N|k)|k) \end{bmatrix} \triangleq \begin{bmatrix} y^{s,\mathrm{ol}}(k|k) \\ y^{s,\mathrm{ol}}(k+1|k) \\ \vdots \\ y^{s,\mathrm{ol}}(k+N|k) \end{bmatrix},$$

$$\tilde{Y}_{N+1}^{s,\mathrm{ol}}(k|k-1) \triangleq \begin{bmatrix} y^{s,\mathrm{ol}}((k|k)|k-1) \\ y^{s,\mathrm{ol}}((k+1|k)|k-1) \\ \vdots \\ y^{s,\mathrm{ol}}((k+N|k)|k-1) \end{bmatrix}.$$

Table 2.1 Analogy of (2.77) and (2.78) to (2.47) and (2.48).

Item in (2.77) and (2.78)	Item in (2.47) and (2.48)
$\tilde{Y}_{N+1}^{s,\mathrm{fr}}(k+1)$	$x(k+1)$
$\tilde{Y}_{N+1}^{s,\mathrm{fr}}(k)$	$x(k)$
M^s	Φ
$M^s \begin{bmatrix} 0 \\ S_1^{u,s} \\ \vdots \\ S_N^{u,s} \end{bmatrix} \Delta u(k) + \begin{bmatrix} 0 \\ S_1^{f,s} \\ \vdots \\ S_N^{f,s} \end{bmatrix} \Delta f(k+1)$	$Bu(k)$
$y^s(k)$	$y(k)$
$\tilde{H}$	H

The ssKF is

$$\begin{aligned}\tilde{Y}^{s,\text{ol}}_{N+1}(k|k) \\ = M^s\left\{\tilde{Y}^{s,\text{ol}}_{N+1}(k-1|k-1) + \begin{bmatrix} 0 \\ S^{u,s}_1 \\ \vdots \\ S^{u,s}_N \end{bmatrix} \Delta u(k-1)\right\} + \begin{bmatrix} 0 \\ S^{f,s}_1 \\ \vdots \\ S^{f,s}_N \end{bmatrix} \Delta f(k) + \tilde{K}\varepsilon^s(k)\end{aligned} \tag{2.79}$$

$$\varepsilon^s(k) = y^s(k) - y^{s,\text{ol}}(k|k-1) - S^{u,s}_1 \Delta u(k-1), \tag{2.80}$$

where $\tilde{K} = \Sigma\tilde{H}^T(\tilde{H}\Sigma\tilde{H}^T + R)^{-1}$, and $\varepsilon^s(k)$ is calculated by $\varepsilon^s(k) = y^s(k) - \tilde{H}\tilde{Y}^{s,\text{ol}}_{N+1}(k|k-1) = y^s(k) - y^{s,\text{ol}}((k|k)|k-1)$. Take the following part of (2.79):

$$\begin{aligned}Y^{s,\text{ol}}_N(k|k) \\ = M^s\left\{Y^{s,\text{ol}}_N(k-1|k-1) + \begin{bmatrix} S^{u,s}_1 \\ S^{u,s}_2 \\ \vdots \\ S^{u,s}_N \end{bmatrix} \Delta u(k-1)\right\} + \begin{bmatrix} S^{f,s}_1 \\ S^{f,s}_2 \\ \vdots \\ S^{f,s}_N \end{bmatrix} \Delta f(k) + K_1\varepsilon^s(k),\end{aligned} \tag{2.81}$$

where

$$M^s = \begin{bmatrix} 0 & I \\ 0 & [0 \cdots 0 \; I] \end{bmatrix}, \quad Y^{s,\text{ol}}_N(k|k) \triangleq \begin{bmatrix} y^{s,\text{ol}}(k+1|k) \\ y^{s,\text{ol}}(k+2|k) \\ \vdots \\ y^{s,\text{ol}}(k+N|k) \end{bmatrix}, \quad \tilde{K} = \begin{bmatrix} K_0 \\ K_1 \end{bmatrix}.$$

In a word, ol prediction is composed of (2.80)–(2.81).

Remark 2.4 By taking the following state-space model:

$$Y^{s,\text{fr}}_N(k+1) = M^s\left\{Y^{s,\text{fr}}_N(k) + \begin{bmatrix} S^{u,s}_1 \\ S^{u,s}_2 \\ \vdots \\ S^{u,s}_N \end{bmatrix} \Delta u(k)\right\} + \begin{bmatrix} S^{f,s}_1 \\ S^{f,s}_2 \\ \vdots \\ S^{f,s}_N \end{bmatrix} \Delta f(k+1), \tag{2.82}$$

$$y^s(k) = HY^{s,\text{fr}}_N(k) + y^{s,\text{fr}}(k|k-1) + S^{u,s}_1 \Delta u(k-1), \tag{2.83}$$

where $H = 0$, we can also obtain (2.80)–(2.81). However, since $H = 0$, the detectability does not hold. Equation (2.82), without f, has been given in e.g. Lee and Xiao [2000].

In Lee et al. [1994] and Lundström et al. [1995], it is equivalent to take the state as $\tilde{Y}^{s,\text{fr}}_N(k)$ and directly take (2.79) and (2.80) ($N+1$ being replaced by N) as ol prediction; this approach is inconsistent with ol prediction in Chapter 4.

2.3.2.2 Open-Loop Prediction of Integral CV

Denote

$$\tilde{Y}_{N+2}^{r,\mathrm{fr}}(k) \triangleq \begin{bmatrix} y^{r,\mathrm{fr}}(k|k) \\ y^{r,\mathrm{fr}}(k+1|k) \\ \vdots \\ y^{r,\mathrm{fr}}(k+N+1|k) \end{bmatrix} \triangleq \begin{bmatrix} y^{r,\mathrm{fr}}(k|k) \\ Y_{N+1}^{r,\mathrm{fr}}(k) \end{bmatrix} = \begin{bmatrix} \tilde{Y}_{N+1}^{r,\mathrm{fr}}(k) \\ y^{r,\mathrm{fr}}(k+N+1|k) \end{bmatrix}.$$

Applying (2.74) yields

$$\tilde{Y}_{N+2}^{r,\mathrm{fr}}(k+1) = M^r \left\{ \tilde{Y}_{N+2}^{r,\mathrm{fr}}(k) + \begin{bmatrix} 0 \\ S_1^{u,r} \\ \vdots \\ S_{N+1}^{u,r} \end{bmatrix} \Delta u(k) \right\} + \begin{bmatrix} 0 \\ S_1^{f,r} \\ \vdots \\ S_{N+1}^{f,r} \end{bmatrix} \Delta f(k+1), \tag{2.84}$$

$$y^r(k) = \tilde{H}\tilde{Y}_{N+2}^{r,\mathrm{fr}}(k), \tag{2.85}$$

where $M^r = \begin{bmatrix} 0 & I \\ 0 & \begin{bmatrix} 0 & \cdots & 0 & -I & 2I \end{bmatrix} \end{bmatrix}$, $\tilde{H} = \begin{bmatrix} I & 0 & \cdots & 0 \end{bmatrix}$. Analogy of (2.84) and (2.85) to (2.47) and (2.48) is shown in Table 2.2.

For (2.84) and (2.85), apply ssKF in Theorem 2.1. Similarly to sCV, denote the state estimation as

$$Y_{N+1}^{r,\mathrm{ol}}(k|k) \triangleq \begin{bmatrix} y^{r,\mathrm{ol}}((k+1|k)|k) \\ y^{r,\mathrm{ol}}((k+2|k)|k) \\ \vdots \\ y^{r,\mathrm{ol}}((k+N+1|k)|k) \end{bmatrix} \triangleq \begin{bmatrix} y^{r,\mathrm{ol}}(k+1|k) \\ y^{r,\mathrm{ol}}(k+2|k) \\ \vdots \\ y^{r,\mathrm{ol}}(k+N+1|k) \end{bmatrix}.$$

Table 2.2 Analogy of (2.84) and (2.85) to (2.47) and (2.48).

Item in (2.84) and (2.85)	Item in (2.47) and (2.48)
$\tilde{Y}_{N+2}^{r,\mathrm{fr}}(k+1)$	$x(k+1)$
$\tilde{Y}_{N+2}^{r,\mathrm{fr}}(k)$	$x(k)$
M^r	Φ
$M^r \begin{bmatrix} 0 \\ S_1^{u,r} \\ \vdots \\ S_{N+1}^{u,r} \end{bmatrix} \Delta u(k) + \begin{bmatrix} 0 \\ S_1^{f,r} \\ \vdots \\ S_{N+1}^{f,r} \end{bmatrix} \Delta f(k+1)$	$Bu(k)$
$y^r(k)$	$y(k)$
$\tilde{H}$	H

The ssKF is

$$Y_{N+1}^{r,\mathrm{ol}}(k|k) = M^r \left\{ Y_{N+1}^{r,\mathrm{ol}}(k-1|k-1) + \begin{bmatrix} S_1^{u,r} \\ S_2^{u,r} \\ \vdots \\ S_{N+1}^{u,r} \end{bmatrix} \Delta u(k-1) \right\} + \begin{bmatrix} S_1^{f,r} \\ S_2^{f,r} \\ \vdots \\ S_{N+1}^{f,r} \end{bmatrix} \Delta f(k) + K_1 \varepsilon^r(k) \tag{2.86}$$

$$\varepsilon^r(k) = y^r(k) - y^{r,\mathrm{ol}}(k|k-1) - S_1^{u,r} \Delta u(k-1). \tag{2.87}$$

Remark 2.5 In Lee et al. [1994] and Lundström et al. [1995], it is equivalent to take the state as $\tilde{Y}_N^{s,\mathrm{fr}}(k)$ and directly apply Theorem 2.1 to obtain ol prediction $\tilde{Y}_N^{r,\mathrm{ol}}(k)$ ($\tilde{Y}_N^{r,\mathrm{ol}}(k)$ being no longer divided into two parts); this approach is inconsistent with ol prediction in Chapter 5.

Lee and Xiao [2000] has given the following equation:

$$Y_{N+1}^{r,\mathrm{fr}}(k+1) = M^r \left\{ Y_{N+1}^{r,\mathrm{fr}}(k) + \begin{bmatrix} S_1^{u,r} \\ S_2^{u,r} \\ \vdots \\ S_{N+1}^{u,r} \end{bmatrix} \Delta u(k) \right\}. \tag{2.88}$$

Same as sCV, one cannot use (2.88) to derive ssKF.

For an engineering application, as long as the model horizon N is strictly greater than TTSS (the maximum TTSS among all sCVs and iCVs), in (2.86) and (2.87), $N+1$ can also be replaced with N and ol prediction for sCV is a special case of that for iCV.

3

Steady-State Target Calculation

In practical applications, real-time optimization (RTO), as a high layer, gives the ideal value of some CVs and MVs. Another layer, higher than RTO, may have more broad-sighted scheduling/optimization. MPC itself usually contains three modules, i.e., prediction, steady-state target calculation (SSTC), and dynamic control (DC). SSTC calculates ss (the steady-state targets) of MV, state or CV in each control interval, taking into account the ideal value provided by RTO. Multivariable predictive controller (MPC) containing SSTC and DC is called two-layered MPC, since SSTC likes an upper layer of DC. In the context of two-layered MPC, the ideal value given by RTO is renamed the external target (ET), the ideal target, or the expected target. SSTC and RTO not only adopt different models, but also have different execution periods. The period of RTO is variable, being much larger than that of SSTC.

SSTC is necessary because the disturbances or the operator interventions, during any control period, may change the optimum ss. The main role of SSTC is to track the result of RTO and to optimize for economic purposes. MPC software package with economic optimization can automatically find, in the neighborhood the equilibrium point, the optimal setpoint for DC.

SSTC calculates the locally optimal ss. It combines the steady-state mathematical model of the process (being consistent with the dynamic model used by MPC), considers the input and output constraints, defines the objective function according to the specific requirement, and finally forms the linear programming (LP) or the quadratic programming (QP). For implementing SSTC, it is necessary to categorize MV into the cost MV and minimum-move MV and CV into sCV (the stable CV) and iCV (the integral CV).

Notations of this chapter: $y \in \mathbb{R}^{n_y}$ ($u \in \mathbb{R}^{n_u}, f \in \mathbb{R}^{n_f}$) denotes CV (MV, DV). y_{ss} (u_{ss}, f_{ss}) denotes ss of CV (MV, DV). y_t (u_t) denotes CVET (MVET). $\|x\|_Q^2 \triangleq x^T Q x$.

3.1 RTO and External Target

The triplet {prediction, SSTC, dynamic control} is called MPC. Here, we briefly discuss the relationship between RTO and MPC. The mature RTO technology mainly adopts the steady-state first-principle model of the process. Some parameters of this model need to

Model Predictive Control, First Edition. Baocang Ding and Yuanqing Yang.

be updated before each RTO implementation. The basic paradigm for update is using the measured data to match the model with the true process. RTO is a constrained nonlinear optimization problem whose constraints are physical and related to the current operating point. For example, the physical constraints of variables in SSTC may be reflected in the optimization problem of RTO, which can ensure that the optimum of RTO is feasible for SSTC under the current operation.

However, the aims of RTO and SSTC are different. Each input of RTO is unnecessarily the input of MPC; each output of RTO is unnecessarily the output of MPC. RTO adopts an objective function chosen by the user. The result of RTO is sent to SSTC, so that some variables of SSTC have ETs. Usually, ET is close to the constraint boundary of the corresponding variable, because it is at the boundary where the economic benefit gets the biggest.

Since ET is often close to the constraint boundary, the most suitable method for tracking RTO results is MPC. MPC is the most suitable method for the constrained multivariable control. ET requires the controller to be able to track it while satisfying various constraints. Other control strategies (e.g., PID) are difficult to handle the constraints. SSTC lies between RTO and DC, being serve as a coordinator. If it tracks ET directly in DC, then SSTC can be simpler (without tracking ET). Besides tracking ET, SSTC has more roles, such as handling sps for the variables without ET and relaxing unsatisfiable constraints. SSTC gives several answers, such as where sp of dynamic control comes from and how ss is feasible.

Moreover, ET can come not only from RTO, but also from other channels, as long as ET is confirmed. Some variables can have ideal rest value (IRV), which is a kind of ET. ET is for either CV or MV. Since all MVs do not have ET, CVET cannot be determined from MVET. Even if MVET and CVET are inconsistent (i.e., not satisfying the steady-state model for SSTC), SSTC still handles.

3.2 Economic Optimization and Target Tracking Problem

For ol sCV, consider the following steady-state model:

$$\nabla y(\infty) = G^u \nabla u(\infty) + G^f \nabla f(\infty), \tag{3.1}$$

where G^u and G^f are steady-state gain matrices, ∞ denote the steady state. ∇ is the difference relative to the equilibrium point. Since DV is a time series, and SSTC runs in every control period, the result obtained by SSTC is also time-varying, thus (3.1) is rewritten as

$$\nabla y_{ss}(k) = G^u \nabla u_{ss}(k) + G^f \nabla f(k), \tag{3.2}$$

where it is assumed that the disturbance f does not change in the future.

Define $y_{ss}^{ol}(k)$ as ol steady-state prediction of CV when the current and future DV and MV no longer change. Thus,

$$\nabla y_{ss}^{ol}(k) = G^u \nabla u(k-1) + G^f \nabla f(k-1). \tag{3.3}$$

Subtracting (3.2) by (3.3), yields

$$\delta y_{ss}(k) = G^u \delta u_{ss}(k) + G^f \delta f_{ss}(k), \tag{3.4}$$

where $\delta y_{ss}(k) = y_{ss}(k) - y_{ss}^{ol}(k)$, $\delta u_{ss}(k) = u_{ss}(k) - u(k-1)$, $\delta f_{ss}(k) = f(k) - f(k-1)$. Taking the difference on both sides of (3.2), yields

$$\Delta y_{ss}(k) = G^u \Delta u_{ss}(k) + G^f \Delta f_{ss}(k), \tag{3.5}$$

where

$$\Delta u_{ss}(k) = u_{ss}(k) - u_{ss}(k-1), \tag{3.6}$$

$$\Delta y_{ss}(k) = y_{ss}(k) - y_{ss}(k-1), \tag{3.7}$$

$$\Delta f_{ss}(k) = f(k) - f(k-1). \tag{3.8}$$

3.2.1 Economic Optimization

3.2.1.1 Optimization Problem

In economic optimization, the design of the objective function should directly reflect the economic loss of the production process. The usual objective function (e.g., see [Kassmann et al., 2000, Nikandrov and Swartz, 2009]) is

$$J = \alpha^{\mathrm{T}} \Delta u_{ss}(k) + \beta^{\mathrm{T}} \Delta y_{ss}(k), \tag{3.9}$$

which reflects the benefit or cost produced by changes in MV and CV. Since $\Delta u_{ss}(k)$ and $\Delta y_{ss}(k)$ are linearly coupled, the increments of MV and CV in (3.9) can be unified as the increment of MV.

For engineering problems, a primary task is to standardize the benefits or costs produced by the unitary change of MV, and then use the standardized parameter h to represent the benefits or costs of all MVs. The symbol $\pm$ distinguishes between costs and benefits, i.e., $+$ stands for the cost and $-$ for the benefit. Therefore, the objective function should be minimized. If an increment of an MV can affect the benefits or costs of the production process, then this MV is called the cost MV in SSTC. An MV, being different from the cost MV, is a minimum-move MV. In summary, the objective function that reflects the economic loss becomes

$$J = h\Delta u_{ss}(k), \tag{3.10}$$

where $h = [\, h_1 \;\, h_2 \;\, \cdots \;\, h_{n_u} \,]$. h is the cost coefficient vector constructed based on the standardized benefit or cost of each MV. The objective function reflects the change in the benefit or cost that occurs after change of MV.

SSTC considers the steady-state constraints on MV and CV, i.e.,

$$\begin{cases} u_{\min} \le u_{ss}(k) \le u_{\max} \\ y_{\min} \le y_{ss}(k) \le y_{\max} \end{cases}, \tag{3.11}$$

where $u_{\min}$ and $u_{\max}$ are the lower and the upper bounds of MV, respectively; $y_{\min}$ and $y_{\max}$ are the lower and the upper bounds of CV, respectively. Equation (3.11) is rewritten as

$$\begin{cases} u_{\min} \le u_{ss}(k-1) + \Delta u_{ss}(k) \le u_{\max} \\ y_{\min} \le y_{ss}(k-1) + \Delta y_{ss}(k) \le y_{\max} \end{cases}. \tag{3.12}$$

Therefore, the economic optimization problem can be described as an LP, i.e.,

$$\min_{\Delta u_{ss}(k)} J = h\Delta u_{ss}(k), \text{ s.t. } (3.12), \tag{3.13}$$

where $\Delta y_{ss}(k)$ is substituted by (3.5).

SSTC economic optimization problem can also be described as a QP. If the lower bound, $J_{\min}$, of J is known in advance, the objective function can be rewritten as $(h\Delta u_{ss}(k) - J_{\min})^2$, so that QP is executed in the direction of minimum economic loss; since the descending directions of LP and QP objective functions are consistent, the optimal solutions to both LP and QP must be equal.

3.2.1.2 Minimum-Move Problem

In an industrial process, for some MV, its change may have little or no impact on the economic loss. If this MV is taken as cost MV, then its cost coefficient will be a small value. Even if this MV varies significantly, its effect on the economic loss can be ignored. For this MV, we choose to punish its variation in the objective function; in this situation, it is called the minimum-move MV. Like the cost MV, the minimum-move MV also has its cost coefficient, which is always positive. In real applications, it is necessary to manually classify each MV as a cost MV or a minimum-move MV. Assume, w.l.o.g., that the first n_{mc} MVs are the cost MVs, and the last $n_u - n_{mc}$ the minimum-move MVs. With the minimum-move MVs included, the objective function of SSTC becomes

$$J = \sum_{i=1}^{n_{mc}} h_i \Delta u_{i,ss}(k) + \sum_{i=n_{mc}+1}^{n_u} h_i \left|\Delta u_{i,ss}(k)\right|. \tag{3.14}$$

The first part of J reflects the economic loss. The second part of J reflects the minimum-move cost. If the minimum-move MV significantly affects the economic loss, then different minimum-move cost coefficients produce different economic losses; in this situation, one can first minimize economic loss, then fix the cost MVs, and finally optimize minimum-move cost to obtain the minimum-move MVs.

The objective function (3.14) contains the absolute operator. By introducing the auxiliary variables, the absolute operator can be removed. Let

$$\left|\Delta u_{i,ss}(k)\right| \le R_i(k), \quad n_{mc} + 1 \le i \le n_u, \tag{3.15}$$

which is equivalent to

$$-R_i(k) \le \Delta u_{i,ss}(k) \le R_i(k), \quad n_{mc} + 1 \le i \le n_u. \tag{3.16}$$

Then, the objective function (3.14) is rewritten as

$$J = \sum_{i=1}^{n_{mc}} h_i \Delta u_{i,ss}(k) + \sum_{i=n_{mc}+1}^{n_u} h_i R_i(k). \tag{3.17}$$

The economic optimization problem can be summarized as

$$\min_{\Delta u_{ss}(k), R_{n_{mc}+1}(k), \cdots, R_{n_u}(k)} J, \text{ s.t. } (3.12), (3.16), \tag{3.18}$$

where $\Delta y_{ss}(k)$ is substituted by (3.5).

Example 3.1 The heavy oil fractionator has three product draws and three circulating refluxes, as shown in Figure 3.1.

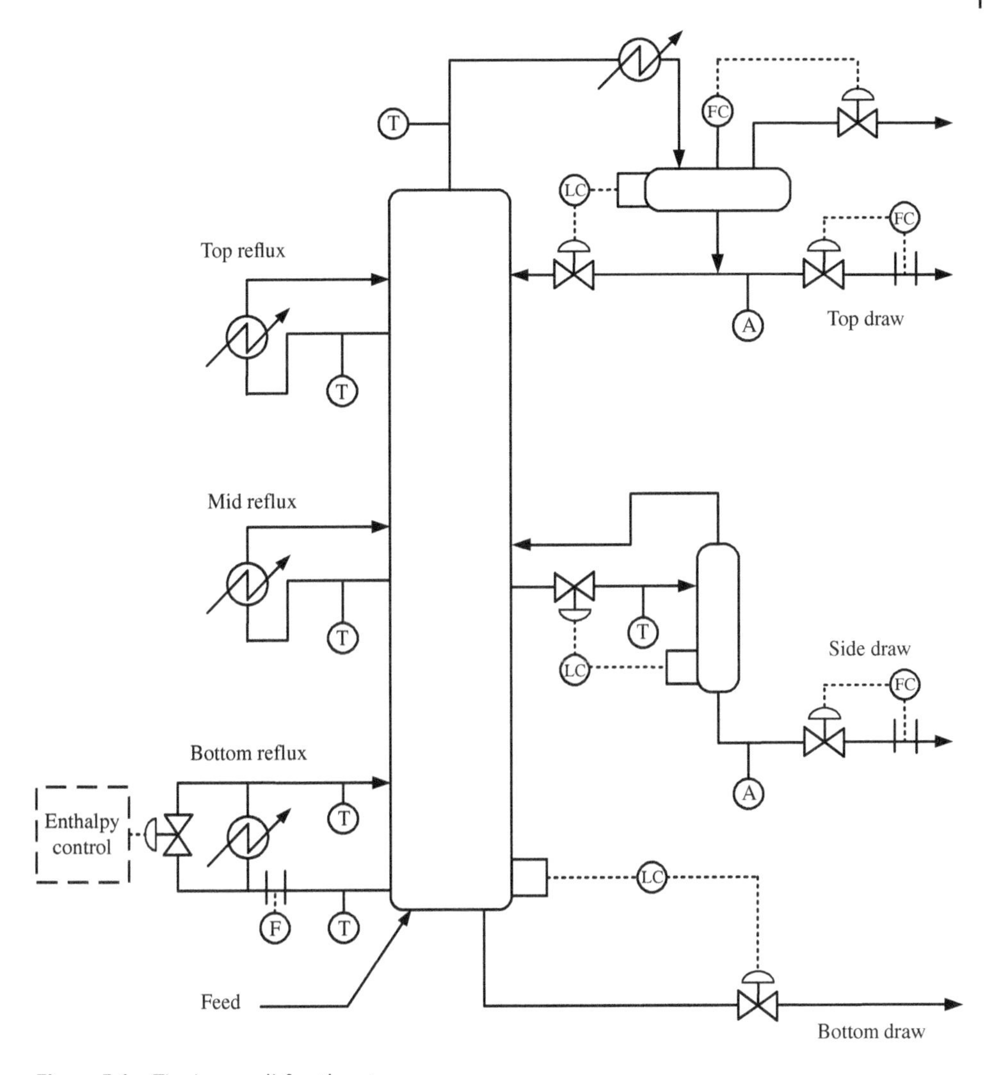

Figure 3.1 The heavy oil fractionator.

The continuous-time transfer function matrix of the heavy oil fractionator around the equilibrium is

$$G^u(s) = \begin{bmatrix} \frac{4.05e^{-27s}}{50s+1} & \frac{1.77e^{-28s}}{60s+1} & \frac{5.88e^{-27s}}{50s+1} \\ \frac{5.39e^{-18s}}{50s+1} & \frac{5.72e^{-14s}}{60s+1} & \frac{6.90e^{-15s}}{40s+1} \\ \frac{4.38e^{-20s}}{33s+1} & \frac{4.42e^{-22s}}{44s+1} & \frac{7.20}{19s+1} \end{bmatrix}, G^f(s) = \begin{bmatrix} \frac{1.20e^{-27s}}{45s+1} & \frac{1.44e^{-27s}}{40s+1} \\ \frac{1.52e^{-15s}}{25s+1} & \frac{1.83e^{-15s}}{20s+1} \\ \frac{1.14}{27s+1} & \frac{1.26}{32s+1} \end{bmatrix},$$

where MVs include the top draw u_1, the side draw u_2, and the bottom reflux u_3; CVs include the top concentration y_1, the side concentration y_2, and the bottom reflux temperature y_3; DVs include the mid reflux f_1 and the top reflux f_2. The system variables have been normalized, i.e., MVs take their values in $[-0.5, 0.5]$, and CVs in $[-0.5, 0.5]$. In order to observe the time-varying trajectories for ss of MVs and CVs, the absolute of single-step steady-state MV increment is bounded by 0.2.

It is easy to obtain the steady-state model as (independent of sampling period)

$$\Delta y_{ss}(k) = \begin{bmatrix} 4.05 & 1.77 & 5.88 \\ 5.39 & 5.72 & 6.90 \\ 4.38 & 4.42 & 7.20 \end{bmatrix} \Delta u_{ss}(k) + \begin{bmatrix} 1.20 & 1.44 \\ 1.52 & 1.83 \\ 1.14 & 1.26 \end{bmatrix} \Delta f_{ss}(k).$$

One can set the cost coefficients of MVs in accordance with the features of fractionator. Let the cost coefficients of u_1, u_2 and u_3 be -2, -1 and $+1$, respectively, because the top and side products of the fractionator are important. It tends to increase the economic benefits by increasing the top and side products, and increase the operating loss by increasing the bottom reflux.

Assume that, at $k = 1$, the process lies at the steady state, i.e., MVs at (0,0, 0) and CVs at (0,0, 0). Let us show the features of SSTC in several cases. In the first case, there is zero disturbance. The time-varying trajectories of ss are obtained through MATLAB. Due to the constraints on the single-step steady-state MV increment, ss gradually stops to vary, as being shown in Figure 3.2 (LHS). The final MVss are $\{0.5, 0.2108, -0.4929\}$, and the final CVss $\{-0.5, 0.5, -0.4269\}$; some final CVss and final MVss reach to their bounds. For the basic SSTC (i.e., all MVs are the cost MVs, and all CVs are sCVs), the number of the boundary-stayed CVss equals to that of the boundary-stayed MVss. Here, what is the benefit

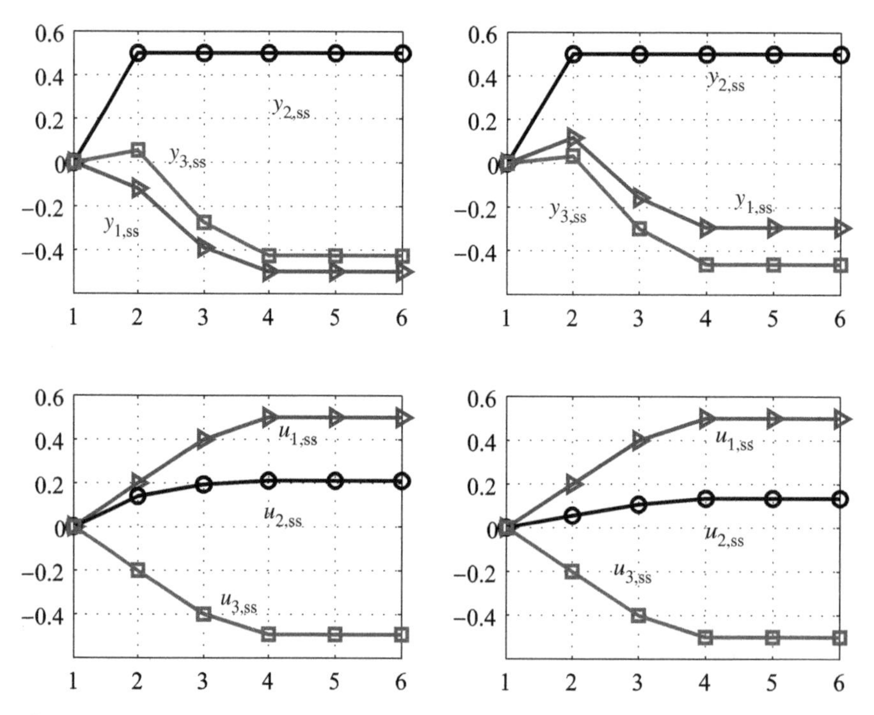

Figure 3.2 Time-varying trajectories of ss (LHS without, but RHS with, disturbance).

of SSTC? The answer is in comparing the values of objective functions before and after SSTC. Before SSTC, the value of the objective function is 0, and after SSTC, −1.7037. Clearly, the benefit of SSTC is 1.7037.

In the second case, there is a nonzero disturbance. Assume that $f_1 = 0.2$ and $f_2 = 0.1$. The time-varying trajectories of ss are shown in Figure 3.2 (RHS). The final MVss are $\{0.5, 0.1343, -0.5\}$, and the final CVss $\{-0.2933, 0.5, -0.4625\}$. The objective function has changed by −1.6343 via SSTC. The disturbance changes ss, and the objective function is slightly worse than the first case, i.e., due to existence of disturbance, the plant operation loses certain benefits (increases certain cost).

Next, choose a minimum-move MV for the first case (no disturbance). Let u_1 and u_2 be the cost MVs, and u_3 the minimum-move MV. Take the cost coefficients of u_1, u_2, and u_3 as −2, −1 and +2, respectively. The time-varying trajectories of ss are shown in Figure 3.3 (LHS). The final MVss are $\{0.5, -0.1113, -0.2259\}$, and the final CVss $\{0.5, 0.5, 0.0719\}$. Since the minimum-move cost coefficient is small, the minimum-move MV varies the same as when it is a cost MV. Then, change the minimum-move cost coefficient of u_3 as +10. The time-varying trajectories of ss are shown in Figure 3.3 (RHS). The final MVss are $\{0.1449, -0.0492, 0\}$, and the final CVss $\{0.5, 0.5, 0.4175\}$. In this situation, since the effect of minimum-move MV increment becomes significant, SSTC zeros the minimum-move MV increment.

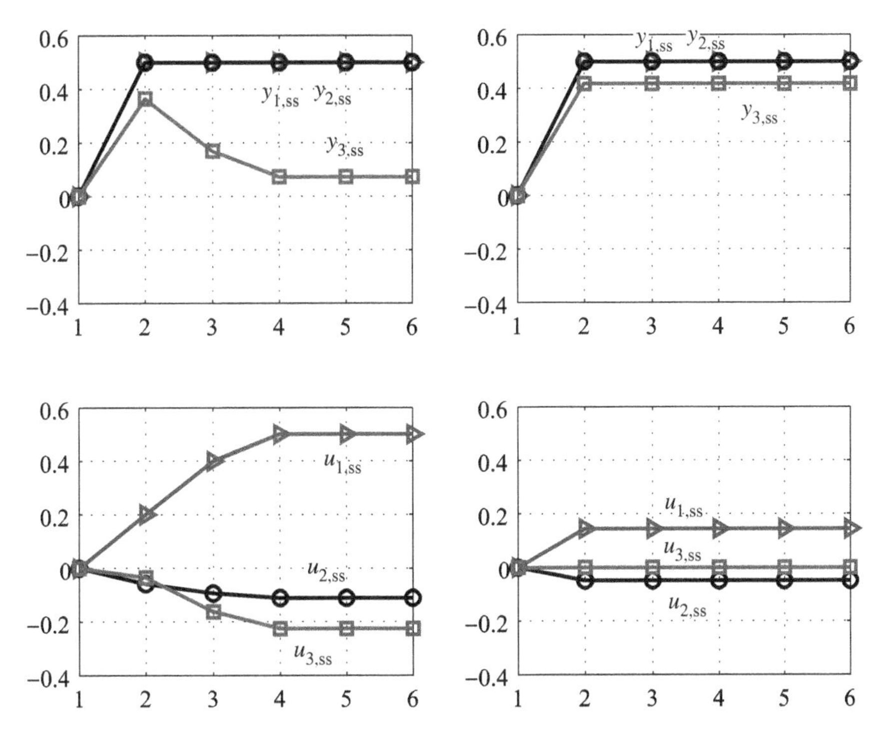

Figure 3.3 The time-varying trajectories of ss when there is a minimum-move MV (on LHS, the cost coefficient of the minimum-move MV is small, and on RHS, large).

3.2.2 Target Tracking Problem

In addition to the economic optimization, SSTC can track ETs, u_t and y_t (t standing for target), which usually comes from the high-layered RTO. Since the model and execution period of RTO are quite different from MPC (e.g., RTO may not consider the system disturbance, or it may invoke some simplifications), u_t and y_t, as sps of MPC, are usually not sent directly to DC layer. A good idea is to apply both current system information and linear steady-state model, to recalculate sps of DC. Thus, the objective function for the target tracking problem is chosen as

$$J = \|y_{ss}(k) - y_t\|_Q^2 + \|u_{ss}(k) - u_t\|_R^2, \tag{3.19}$$

where R and Q are weight coefficients.

In engineering applications, we often expect that ss obtained from the target tracking problem is highly consistent with $\{u_t, y_t\}$. This can be achieved by imposing some range restrictions on $y_{ss}(k)$ and $u_{ss}(k)$. Impose

$$u_{t\,\min} \le u_{ss}(k) \le u_{t\,\max}, \tag{3.20}$$

where $u_{t\,\min}$ and $u_{t\,\max}$ can be regarded as the expected lower and upper bounds of MVss. Similarly, impose

$$y_{t\,\min} \le y_{ss}(k) \le y_{t\,\max}, \tag{3.21}$$

where $y_{t\,\min}$ and $y_{t\,\max}$ can be regarded as the expected lower and upper bounds of CVss. Then, the target tracking problem of SSTC is described as

$$\min_{\Delta u_{ss}(k)} J, \text{ s.t. } \begin{cases} u_{t\,\min} \le u_{ss}(k-1) + \Delta u_{ss}(k) \le u_{t\,\max}, \\ y_{t\,\min} \le y_{ss}(k-1) + \Delta y_{ss}(k) \le y_{t\,\max}, \end{cases} \tag{3.22}$$

where $\Delta y_{ss}(k)$ is substituted by (3.5).

The target tracking problem of SSTC aims at tracking the results of RTO. Sometimes, the process or operator may need to maintain some MVs and CVs at their fixed values, which are referred to as IRVs in MPC engineering technology.

In each control period, for tracking either the results of RTO or IRVs, it is necessary to solve (3.22). Hence, one may want to specify u_t and y_t, which is enough for engineering applications.

3.3 Judging Feasibility and Adjusting Soft Constraint

SSTC in Section 3.2 applies in the situation when the feasible solution exists. If the optimization is infeasible, the previous SSTC does not yield ss. A feasible optimization means that it has at least one solution. Any set of decision variables, satisfying all the constraints of the optimization problem, is called a feasible solution of this optimization problem, and the set including all the feasible solutions is called the feasible region of the optimization problem. For an industrially applied optimal controller, it should try to avoid interruptions caused by infeasible optimizations. Therefore, an industrial optimization method needs to complete its execution by handling the infeasibility, so as to ensure the sustainability.

SSTC can be artificially decomposed into two stages, i.e.,

1) feasibility stage: adjusting some constraints so that feasible region is nonempty;
2) economic optimization stage: searching the optimum of economic objective function within the feasible region.

First, determine whether or not the region being formulated by the constraints exists; if it does, then search for the optimum within it; otherwise, adjust the soft constraints to obtain a feasible region, and then search. This method is called "two-stage." In MPC engineering, the constraints that can be adjusted are called the soft constraints, and those that cannot, the hard constraints. Adjusting the soft constraints (i.e., softening constraints) is an effective way to tackle infeasibility.

The mathematics for adjusting the soft constraints is essentially the optimization, whose decision variables are called the slack variables. Generally, in industrial MPC software, each CV has four kinds of limits: high limit (HL), low limit (LL), high high limit (HHL), and low low limit (LLL). These limits reflect the normal operating range of CV and the maximal adjustable engineering range. HL and LL are referred to as the operating limits, and HHL and LLL, the engineering limits. Engineering limits are the hard constraints of CV that cannot be violated.

In the following, without the minimum-move MV, let us first propose the weight method, and then the priority-rank method. For the case with the minimum-move MV, it needs an appropriate extension.

3.3.1 Weight Method

There are two problems. One is judging feasibility, i.e., deciding whether or not the economic optimization is feasible for the steady-state model and constraints. The other is adjusting the soft constraints, i.e., when it is infeasible, relaxing the operating constraint bounds to retrieve feasibility. By an appropriate handler, the first problem can be merged into the second. Usually, by adjusting the soft constraints, the feasibility can be judged by the resultant slack variables.

3.3.1.1 An Illustrative Example

First, illustrate the feasibility. Taking the two-dimensional space as an example. In this space, each constraint bound can be drawn as a straight line; one side of the line is feasible. Hence, multiple constraint bounds can be drawn as multiple lines. If the feasible sides of all the straight lines have a common non-empty area, then there is at least one feasible solution. As shown by the shadow in Figure 3.4, the feasible sides of the five constraints have a pentagonal feasible region. Imagine that, in Figure 3.4, the feasible side of one constraint bound is the opposite, then the feasible region no longer exists.

In the following, consider a model with one input and two outputs. Let us illustrate how to adjust the soft constraints and judge feasibility. Assume that the steady-state model is

$$\begin{cases} \Delta y_{1,ss}(k) = G_1 \Delta u_{ss}(k) \\ \Delta y_{2,ss}(k) = G_2 \Delta u_{ss}(k) \end{cases}, \tag{3.23}$$

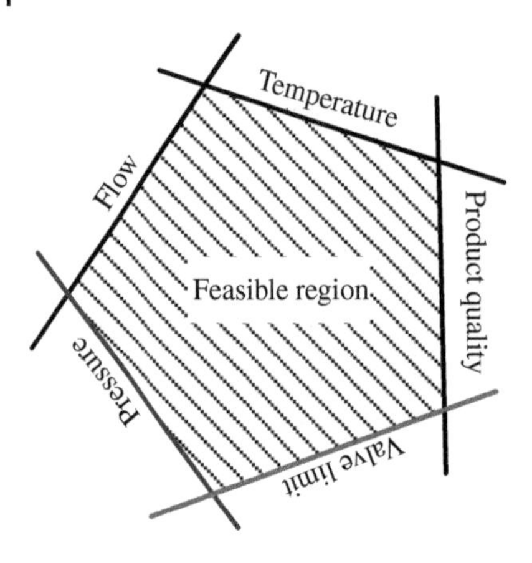

Figure 3.4 Feasible region surrounded by multiple constraints.

and the constraints are

$$\begin{cases} y_{1LL} \leq y_{1,ss}(k-1) + G_1 \Delta u_{ss}(k) \leq y_{1HL} \\ y_{2LL} \leq y_{2,ss}(k-1) + G_2 \Delta u_{ss}(k) \leq y_{2HL} \end{cases}. \tag{3.24}$$

Since there is only one MV, (3.24) is transformed into

$$\begin{cases} \Delta u_{ss}(k) \leq \mathcal{B} \\ \mathcal{A} \leq \Delta u_{ss}(k) \end{cases}, \tag{3.25}$$

or

$$\begin{cases} \Delta u_{ss}(k) \leq \mathcal{A} \\ \mathcal{B} \leq \Delta u_{ss}(k) \end{cases}, \tag{3.26}$$

where $\mathcal{A}$, $\mathcal{B}$ are reals, with $\mathcal{A} < \mathcal{B}$. Obviously, the constraint (3.25) surrounds a feasible region, while (3.26) does not. Introduce two slack variables $\varepsilon_1 \geq 0$ and $\varepsilon_2 \geq 0$. Since the slack variables are given for the soft constraints of CV, let us introduce the associated parameters $\alpha \geq 0$ and $\beta \geq 0$ for adjusting soft constraints. Then, (3.25)–(3.26) become

$$\begin{cases} \Delta u_{ss}(k) \leq \mathcal{B} + \alpha\varepsilon_1 \\ \mathcal{A} - \beta\varepsilon_2 \leq \Delta u_{ss}(k) \end{cases}, \tag{3.27}$$

$$\begin{cases} \Delta u_{ss}(k) \leq \mathcal{A} + \alpha\varepsilon_1 \\ \mathcal{B} - \beta\varepsilon_2 \leq \Delta u_{ss}(k) \end{cases}. \tag{3.28}$$

By adjusting the soft constraints, it means tuning the constraint as feasible by the minimum slack variables. Hence, the adjustment of the constraint bound can be achieved through optimization. For example, the objective function can be written as

$$J = \lambda\varepsilon_1 + \gamma\varepsilon_2, \tag{3.29}$$

where λ and γ are the positive weights. Since (3.27)–(3.28) include $\Delta u_{ss}(k)$, which is unknown for adjusting the soft constraints, we will remove $\Delta u_{ss}(k)$. In (3.27), adding the two inequalities removes $\Delta u_{ss}(k)$, i.e.,

$$\alpha\varepsilon_1 + \beta\varepsilon_2 \geq \mathcal{A} - \mathcal{B}. \tag{3.30}$$

Similarly, for (3.28),

$$\alpha\varepsilon_1 + \beta\varepsilon_2 \geq \mathcal{B} - \mathcal{A}. \tag{3.31}$$

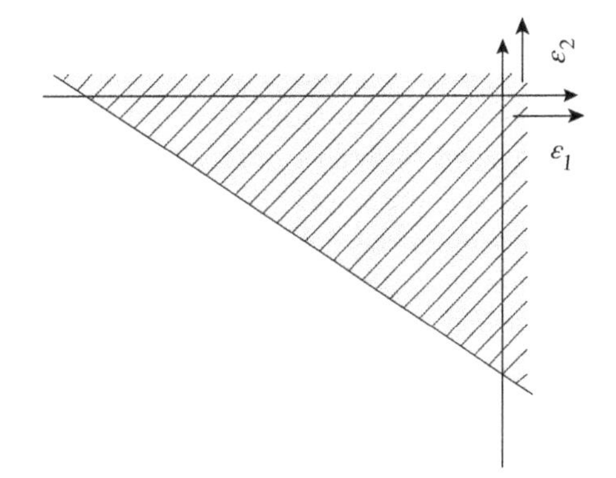

Figure 3.5 The feasible region for adjusting the soft constraints, but there is no adjustment.

For the case (3.25), the optimization problem for adjusting the soft constraints can be described as

$$\min_{\varepsilon_1,\varepsilon_2} J, \text{ s.t. } \alpha\varepsilon_1 + \beta\varepsilon_2 \geq \mathcal{A} - \mathcal{B}\ ,\ \varepsilon_1 \geq 0\ ,\ \varepsilon_2 \geq 0, \tag{3.32}$$

which is an LP. Since there are only two decision variables in (3.32), one can directly use the geometric method to solve (3.32). Because $\alpha \geq 0, \beta \geq 0$ and $\mathcal{A} < \mathcal{B}$, the intersection points of the line $\alpha\varepsilon_1 + \beta\varepsilon_2 = \mathcal{A} - \mathcal{B}$ with both ε_1-axis and ε_2-axis, are both negative. The constraint $\alpha\varepsilon_1 + \beta\varepsilon_2 \geq \mathcal{A} - \mathcal{B}$ means that only the region above the line $\alpha\varepsilon_1 + \beta\varepsilon_2 = \mathcal{A} - \mathcal{B}$, shown as the shadow in Figure 3.5, can be feasible; considering the shadow together with $\varepsilon_1 \geq 0$ and $\varepsilon_2 \geq 0$, the feasible region is the whole first quadrant. Since the optimization for adjusting the soft constraints will obtain ε_1 and ε_2 so that $\lambda\varepsilon_1 + \gamma\varepsilon_2$ has the minimum, it is shown that the optimum is the origin, i.e., $\varepsilon_1 = 0$ and $\varepsilon_2 = 0$. The details are shown in Figure 3.5.

For the case (3.26), the optimization for adjusting the soft constraints can be described as

$$\min_{\varepsilon_1,\varepsilon_2} J, \text{ s.t. } \alpha\varepsilon_1 + \beta\varepsilon_2 \geq \mathcal{B} - \mathcal{A}\ ,\ \varepsilon_1 \geq 0,\ \varepsilon_2 \geq 0. \tag{3.33}$$

The analysis is the same as (3.32), so is not repeated. The feasible region, shown as the shadow in Figure 3.6, is surrounded by three lines ($\alpha\varepsilon_1 + \beta\varepsilon_2 = \mathcal{B} - \mathcal{A}$, $\varepsilon_1 = 0$, $\varepsilon_2 = 0$).

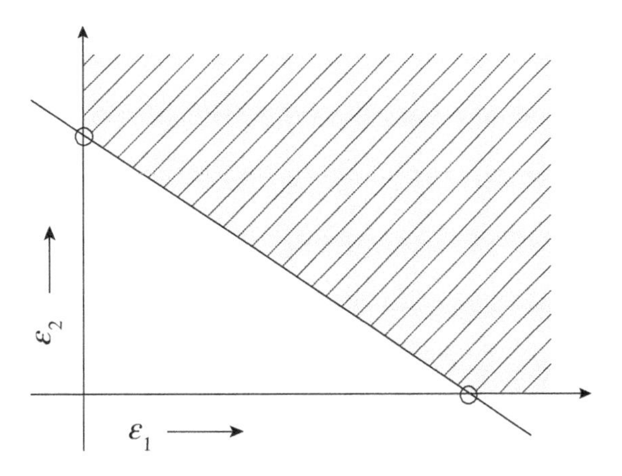

Figure 3.6 The feasible region for adjusting the soft constraints, and there is nonzero adjustment.

According to the feature of LP, the optimum of the problem (3.33) must be one of the intersection points of the line $\alpha\varepsilon_1 + \beta\varepsilon_2 = \mathcal{B} - \mathcal{A}$ with both ε_1-axis and ε_2-axis. Which of the two points being the optimum depends on the values λ and γ.

In general, the problem of judging the feasibility of SSTC can be merged into that of adjusting the soft constraints. When the optimum of the slack variables are zeros, it implies that the constraints need not be softened, and there exists a feasible region; when there is a nonzero slack variable, it implies that some constraint needs to be softened, and the nonzero slack variables adjust the soft constraints in order to retrieve a feasible region; if no slack variable solution exists, it implies that the feasible region cannot be retrieved by adjusting the soft constraints.

3.3.1.2 Weight Method

We will standardize the constraints. If it does not need to soften the constraints, then it should satisfy

$$\begin{cases} \Delta y_{ss}(k) = G^u \Delta u_{ss}(k) + G^f \Delta f_{ss}(k) \\ u_{LL} \le u_{ss}(k-1) + \Delta u_{ss}(k) \le u_{HL} \\ y_{LL} \le y_{ss}(k-1) + \Delta y_{ss}(k) \le y_{HL} \end{cases}. \tag{3.34}$$

Introducing the slack variables for CV operating constraints, and invoking the equation of $\Delta y_{ss}(k)$, yields

$$\begin{cases} u_{LL} \le u_{ss}(k-1) + \Delta u_{ss}(k) \le u_{HL} \\ y_{LL} - \varepsilon_2 \le y_{ss}(k-1) + G^u \Delta u_{ss}(k) + G^f \Delta f_{ss}(k) \le y_{HL} + \varepsilon_1 \\ y_{LLL} \le y_{ss}(k-1) + G^u \Delta u_{ss}(k) + G^f \Delta f_{ss}(k) \le y_{HHL} \end{cases}, \tag{3.35}$$

where $\varepsilon_1, \varepsilon_2$ are the slack variables for CV constraints, and the constraints without slack variables are hard.

Usually, for adjusting the soft constraints, the operator's intentions can be reflected through weight. Extending the objective function in (3.29) to the general weight form, the following LP problem is obtained:

$$\min_{\varepsilon_1, \varepsilon_2, \Delta u_{ss}(k)} J = w_1 \varepsilon_1 + w_2 \varepsilon_2, \text{ s.t. } (3.35), \tag{3.36}$$

where all the elements of $w_1 = [w_{11}, w_{12}, \dots, w_{1n_y}]$ and $w_2 = [w_{21}, w_{22}, \dots, w_{2n_y}]$ are positive. By solving LP problem (3.36), if it is feasible, one obtains a set of slack variables $\varepsilon_1, \varepsilon_2$.

Adjusting the soft constraints can also be achieved by solving the following QP:

$$\min_{\varepsilon_1, \varepsilon_2, \Delta u_{ss}(k)} J = \varepsilon_1^T W_1^2 \varepsilon_1 + \varepsilon_2^T W_2^2 \varepsilon_2, \text{ s.t. } (3.35), \tag{3.37}$$

where $W_1 = \text{diag}\{w_{11}, w_{12}, \dots, w_{1n_y}\}$ and $W_2 = \text{diag}\{w_{21}, w_{22}, \dots, w_{2n_y}\}$ are both positive-definite.

Notice that although both LP and QP can adjust the soft constraints, their results are different in general. In solving LP, in general, all the active constraints (i.e., constraints affecting the optimization, also known as the effective constraints) are unnecessarily relaxed; in QP, they are often relaxed.

3.3.2 Priority-Rank Method

In industrial production, the importance of each constraint may have a different level. An important constraint should be satisfied in priority. Applying the aforementioned weight method, sometimes both LP and QP relax several soft constraints. Although weight method reflects the importance of soft constraints to some extent, the priority of the soft constraints cannot be fully reflected. The so-called priority-rank method refers to relaxing the soft constraints in a specific order (see e.g., [Kerrigan and Maciejowski, 2002, Qin and Badgwell, 2003]). The priority rank of a soft constraint indicates the importance level of this constraint. The higher the priority rank, the more important the constraint. Assign a rank number for each priority rank, being called the priority number; the smaller the rank number, the higher the priority rank. The rank numbers of different soft constraints can be either equal or unequal, while the upper and lower limits of the same variable can be assigned with different rank numbers.

The soft constraints in process control are usually associated with the production requirements, which can be sorted in order from high importance to low importance, e.g., from safety, environmental protection, iCV bound, model validity zone, product quality, and economic profit. Safety is more important than every other considerations. Sustainable development has become the theme of current society, so the constraints relating to environmental protection are also the primary concerns. The importance of iCV is related to the stability of MPC system, and violating the constraints of iCV is very dangerous. Thus, it is necessary to assign iCV with a high priority rank. Model validity zone involves accuracy of the running model (e.g., the linear model associated with an equilibrium point). Thus, in order to ensure the model accuracy, the importance of model validity zone may be the most important after iCV. The requirements for ensuring the product quality should be met as much as possible, but there should exist some flexibility (e.g., allowing a fluctuation within a certain range). MVET and CVET are usually more "soft" constraints.

Remark 3.1 In the priority-rank method, there are several usages of the term "constraint," so it is necessary to clarify them. The soft constraints are adhere to each priority rank, and are determined along with the priority rank. The original hard constraints are inherent for SSTC, which are not allowed softening and must be satisfied for each priority rank. The constraints of a priority rank include all the constraints to be satisfied in this rank, including the original hard constraints, the soft-2-hard (relaxed soft) constraints, and the original (unrelaxed) soft constraints.

Assume that there are totally r_0 ranks. With the priority rank being determined, the constraints for each rank are formally denoted as

$$\begin{cases} \Theta^r \varepsilon^r = \vartheta^r \\ \Omega^r \varepsilon^r \leq \omega^r \end{cases}, \quad r = 1, \ldots, r_0, \tag{3.38}$$

where the superscript r indicates the rank number. Assume that (3.38) contains $\varepsilon^r \geq 0$ and the only unknown variable is ε^r, so (3.38) has contained all the hard constraints and the related soft constraints (may be relaxed). There are two methods for judging feasibility and adjusting the soft constraints, i.e., ascending-number method and descending-number method.

3.3.2.1 Ascending-Number Method

The algorithm for judging feasibility and adjusting the soft constraints, based on the ascending number, is as follows.

Step 1. Assign a rank number and a weight coefficient for each soft constraint.
Step 2. Judge feasibility and adjust the soft constraints, for each rank number, from low to high.

For the rank r, take all the soft constraints (which may have been relaxed), with the rank number being less than r, as the hard constraints. At the rank r, after judging feasibility and adjusting the soft constraints, consider the rank $r+1$. Continue until all r_0 ranks have been judged for feasibility and adjusted for soft constraints.

For Step 2, we give an additional explanation. Consider the rank number 1. We should consider the original hard constraints, and the soft constraint for rank 1. The optimization problem for the rank 1 is

$$\min_{\varepsilon^1} J = w_L^1 \varepsilon^1, \text{ s.t. (3.38)}, \ r = 1, \tag{3.39}$$

or

$$\min_{\varepsilon^1} J = (\varepsilon^1)^{\mathrm{T}} (W_Q^1)^2 \varepsilon^1, \text{ s.t. (3.38)}, \ r = 1, \tag{3.40}$$

where w_L^1 is the weight coefficient vector with the positive elements, and W_Q^1 is the positive-definite weight matrix. The necessary and sufficient condition for the feasibility of the optimization problem for the rank 1 is that the original hard constraints is compatible.

Consider the rank number 2. We should consider the original hard constraints, the soft constraints for rank 2, and the soft constraints for rank 1 (which may have been relaxed) as hard, so as to collect the constraints for rank 2. The optimization problem for rank 2 is

$$\min_{\varepsilon^2} J = w_L^2 \varepsilon^2, \text{ s.t. (3.38)}, \ r = 2, \tag{3.41}$$

or

$$\min_{\varepsilon^2} J = (\varepsilon^2)^{\mathrm{T}} (W_Q^2)^2 \varepsilon^2, \text{ s.t. (3.38)}, \ r = 2, \tag{3.42}$$

where w_L^2 is the weight coefficient vector with the positive elements, and W_Q^2 is the positive-definite weight matrix. The necessary and sufficient condition for the feasibility of the optimization problem for the rank 2 is that the optimization problem for the rank 1 is feasible.

The rank number in $\{3, \dots, r_0\}$ is similar to the rank 2, which is not repeated. The ascending-number method is recursively feasible, i.e., the necessary and sufficient condition for its feasibility is that the optimization problem for the rank 1 is feasible. For the optimum of the rth rank, we analyze as follows. If $J = 0$, it implies that the region surrounded by the soft and hard constraints for rank r is nonempty; if $J \neq 0$, it implies empty, but the region, after relaxing the soft constraints for rank r, is again nonempty.

In a word, the so-called ascending-number method is maneuvering the soft constraints from "soft" to "hard" in the order of rank number from low to high.

3.3.2.2 Descending-Number Method

Contrary to the ascending-number, judging and adjusting in the descending-number method is in the order of rank numbers from high to low. The descending-number method considers all the constraints at the beginning. The algorithm for judging feasibility and adjusting the soft constraints, based on the descending number, is as follows.

Step 1. Set a rank number and a weight coefficient for each soft constraint.
Step 2. Judge feasibility and adjust the soft constraints, for each rank number, from high to low.

At the rank r, if the optimization is infeasible, relax each soft constraint for the rank r to its allowable limit (e.g., to the engineering limit), then consider the rank $r-1$; if the optimization is feasible, the soft constraints for the rank r are relaxed by the optimal slack variables, and there is no further optimization for the rank number less than r. For the rank r, take all the soft constraints (which may have been relaxed) for the rank number unequal to r, as hard constraints.

For Step 2, we give a further explanation. Consider the rank number r_0. Consider all the ranks of soft constraints and the original hard constraints so as to collect the constraints for rank r_0. The optimization problem for the rank r_0 is

$$\min_{\varepsilon^{r_0}} J = w_L^{r_0}\varepsilon^{r_0}, \text{ s.t. } (3.38),\ r = r_0, \tag{3.43}$$

or

$$\min_{\varepsilon^{r_0}} J = (\varepsilon^{r_0})^{\mathrm{T}}(W_Q^{r_0})^2\varepsilon^{r_0}, \text{ s.t. } (3.38),\ r = r_0, \tag{3.44}$$

where $w_L^{r_0}$ is the weight coefficient vector with the positive elements, and $W_Q^{r_0}$ is the positive-definite weight matrix. We only consider relaxing the soft constraints for rank r_0, while the other ranks of soft constraints are treated as hard.

If the optimization of the rank r_0 is infeasible, then relax the soft constraints for rank r_0 to their allowable limits, and then consider the optimization for rank r_0-1. Consider all the ranks of soft constraints (which may have been relaxed) and the original hard constraints so as to collect the constraints for the rank r_0-1. The optimization problem for the rank r_0-1 is

$$\min_{\varepsilon^{r_0-1}} J = w_L^{r_0-1}\varepsilon^{r_0-1}, \text{ s.t. } (3.38),\ r = r_0-1, \tag{3.45}$$

or

$$\min_{\varepsilon^{r_0-1}} J = (\varepsilon^{r_0-1})^{\mathrm{T}}(W_Q^{r_0-1})^2\varepsilon^{r_0-1}, \text{ s.t. } (3.38),\ r = r_0-1, \tag{3.46}$$

where $w_L^{r_0-1}$ is the weight coefficient vector with the positive elements, and $W_Q^{r_0-1}$ is the positive-definite weight matrix. We only consider relaxing the soft constraints for rank r_0-1, while the other ranks of soft constraints are treated as hard.

The rank $\{r_0-2,\ldots,1\}$ is similar to the rank r_0-1, which is not repeated. For the optimum of the rth rank, we analyze as follows. If $J=0$, it implies that all the constraints (including the relaxed soft constraints, the non-relaxed soft constraints, and the original hard constraints) surround a nonempty feasible region. If $J \neq 0$, it implies empty, but the region, after relaxing the soft constraints for rank r, is again nonempty. When the descending-number method is carried out for rank 1, and the optimization is infeasible, then it implies that relaxing all the soft constraints within their allowable limits cannot retrieve a nonempty feasible region.

In a word, the so-called descending-number method is manoeuvering the soft constraints from "hard" to "soft" in the order of rank numbers from high to low.

Remark 3.2 Note that a premise of (3.39)–(3.46) is that the slack variables of the rank are the only unknown variables (decision variables); if there are other unknown variables, their hard constraints should be imposed. In addition, for each rank r, the constraints (3.38) of the ascending-number and descending-number methods are different, and the parameters in the constraints are recursively updated rather than being determined, once and for all, after assigning the rank numbers (Step 1). In general, it unnecessarily reduces the computational burden by eliminating the unknown variables other than the slack variables. Hence, such an elimination may be unnecessary, and (3.39)–(3.46) only represent the method in theory.

Example 3.2 We still take the Shell benchmark problem as an example. Each MV is constrained in $[-0.5, 0.5]$; each CV operating constraint is $[-0.5, 0.5]$, and its engineering constraint $[-0.6, 0.6]$; the current MVss is [0,0, 0], the current CVss is [0,0, 0], and DV is $[-1.7, -1.5]$. Using CV operating constraints as the hard constraints for SSTC, it is known that optimization is infeasible due to the large disturbance.

1) Weight method

To illustrate the impact of weight on adjusting, seven sets of weights are chosen (the weights for the upper and lower bounds of each CV are the same) to indicate the importance of each CV constraint.

First, apply LP, and the result is shown in Table 3.1. Second, apply QP, and the result is shown in Table 3.2.

It is shown that, although the LP weight method can adjust CV constraints so that SSTC retrieves feasible, the adjustments cannot fully reflect the inclination of the operator who sets the weights. The reason is that the adjustments are determined by both the descent direction formed by the weight coefficients and the vertices intersected by multiple constraint lines. The QP weight method cannot either, but it differs from LP in that more bounds of CV constraints are relaxed.

Table 3.1 Lookup table for adjusting the soft constraints and choosing weight coefficients (LP).

Weight coefficient	Relaxed bound	Slack variable ε_1 or ε_2
1,1,1	upper bound	0.0000, 0.0000, 0.0172
1,2,3	lower bound	0.0344, 0.0000, 0.0000
1,3,2	upper bound	0.0000, 0.0000, 0.0172
3,1,2	lower bound	0.0000, 0.0278, 0.0000
2,1,3	lower bound	0.0000, 0.0278, 0.0000
3,2,1	upper bound	0.0000, 0.0000, 0.0172
2,3,1	upper bound	0.0000, 0.0000, 0.0172

Table 3.2 Lookup table for adjusting the soft constraints and choosing weight coefficients (QP).

Weight coefficient	Slack variable $\varepsilon_1, \varepsilon_2$
1,1,1	[0.0000, 0.0000, 0.0105], [0.0053, 0.0065, 0.0000]
1,2,3	[0.0000, 0.0000, 0.0074], [0.0111, 0.0069, 0.0000]
1,3,2	[0.0000, 0.0000, 0.0098], [0.0098, 0.0040, 0.0000]
3,1,2	[0.0000, 0.0000, 0.0089], [0.0030, 0.0110, 0.0000]
2,1,3	[0.0000, 0.0000, 0.0068], [0.0051, 0.0126, 0.0000]
3,2,1	[0.0000, 0.0000, 0.0135], [0.0022, 0.0042, 0.0000]
2,3,1	[0.0000, 0.0000, 0.0137], [0.0034, 0.0028, 0.0000]

2) Priority-rank method

Let us set the rank number for CV1, CV2, and CV3 as 1, 2, and 3, respectively. For each CV, for both upper and lower constraint bounds, set the same weight.

First, choose the ascending-number method. The LP result is as follows. Judge the feasibility for the rank 1, and the constraints are feasible; judge the feasibility for the rank 2, and the constraints are feasible; judge the feasibility for the rank 3, and the constraints are infeasible. Adjust the upper bound of CV3, and the adjustment is $\varepsilon_{1,3} = 0.0172$. QP is the same as LP, and results are the same (this phenomenon being only a special case).

Second, choose the descending-number method. The LP result is as follows. Judge the feasibility for the rank 3 (it is unnecessary to adjust the soft constraints for ranks 1 and 2, so the problem here is completely equivalent to the ascending-number method), the constraints are incompatible and $\varepsilon_{1,3} = 0.0172$ is obtained. QP is the same as LP.

It is shown that, based on the priority rank, adjusting the soft constraints can fully reflect the operator's intention.

3.3.3 Compromise Between Adjusting Soft Constraints and Economic Optimization

SSTC can be divided into two stages, i.e., feasibility and economic optimization. In Sections 3.3.1 and 3.3.2, we introduce the weight method for judging feasibility and adjusting the soft constraints, and based on this method, further the priority-rank method. Section 3.2.1 introduces the economic optimization.

The essential of the two-stage method is adjusting the soft constraints in order to construct a feasible region when it does not exist. This is standard for SSTC to handle constraint incompatibility. Some questions may arise. When adjusting the soft constraints, can the economic loss be considered? In reconstructing the feasible region, can the descent direction of the economic loss be considered? This question can be further extended. Even if the original optimization problem is already feasible, can the feasible region be adjusted appropriately to further reduce the economic loss? The answer is positive. This is a compromise between adjusting the soft constraints and the economic optimization.

In the following, we provide a description of the compromise problem between adjusting the soft constraints and the economic optimization, including LP and QP. LP optimization problem is described as

$$\min_{\Delta u_{ss}(k),\varepsilon} J = h\Delta u_{ss}(k) + w_L\varepsilon, \quad \text{s.t. } \Psi \begin{bmatrix} \Delta u_{ss}(k) \\ \varepsilon \end{bmatrix} \leq \psi, \tag{3.47}$$

where Ψ and ψ are parameters related to $\Delta u_{ss}(k)$ and ε. Similarly, QP optimization problem is described as

$$\min_{\Delta u_{ss}(k),\varepsilon} J = (h\Delta u_{ss}(k) - J_{\min})^2 + \varepsilon^{\mathrm{T}} W_Q^2 \varepsilon, \quad \text{s.t. } \Psi \begin{bmatrix} \Delta u_{ss}(k) \\ \varepsilon \end{bmatrix} \leq \psi. \tag{3.48}$$

The decision variables of the optimization problems include $\Delta u_{ss}(k)$ and ε. The first half of J reflects the economic loss. If relaxing the soft constraints leads to an increase in the economic benefit, then the constraint bounds will be adjusted accordingly.

4

Two-Layered DMC for Stable Processes

New notations:

M:	control horizon;
N:	model horizon, length of the step response coefficients;
P:	prediction horizon;
P':	steps of future f which can be estimated;
f:	in $\mathbb{R}^{n_f}$, measurable disturbance variable (DV);
f_{eq}:	value of f at the equilibrium;
∇f:	$f - f_{\text{eq}}$;
$f(k+j\|k)$:	$j = 0, 1, \dots, P'$, current and future estimation of f at time k;
u:	in $\mathbb{R}^{n_u}$, manipulated variable (MV);
u_{eq}:	value of u at the equilibrium;
∇u:	$u - u_{\text{eq}}$;
$\Delta u(k)$:	$u(k) - u(k-1)$, increment of u at time k;
$\underline{u}$:	lower bound of u;
$\bar{u}$:	upper bound of u;
$\Delta\bar{u}$:	the maximum allowable absolute change rate of u;
$u_t^{(\text{sm})}(k)$:	ET of u, where "sm" means "some";
$u_{i,t}(k)$:	ET of u_i;
$u_{i,\text{ss,range}}$:	expected allowable range of $u_{i,\text{ss}}$ around $u_{i,t}$;
$\delta u_{\text{ss}}(k)$:	$u_{\text{ss}}(k) - u(k-1)$, or equivalently $\sum_{i=0}^{M-1} \Delta u(k+i\|k)$, steady-state target increment of u at time k;
$\delta\bar{u}_{\text{ss}}$:	the maximum allowable absolute steady-state change rate of u, the bound on absolute of single-step steady-state u increment;
y:	in $\mathbb{R}^{n_y}$, controlled variable (CV);
y_{eq}:	value of y at the equilibrium;
∇y:	$y - y_{\text{eq}}$;
$\underline{y}_0$:	LL of y;
$\bar{y}_0$:	HL of y;
$\underline{y}_{0,h}$:	LLL of y;
$\bar{y}_{0,h}$:	HHL of y;

Model Predictive Control, First Edition. Baocang Ding and Yuanqing Yang.

$y_{ss}^{ol}(k)$: ol steady-state prediction of y at time k;
$y_{ss}(k)$: ss of y at time k;
$y^{ol}(k+n|k)$: $n = 1, 2, \ldots, N$, ol dynamic prediction of y at time k;
$y^{fr}(k+p|k)$: free dynamic prediction of y at time k;
$y_t^{(sm)}(k)$: ET of y, where "sm" means "some";
$y_{j,t}(k)$: ET of y_j;
$y_{j,ss,range}$: expected allowable range of $y_{j,ss}$ around $y_{j,t}$;
$\delta y_{ss}(k)$: $y_{ss}(k) - y_{ss}^{ol}(k)$, steady-state target increment of y at time k;
$\delta \bar{y}_{ss}$: the maximum allowable absolute steady-state change rate of y;
$\xi(k+i|k)$: prediction of $\xi(k+i)$ at time k;
$\|\xi\|_Q^2$: $\xi^T Q \xi$;
$\mathbb{N}_u$: $\{1, 2, \ldots, n_u\}$, set of indices of u;
$\mathbb{N}_y$: $\{1, 2, \ldots, n_y\}$, set of indices of y;
S_i^f: step-response coefficient matrix of f, $S_{N+i}^f = S_N^f$ for all $i \geq 0$;
S_i^u: step-response coefficient matrix of u, $S_{N+i}^u = S_N^u$ for all $i \geq 0$;
S_N^f: steady-state gain matrix from f to y;
S_N^u: steady-state gain matrix from u to y;
$\mathcal{I}_t$: index set of u which has ET;
$\mathcal{O}_t$: index set of y which has ET;
$|\mathcal{S}|$: cardinality of (number of elements in) the set $\mathcal{S}$.

Given some CVETs and MVETs, the two-layered MPC in this chapter finds the real-time control move satisfying MV rate, MV magnitude and CV magnitude constraints. The expected properties of this two-layered MPC include fast control, smooth control, closed-loop stability and offset-free.

The two-layered MPC consists of three modules, i.e., the open-loop prediction module, the steady-state target calculation (SSTC) module, and the dynamic calculation (DC) module. In ol prediction module, which can be conceptually represented as

$$\begin{aligned}&\{y_{ss}^{ol}(k), \{y^{ol}(k+n|k),\ n = 1, 2, \ldots, N\}\}\\&\quad = \Phi_{olp}\left(y(k), \{y^{ol}(k+n-1|k-1),\ n = 1, 2, \ldots, N\}, \Delta u(k-1),\right.\\&\qquad \left.\Delta f(k), \Delta f(k+1|k), \ldots, \Delta f(k+P'|k)\right),\end{aligned} \tag{4.1}$$

the future values of CV are predicted based on the current CV measurements, assuming that the future MV keeps unchanged. In SSTC, which can be conceptually represented as

$$\left\{\delta u_{ss}(k), y_{ss}(k)\right\} = \Phi_{SSTC}\left(u(k-1), y_{ss}(k-1), y_{ss}^{ol}(k), u_t^{(sm)}(k), y_t^{(sm)}(k)\right), \tag{4.2}$$

some constrained LPs or constrained QPs are solved, which obtains MVss and CVss. In DC, which can be conceptually represented as

$$\Delta u(k|k) = \Phi_{DC}\left(u(k-1), \delta u_{ss}(k), \{y^{ol}(k+p|k),\ p = 1, 2, \ldots, P\}, y_{ss}(k)\right), \tag{4.3}$$

ss from SSTC are tracked by optimizing the future MV moves.

4.1 Open-Loop Prediction Module

Consider the following FSR model (equivalent for all $N' \geq N$):

$$\begin{aligned}\nabla y(k) = &\sum_{i=1}^{N'-1} S_i^u \Delta u(k-i) + S_{N'}^u \nabla u(k-N') \\ &+ \sum_{i=1}^{N'-1} S_i^f \Delta f(k-i) + S_{N'}^f \nabla f(k-N'),\end{aligned} \tag{4.4}$$

where the step-response coefficient matrices satisfy $S_{N+i}^u = S_N^u$ and $S_{N+i}^f = S_N^f$ for all $i \geq 0$. Some predictions based on (4.4) are

$$\begin{aligned}\nabla y(k+p|k-1) = &\sum_{i=1}^{N'-1} S_i^u \Delta u(k-i+p|k-1) + S_{N'}^u \nabla u(k-N'+p|k-1) \\ &+ \sum_{i=1}^{N'-1} S_i^f \Delta f(k-i+p|k-1) + S_{N'}^f \nabla f(k-N'+p|k-1),\end{aligned}$$

$$\begin{aligned}\nabla y(k+p|k) = &\sum_{i=1}^{N'-1} S_i^u \Delta u(k-i+p|k) + S_{N'}^u \nabla u(k-N'+p|k) \\ &+ \sum_{i=1}^{N'\ 1} S_i^f \Delta f(k-i+p|k) + S_{N'}^f \nabla f(k-N'+p|k),\end{aligned}$$

where $\Delta u(h|\ell) = \Delta u(h)$ for all $h < \ell$, $\nabla u(h|\ell) = \nabla u(h)$ for all $h < \ell$. Assume that $f(k)$ is measurable at time k, and $\Delta f(h|\ell) = \begin{cases} \Delta f(h), & h \leq \ell \\ 0, & h > \ell \end{cases}$ which means that f keeps unchanged in the future. Let $y^{\mathrm{fr}}(k+p|k)$ be the prediction based on (4.4) by taking $\Delta u(k+i-1|k) = 0$ for $1 \leq i \leq p$. Applying the above predictions yields

$$y^{\mathrm{fr}}(k+p|k) - y^{\mathrm{fr}}(k+p|k-1) = S_{p+1}^u \Delta u(k-1) + S_p^f \Delta f(k), \tag{4.5}$$

$$y(k+p|k) - y^{\mathrm{fr}}(k+p|k) = \sum_{i=1}^{p} S_i^u \Delta u(k+p-i|k). \tag{4.6}$$

In the aforementioned, the prediction step p satisfies $p > 0$; by taking $S_0^f = 0$, it is allowed to take $p = 0$. For $p = 0$, notice that $y^{\mathrm{fr}}(k|k) = y(k|k) \neq y(k)$.

In predicting $y^{\mathrm{fr}}(k+p|k)$, the current measurement $y(k)$ is not applied, so there is no feedback correction based on $y(k)$. In the following, we introduce the feedback correction, then replace $y^{\mathrm{fr}}(k+p|k)$ by $y^{\mathrm{ol}}(k+p|k)$. Let us denote

$$Y_N^{\mathrm{ol}}(k|k) = \begin{bmatrix} y^{\mathrm{ol}}(k+1|k) \\ y^{\mathrm{ol}}(k+2|k) \\ \vdots \\ y^{\mathrm{ol}}(k+N|k) \end{bmatrix}.$$

The ol predictions are based on (4.5), changing the superscript fr as ol. At time $k > 0$, after $\Delta u(k-1)$ is measured, it yields

$$y^{\mathrm{ol}}(k|k) = y^{\mathrm{ol}}(k|k-1) + S_1^u \Delta u(k-1).$$

Applying the measured $y(k)$ yields

$$\epsilon(k) = y(k) - y^{\rm ol}(k|k) = y(k) - y^{\rm ol}(k|k-1) - S_1^u \Delta u(k-1). \tag{4.7}$$

$\epsilon(k)$, named prediction error, represents the influence of model-plant mismatch on the prediction of y. $\epsilon_j(k)$ is added to each future prediction of y_j, so that

$$Y_N^{\rm ol}(k|k) = \mathcal{M}\left\{Y_N^{\rm ol}(k-1|k-1) + \begin{bmatrix} S_1^u \\ S_2^u \\ \vdots \\ S_N^u \end{bmatrix} \Delta u(k-1)\right\} + \begin{bmatrix} S_1^f \\ S_2^f \\ \vdots \\ S_N^f \end{bmatrix} \Delta f(k) + \begin{bmatrix} \epsilon(k) \\ \epsilon(k) \\ \vdots \\ \epsilon(k) \end{bmatrix}, \tag{4.8}$$

where $\mathcal{M} = \begin{bmatrix} 0 & I & 0 & \dots & 0 \\ \vdots & \ddots & \ddots & \ddots & \vdots \\ 0 & \dots & 0 & I & 0 \\ 0 & \dots & 0 & 0 & I \\ 0 & \dots & 0 & 0 & I \end{bmatrix}$, with $Y_N^{\rm ol}(0|0) = \begin{bmatrix} y(0) \\ y(0) \\ \vdots \\ y(0) \end{bmatrix} + \begin{bmatrix} S_1^f \\ S_2^f \\ \vdots \\ S_N^f \end{bmatrix} \Delta f(0)$ being the initial value.

Equation (4.8) is called ol dynamic prediction equation. The ol steady-state prediction is

$$y_{\rm ss}^{\rm ol}(k) = y^{\rm ol}(k+N+i|k), \quad i \geq 0. \tag{4.9}$$

Note that in deducing (4.8) we have applied $y^{\rm ol}(k+N|k-1) = y^{\rm ol}(k+N-1|k-1)$, which is consistent with (4.9).

Remark 4.1 Assume that we can estimate the future values of f, denoted as

$$\Delta\tilde{f}(k+1|k) = \begin{bmatrix} \Delta f(k+1|k) \\ \Delta f(k+2|k) \\ \vdots \\ \Delta f(k+P'|k) \end{bmatrix}.$$

Then, we can replace $\{N, Y_N^{\rm ol}(k|k)\}$ in the following:

$$\begin{aligned} N &\rightarrow N + P', \\ Y_N^{\rm ol}(k|k) &\rightarrow Y_{N+P'}^{\rm ol}(k|k) + \mathcal{S}^f \Delta\tilde{f}(k+1|k), \end{aligned}$$

where

$$\mathcal{S}^f = \begin{bmatrix} 0 & 0 & \cdots & 0 \\ S_1^f & 0 & \cdots & 0 \\ S_2^f & S_1^f & \ddots & \vdots \\ \vdots & \ddots & \ddots & 0 \\ S_{P'}^f & \cdots & S_2^f & S_1^f \\ \vdots & & \vdots & \vdots \\ S_{N+P'-1}^f & \cdots & S_{N+1}^f & S_N^f \end{bmatrix}.$$

4.2 Steady-State Target Calculation Module

SSTC optimizes $\delta u_{ss}(k)$ and $y_{ss}(k)$, which satisfy the steady-state CV prediction equation

$$y_{ss}(k) = S_N^u \delta u_{ss}(k) + y_{ss}^{ol}(k). \tag{4.10}$$

This equation is obtained from (4.6), changing the superscript fr as ol.

SSTC receives ETs from the real-time optimization (RTO), process operator or control engineer. An ET is an expected CVss or MVss. It is not necessary that every MV and CV has its ET. In SSTC, each scalar constraint in LP and QP is either hard or soft. The hard constraints cannot be violated. For the feasibility of SSTC, any soft constraint can be relaxed.

4.2.1 Hard and Soft Constraints

Each MVss has a hard magnitude constraint, i.e.,

$$\underline{u} \leq u_{ss}(k) \leq \bar{u}. \tag{4.11}$$

If, in DC, there is MV rate constraint $|\Delta u(k+j|k)| \leq \Delta\bar{u}$ $(0 \leq j \leq M-1)$, then in order to be consistent with DC, the MV target should satisfy

$$|\delta u_{ss}(k)| \leq M\Delta\bar{u}. \tag{4.12}$$

Usually, an additional constraint is imposed on $\delta u_{ss}(k)$, i.e.,

$$|\delta u_{ss}(k)| \leq \delta\bar{u}_{ss}. \tag{4.13}$$

Let us combine (4.12) and (4.13) to yield

$$|\delta u_{ss}(k)| \leq \delta\bar{u}'_{ss}, \tag{4.14}$$

where

$$\delta\bar{u}'_{ss} = \min\{M\Delta\bar{u}, \delta\bar{u}_{ss}\}.$$

If (4.14) conflicts with (4.11), then (4.11) has to be satisfied (i.e., (4.14) is somewhat not always hard). Hence, (4.11) and (4.14) are combined to yield the hard constraint

$$\underline{u}''(k) \leq \delta u_{ss}(k) \leq \bar{u}''(k), \tag{4.15}$$

where

$$\underline{u}''(k) = \min\{\max\{\underline{u} - u(k-1), -\delta\bar{u}'_{ss}\}, \bar{u} - u(k-1)\},$$

$$\bar{u}''(k) = \max\{\min\{\bar{u} - u(k-1), \delta\bar{u}'_{ss}\}, \underline{u} - u(k-1)\}.$$

If $i \in \mathcal{I}_t \subseteq \mathbb{N}_u$, then $u_{i,ss}(k)$ is expected to satisfy

$$u_{i,t}(k) - \frac{1}{2}u_{i,ss,range} \leq u_{i,ss}(k) \leq u_{i,t}(k) + \frac{1}{2}u_{i,ss,range}. \tag{4.16}$$

Since (4.15) has to be satisfied, (4.16) is revised as the soft constraint

$$\underline{u}_{i,ss}(k) \leq \delta u_{i,ss}(k) \leq \bar{u}_{i,ss}(k), \quad i \in \mathcal{I}_t, \tag{4.17}$$

where

$$\underline{u}_{i,\text{ss}}(k) = \min\{\max\{u_{i,t}(k) - \frac{1}{2}u_{i,\text{ss,range}} - u_i(k-1), \underline{u}_i''(k)\}, \bar{u}_i''(k)\},$$

$$\bar{u}_{i,\text{ss}}(k) = \max\{\min\{u_{i,t}(k) + \frac{1}{2}u_{i,\text{ss,range}} - u_i(k-1), \bar{u}_i''(k)\}, \underline{u}_i''(k)\}.$$

By adding MVET, the soft constraints for $u_{\text{ss}}(k)$ become

$$\underline{u}_{i,\text{ss}}(k) \le \delta u_{i,\text{ss}}(k) \le \bar{u}_{i,\text{ss}}(k), \quad \delta u_{i,\text{ss}}(k) = u_{i,t}(k) - u_i(k-1), \quad i \in \mathcal{I}_t. \tag{4.18}$$

Each CVss has hard magnitude constraint (engineering limits), i.e.,

$$\underline{y}_{0,h} \le y_{\text{ss}}(k) \le \bar{y}_{0,h}. \tag{4.19}$$

Usually, an additional constraint is imposed on $\delta y_{\text{ss}}(k)$, i.e.,

$$|\delta y_{\text{ss}}(k)| \le \delta \bar{y}_{\text{ss}}. \tag{4.20}$$

Violating (4.20) is not inconsistent with DC. If (4.20) conflicts with (4.19), then (4.19) has to be satisfied (i.e., (4.20) is somewhat not always hard). Hence, by applying (4.10), it is shown that (4.19)–(4.20) are combined to yield the hard constraint

$$\underline{y}_h(k) \le S_N^u \delta u_{\text{ss}}(k) \le \bar{y}_h(k), \tag{4.21}$$

where

$$\underline{y}_h(k) = \min\{\max\{-\delta\bar{y}_{\text{ss}}, \underline{y}_{0,h} - y_{\text{ss}}^{\text{ol}}(k)\}, \bar{y}_{0,h} - y_{\text{ss}}^{\text{ol}}(k)\},$$

$$\bar{y}_h(k) = \max\{\min\{\delta\bar{y}_{\text{ss}}, \bar{y}_{0,h} - y_{\text{ss}}^{\text{ol}}(k)\}, \underline{y}_{0,h} - y_{\text{ss}}^{\text{ol}}(k)\}.$$

Each CVss is expected to satisfy (operating limits)

$$\underline{y}_0 \le y_{\text{ss}}(k) \le \bar{y}_0. \tag{4.22}$$

Since (4.21) has to be satisfied, (4.22) is revised as the soft constraint

$$S_N^u \delta u_{\text{ss}}(k) \le \bar{y}(k), \tag{4.23}$$

$$S_N^u \delta u_{\text{ss}}(k) \ge \underline{y}(k), \tag{4.24}$$

where

$$\underline{y}(k) = \min\{\max\{\underline{y}_0 - y_{\text{ss}}^{\text{ol}}(k), \underline{y}_h(k)\}, \bar{y}_h(k)\},$$

$$\bar{y}(k) = \max\{\min\{\bar{y}_0 - y_{\text{ss}}^{\text{ol}}(k), \bar{y}_h(k)\}, \underline{y}_h(k)\}.$$

If $j \in \mathcal{O}_t \subseteq \mathbb{N}_y$, then $y_{j,\text{ss}}(k)$ is expected to satisfy

$$y_{j,t}(k) - \frac{1}{2}y_{j,\text{ss,range}} \le y_{j,\text{ss}}(k) \le y_{j,t}(k) + \frac{1}{2}y_{j,\text{ss,range}}. \tag{4.25}$$

Since (4.21) has to be satisfied, (4.25) is revised as the soft constraint

$$\underline{y}_{j,\text{ss}}(k) \le S_{j,N}^u \delta u_{\text{ss}}(k) \le \bar{y}_{j,\text{ss}}(k), \quad j \in \mathcal{O}_t, \tag{4.26}$$

where

$$\underline{y}_{j,\mathrm{ss}}(k)=\min\{\max\{y_{j,t}(k)-\frac{1}{2}y_{j,\mathrm{ss,range}}-y^{\mathrm{ol}}_{j,\mathrm{ss}}(k),\underline{y}_{j,h}(k)\},\bar{y}_{j,h}(k)\}, \tag{4.27}$$

$$\bar{y}_{j,\mathrm{ss}}(k)=\max\{\min\{y_{j,t}(k)+\frac{1}{2}y_{j,\mathrm{ss,range}}-y^{\mathrm{ol}}_{j,\mathrm{ss}}(k),\bar{y}_{j,h}(k)\},\underline{y}_{j,h}(k)\}. \tag{4.28}$$

By adding CVET to (4.26), it becomes

$$\underline{y}_{j,\mathrm{ss}}(k)\le S^{u}_{j,N}\delta u_{\mathrm{ss}}(k)\le\bar{y}_{j,\mathrm{ss}}(k),\ \ S^{u}_{j,N}\delta u_{\mathrm{ss}}(k)=y_{j,t}(k)-y^{\mathrm{ol}}_{j,\mathrm{ss}}(k),\ \ j\in\mathcal{O}_{t}. \tag{4.29}$$

4.2.2 Priority Rank of Soft Constraints

The soft constraint set {(4.18), (4.23), (4.24), (4.29)} need to be grouped into a number of (say, r_0) subsets, each corresponding to a priority rank. The soft constraints with lower value of r have higher priority. The soft constraints are handled in the order of priority rank. The following set of notations will be utilized:

$\tilde{\mathcal{O}}_u^{(r)}$: the index for upper bounds of CV in rank r;
$\mathcal{O}_u^{(r)}$: the accumulated index for upper bounds of CV in and before rank r;
$\overline{\mathcal{O}}_u^{(r)}$: the accumulated index for $\varepsilon^* \neq 0$ upper bounds of CV in and before rank r;
$\tilde{\mathcal{O}}_l^{(r)}$: the index for lower bounds of CV in rank r;
$\mathcal{O}_l^{(r)}$: the accumulated index for lower bounds of CV in and before rank r;
$\overline{\mathcal{O}}_l^{(r)}$: the accumulated index for $\varepsilon^* \neq 0$ lower bounds of CV in and before rank r;
$\tilde{\mathbb{O}}^{(r)}$: the index for CVET in rank r;
$\mathbb{O}^{(r)}$: the accumulated index for CVET in and before rank r;
$\tilde{\mathbb{I}}^{(r)}$: the index for MVET in rank r;
$\mathbb{I}^{(r)}$: the accumulated index for MVET in and before rank r;

It satisfies $\mathcal{O}_u^{(r)}=\bigcup_{k=1}^{r}\tilde{\mathcal{O}}_u^{(k)}$, $\mathcal{O}_l^{(r)}=\bigcup_{k=1}^{r}\tilde{\mathcal{O}}_l^{(k)}$, $\mathbb{O}^{(r)}=\bigcup_{k=1}^{r}\tilde{\mathbb{O}}^{(k)}$, $\mathbb{I}^{(r)}=\bigcup_{k=1}^{r}\tilde{\mathbb{I}}^{(k)}$. In rank $r\ (r=1,2,\dots,r_0)$, the following soft constraints are included:

- the upper bounds for some CVs, i.e., (4.23) for $j\in\tilde{\mathcal{O}}_u^{(r)}$. These constraints can be relaxed as

$$S^{u}_{j,N}\delta u_{\mathrm{ss}}(k)\le\bar{y}_j(k)+\varepsilon_{\bar{y},j},\ \ j\in\tilde{\mathcal{O}}_u^{(r)}\backslash\mathbb{O}^{(r-1)}, \tag{4.30}$$

  where $\varepsilon_{\bar{y},j}$ is the nonnegative slack variable;
- the lower bounds for some CVs, i.e., (4.24) for $j\in\tilde{\mathcal{O}}_l^{(r)}$. These constraints can be relaxed as

$$-S^{u}_{j,N}\delta u_{\mathrm{ss}}(k)\le-\underline{y}_j(k)+\varepsilon_{\underline{y},j},\ \ j\in\tilde{\mathcal{O}}_l^{(r)}\backslash\mathbb{O}^{(r-1)}, \tag{4.31}$$

  where $\varepsilon_{\underline{y},j}$ is the nonnegative slack variable;
- the expected value and upper/lower bounds for some CVs, i.e., (4.29) for $j\in\tilde{\mathbb{O}}^{(r)}$. These constraints can be relaxed as

$$\begin{aligned}&-S^{u}_{j,N}\delta u_{\mathrm{ss}}(k)\le-\underline{y}_{j,\mathrm{ss}}(k)+\varepsilon_{\underline{y},j,\mathrm{ss}},\ \ S^{u}_{j,N}\delta u_{\mathrm{ss}}(k)\le\bar{y}_{j,\mathrm{ss}}(k)+\varepsilon_{\bar{y},j,\mathrm{ss}},\\&S^{u}_{j,N}\delta u_{\mathrm{ss}}(k)=y_{j,t}(k)-y^{\mathrm{ol}}_{j,\mathrm{ss}}(k)+\varepsilon_{y,j,t,+}-\varepsilon_{y,j,t,-},\\&j\in\tilde{\mathbb{O}}^{(r)}\backslash\overline{\mathcal{O}}_u^{(r-1)}\backslash\overline{\mathcal{O}}_l^{(r-1)},\end{aligned} \tag{4.32}$$

  where $\varepsilon_{\underline{y},j,\mathrm{ss}}$, $\varepsilon_{\bar{y},j,\mathrm{ss}}$, $\varepsilon_{y,j,t,+}$, $\varepsilon_{y,j,t,-}$ are the nonnegative slack variables;

- the expected value and upper/lower bounds for some MVs, i.e., (4.18) for $i \in \tilde{\mathbb{I}}^{(r)}$. These constraints can be relaxed as

$$
\begin{aligned}
&- \delta u_{i,\mathrm{ss}}(k) \leq -\underline{u}_{i,\mathrm{ss}}(k) + \varepsilon_{\underline{u},i,\mathrm{ss}}, \quad \delta u_{i,\mathrm{ss}}(k) \leq \bar{u}_{i,\mathrm{ss}}(k) + \varepsilon_{\bar{u},i,\mathrm{ss}}, \\
&\delta u_{i,\mathrm{ss}}(k) = u_{i,t}(k) - u_i(k-1) + \varepsilon_{u,i,t,+} - \varepsilon_{u,i,t,-}, \quad i \in \tilde{\mathbb{I}}^{(r)},
\end{aligned} \tag{4.33}
$$

where $\varepsilon_{\underline{u},i,\mathrm{ss}}$, $\varepsilon_{\bar{u},i,\mathrm{ss}}$, $\varepsilon_{u,i,t,+}$, $\varepsilon_{u,i,t,-}$ are the nonnegative slack variables.

Let us express the soft constraints in rank r (i.e., (4.30)–(4.33)) as

$$
\begin{cases}
\tilde{C}^{(r)} \delta u_{\mathrm{ss}}(k) \leq \tilde{c}^{(r)}(k) + \varepsilon^{(r)}(k) \\
\tilde{C}_{\mathrm{eq}}^{(r)} \delta u_{\mathrm{ss}}(k) = \tilde{c}_{\mathrm{eq}}^{(r)}(k) + \varepsilon_{\mathrm{eq}+}^{(r)}(k) - \varepsilon_{\mathrm{eq}-}^{(r)}(k)
\end{cases}, \tag{4.34}
$$

where $\varepsilon^{(r)}(k)$, $\varepsilon_{\mathrm{eq}+}^{(r)}(k)$ and $\varepsilon_{\mathrm{eq}-}^{(r)}(k)$ are nonnegative slack variables, appropriately composed of ε in (4.30)–(4.33). By analogy, the soft constraints in rank $r+1$ are expressed as

$$
\begin{cases}
\tilde{C}^{(r+1)} \delta u_{\mathrm{ss}}(k) \leq \tilde{c}^{(r+1)}(k) + \varepsilon^{(r+1)}(k) \\
\tilde{C}_{\mathrm{eq}}^{(r+1)} \delta u_{\mathrm{ss}}(k) = \tilde{c}_{\mathrm{eq}}^{(r+1)}(k) + \varepsilon_{\mathrm{eq}+}^{(r+1)}(k) - \varepsilon_{\mathrm{eq}-}^{(r+1)}(k)
\end{cases}. \tag{4.35}
$$

4.2.3 Feasibility Stage

In each rank, the following hard constraint has to be satisfied:

$$
\begin{bmatrix} I \\ -I \\ S_N^u \\ -S_N^u \end{bmatrix} \delta u_{\mathrm{ss}}(k) \leq \begin{bmatrix} \bar{u}''(k) \\ -\underline{u}''(k) \\ \bar{y}_h(k) \\ -\underline{y}_h(k) \end{bmatrix}, \tag{4.36}
$$

which combines (4.15) and (4.21). In rank r, (4.34) is handled, which means minimizing/finding/fixing ε in rank r so that (4.34) does not conflict in itself or with the hard constraints. With $\varepsilon = \varepsilon^*$ having been determined, (4.34) is the hard constraint for rank $r+1$. The process, from expecting $\varepsilon = 0$ to fixing $\varepsilon = \varepsilon^*$, is called relaxing or hardening. The constraint (4.34) with fixed $\varepsilon = \varepsilon^*$ is called soft-to-hard constraint. Denote the soft-to-hard constraints, from ranks 1 to r, as

$$
\begin{cases}
C^{(r)} \delta u_{\mathrm{ss}}(k) \leq c^{(r)}(k) \\
C_{\mathrm{eq}}^{(r)} \delta u_{\mathrm{ss}}(k) = c_{\mathrm{eq}}^{(r)}(k)
\end{cases}. \tag{4.37}
$$

For the rank $r+1$, (4.37) is the hard constraint. By applying (4.30)–(4.33), it is easy to show that (4.37) includes

$$
S_{j,N}^u \delta u_{\mathrm{ss}}(k) \leq \bar{y}_j(k) + \varepsilon_{\bar{y},j}, \quad j \in \overline{\mathcal{O}}_u^{(r)}, \tag{4.38}
$$

$$
-S_{j,N}^u \delta u_{\mathrm{ss}}(k) \leq -\underline{y}_j(k) + \varepsilon_{\underline{y},j}, \quad j \in \overline{\mathcal{O}}_l^{(r)}, \tag{4.39}
$$

$$
S_{j,N}^u \delta u_{\mathrm{ss}}(k) \leq \bar{y}_j(k), \quad j \in \mathcal{O}_u^{(r)} \backslash \overline{\mathcal{O}}_u^{(r)}, \tag{4.40}
$$

$$
-S_{j,N}^u \delta u_{\mathrm{ss}}(k) \leq -\underline{y}_j(k), \quad j \in \mathcal{O}_l^{(r)} \backslash \overline{\mathcal{O}}_l^{(r)}, \tag{4.41}
$$

$$
S_{j,N}^u \delta u_{\mathrm{ss}}(k) = y_{j,t}(k) - y_{j,\mathrm{ss}}^{\mathrm{ol}}(k) + \varepsilon_{y,j,t,+} - \varepsilon_{y,j,t,-}, \quad j \in \mathbb{O}^{(r)} \backslash \overline{\mathcal{O}}_u^{(r-1)} \backslash \overline{\mathcal{O}}_l^{(r-1)}, \tag{4.42}
$$

$$
\delta u_{i,\mathrm{ss}}(k) = u_{i,t}(k) - u_i(k-1) + \varepsilon_{u,i,t,+} - \varepsilon_{u,i,t,-}, \quad i \in \mathbb{I}^{(r)}, \tag{4.43}
$$

where $\overline{\mathcal{O}}_u^{(r)} \subseteq \mathcal{O}_u^{(r)}$ and $\overline{\mathcal{O}}_l^{(r)} \subseteq \mathcal{O}_l^{(r)}$ are the index sets with $\varepsilon^* \neq 0$. For the continued optimizations, the equalities in (4.38) and (4.39) will be satisfied. Hence, (4.37) is reduced to the following constraints:

$$\begin{cases} S_{j,N}^u \delta u_{ss}(k) \le \bar{y}_j(k), & j \in \mathcal{O}_u^{(r)} \backslash \overline{\mathcal{O}}_u^{(r)} \backslash \overline{\mathcal{O}}_l^{(r)} \\ -S_{j,N}^u \delta u_{ss}(k) \le -\underline{y}_j(k), & j \in \mathcal{O}_l^{(r)} \backslash \overline{\mathcal{O}}_u^{(r)} \backslash \overline{\mathcal{O}}_l^{(r)} \\ S_{j,N}^u \delta u_{ss}(k) = y_{j,ss}(k) - y_{j,ss}^{ol}(k), & j \in \mathbb{O}^{(r)} \bigcup \overline{\mathcal{O}}_u^{(r)} \bigcup \overline{\mathcal{O}}_l^{(r)} \\ \delta u_{i,ss}(k) = u_{i,ss}(k) - u_i(k-1), & i \in \mathbb{I}^{(r)} \end{cases}, \tag{4.44}$$

where

$$y_{j,ss}(k) = \begin{cases} y_{j,t}(k) + \varepsilon_{y,j,t,+} - \varepsilon_{y,j,t,-}, & j \in \mathbb{O}^{(r)} \backslash \overline{\mathcal{O}}_u^{(r-1)} \backslash \overline{\mathcal{O}}_l^{(r-1)} \\ \bar{y}_j(k) + \varepsilon_{\bar{y},j}, & j \in \overline{\mathcal{O}}_u^{(r)} \\ \underline{y}_j(k) - \varepsilon_{\underline{y},j}, & j \in \overline{\mathcal{O}}_l^{(r)} \end{cases}$$

and $u_{i,ss}(k) = u_{i,t}(k) + \varepsilon_{u,i,t,+} - \varepsilon_{u,i,t,-}$.

Lemma 4.1 With (4.44) satisfied, the hard constraints in (4.36) can be simplified as

$$\begin{cases} \underline{u}_i''(k) \le \delta u_{i,ss}(k) \le \bar{u}_i''(k), & i \in \mathbb{N}_u \backslash \mathbb{I}^{(r)} \\ S_{j,N}^u \delta u_{ss}(k) \le \bar{y}_{j,h}(k), & j \in \mathbb{N}_y \backslash \mathbb{O}^{(r)} \backslash \mathcal{O}_u^{(r)} \backslash \overline{\mathcal{O}}_l^{(r)} \\ -S_{j,N}^u \delta u_{ss}(k) \le -\underline{y}_{j,h}(k), & j \in \mathbb{N}_y \backslash \mathbb{O}^{(r)} \backslash \mathcal{O}_l^{(r)} \backslash \overline{\mathcal{O}}_u^{(r)} \end{cases}. \tag{4.45}$$

In rank $r+1$, either LP or QP is invoked, so as to find ε as small as possible. In each rank, if multiple ε are to be optimized, then they are equally important. For each ε, an equal concern error $\bar{\varepsilon}$ is applied, which is the maximum value of ε, i.e.,

$$\bar{\varepsilon}_{\bar{y},j} = \bar{y}_{j,h}(k) - \bar{y}_j(k), \qquad \bar{\varepsilon}_{\underline{y},j} = \underline{y}_j(k) - \underline{y}_{j,h}(k),$$

$$\bar{\varepsilon}_{\bar{y},j,ss} = \bar{y}_{j,h}(k) - \bar{y}_{j,ss}(k), \qquad \bar{\varepsilon}_{\underline{y},j,ss} = \underline{y}_{j,ss}(k) - \underline{y}_{j,h}(k),$$

$$\bar{\varepsilon}_{y,j,t,+} = \bar{\varepsilon}_{y,j,t,-} = \frac{1}{2}[\bar{y}_{j,h}(k) - \underline{y}_{j,h}(k)],$$

$$\bar{\varepsilon}_{\bar{u},i,ss} = \bar{u}_i''(k) - \bar{u}_{i,ss}(k), \qquad \bar{\varepsilon}_{\underline{u},i,ss} = \underline{u}_{i,ss}(k) - \underline{u}_i''(k),$$

$$\bar{\varepsilon}_{u,i,t,+} = \bar{\varepsilon}_{u,i,t,-} = \frac{1}{2}[\bar{u}_i''(k) - \underline{u}_i''(k)].$$

If a $\bar{\varepsilon} \le 0$, then its corresponding soft constraint cannot be relaxed; in this case, we can remove the corresponding soft constraint, or simply take this $\bar{\varepsilon} = 1$ (it is equivalent to taking any value for this $\bar{\varepsilon}$).

By invoking LP, the optimization problem is

$$\min_{\varepsilon^{(r+1)}(k), \varepsilon_{eq+}^{(r+1)}(k), \varepsilon_{eq-}^{(r+1)}(k), \delta u_{ss}(k)} \left[\sum_{\ell=1}^{d^{(r+1)}} \left(\bar{\varepsilon}_\ell^{(r+1)}\right)^{-1} \varepsilon_\ell^{(r+1)}(k) \right.$$
$$\left. + \sum_{\tau=1}^{d_{eq}^{(r+1)}} \left[\left(\bar{\varepsilon}_{eq+,\tau}^{(r+1)}\right)^{-1} \varepsilon_{eq+,\tau}^{(r+1)}(k) + \left(\bar{\varepsilon}_{eq-,\tau}^{(r+1)}\right)^{-1} \varepsilon_{eq-,\tau}^{(r+1)}(k) \right] \right],$$

$$\text{s.t. (4.44)–(4.45), (4.35) and } \varepsilon \ge 0, \tag{4.46}$$

where the subscript ℓ corresponds to the ℓ-th element of $\varepsilon^{(r+1)}(k)$, and $d^{(r+1)}$ is the dimension of $\varepsilon^{(r+1)}$; the subscript τ corresponds to the τth element of $\varepsilon_{\rm eq}^{(r+1)}(k)$, and $d_{\rm eq}^{(r+1)}$ is the dimension of $\varepsilon_{\rm eq}^{(r+1)}(k)$.

By invoking QP, the optimization problem is

$$\min_{\varepsilon^{(r+1)}(k),\varepsilon_{\rm eq+}^{(r+1)}(k),\varepsilon_{\rm eq-}^{(r+1)}(k),\delta u_{\rm ss}(k)} \left[\sum_{\ell=1}^{d^{(r+1)}} (\overline{\varepsilon}_{\ell}^{(r+1)})^{-2} \varepsilon_{\ell}^{(r+1)}(k)^2 + \sum_{\tau=1}^{d_{\rm eq}^{(r+1)}} \left[(\overline{\varepsilon}_{{\rm eq}+,\tau}^{(r+1)})^{-2} \varepsilon_{{\rm eq}+,\tau}^{(r+1)}(k)^2 + (\overline{\varepsilon}_{{\rm eq}-,\tau}^{(r+1)})^{-2} \varepsilon_{{\rm eq}-,\tau}^{(r+1)}(k)^2 \right] \right],$$

$$\text{s.t. (4.44)–(4.45), (4.35), and } \varepsilon \geq 0. \tag{4.47}$$

4.2.4 Economic Stage

There are two choices for the economic stage. One, being called pure economic, is that the economic optimization has lower priority than any rank r ($r = 1, 2, \dots, r_0$). The other, being called mixed economic, is that the economic optimization has the same priority with the lowest priority rank r_0.

After the feasibility stage, for a part of CVs/MVs, their ss have been given; for the remaining CVs/MVs, their ss can be evaluated by an economic cost. This economic cost measures the benefit/loss by changing $u_{\rm ss}(k)$ and $y_{\rm ss}(k)$, so it can be expressed in $\delta u_{\rm ss}(k)$. Some MVs may be negligible for the economic cost, and some MVs are desired "better not to change"; these MVs are called minimum-move MVs. Denote the index set of minimum-move MV as $\mathcal{I}_{\rm mm}$. For any $i \in \mathcal{I}_{\rm mm}$, minimizing the move of $\delta u_{i,{\rm ss}}(k)$ means penalizing $|\delta u_{i,{\rm ss}}(k)|$. Minimizing $|\delta u_{i,{\rm ss}}(k)|$ is equivalent to minimizing $U_i(k)$ satisfying

$$-U_i(k) \leq \delta u_{i,{\rm ss}}(k) \leq U_i(k). \tag{4.48}$$

The minimum-move penalty term is also added to the economic cost.

Lemma 4.2 After feasibility stage optimizations from rank 1 to rank r_0, all the hard (including soft-to-hard) constraints can be expressed as

$$\begin{cases} \underline{u}_i''(k) \leq \delta u_{i,{\rm ss}}(k) \leq \bar{u}_i''(k), & i \in \mathbb{N}_u \backslash \mathcal{I}_t \\ \underline{y}_j(k) \leq S_{j,N}^u \delta u_{\rm ss}(k) \leq \bar{y}_j(k), & j \in \mathbb{N}_y \backslash \mathcal{O}_t \backslash \overline{\mathcal{O}}_u^{(r_0)} \backslash \overline{\mathcal{O}}_l^{(r_0)} \\ S_{j,N}^u \delta u_{\rm ss}(k) = y_{j,{\rm ss}}(k) - y_{j,{\rm ss}}^{\rm ol}(k), & j \in \mathcal{O}_t \bigcup \overline{\mathcal{O}}_u^{(r_0)} \bigcup \overline{\mathcal{O}}_l^{(r_0)} \\ \delta u_{i,{\rm ss}}(k) = u_{i,{\rm ss}}(k) - u_i(k-1), & i \in \mathcal{I}_t \end{cases}. \tag{4.49}$$

Consider the pure economic. By invoking LP, the optimization problem is

$$\min_{\delta u_{i,{\rm ss}}(k),U_i(k)} \left[\sum_{i \notin \mathcal{I}_{\rm mm} \bigcup \mathcal{I}_t} h_i \delta u_{i,{\rm ss}}(k) + \sum_{i \in \mathcal{I}_{\rm mm}} h_i U_i(k) \right],$$

$$\text{s.t. (4.48), } i \in \mathcal{I}_{\rm mm}\text{; (4.49),}$$

where h_i is the weight coefficient. By invoking QP, the optimization problem is

$$\min_{\delta u_{i,ss}(k),U_i(k)} \left[\left(\sum_{i \notin \mathcal{I}_{mm} \bigcup \mathcal{I}_t} h_i \delta u_{i,ss}(k) - J_{min} \right)^2 + \sum_{i \in \mathcal{I}_{mm}} h_i^2 U_i(k)^2 \right],$$

$$\text{s.t. } (4.48),\ i \in \mathcal{I}_{mm};\ (4.49),$$

where J_{min} is equal to, or smaller than, the minimum of $\sum_{i \notin \mathcal{I}_{mm} \bigcup \mathcal{I}_t} h_i \delta u_{i,ss}(k)$. One can utilize the upper/lower bounds of $\delta u_{i,ss}(k)$, and the signs of h_i, to determine a J_{min}. If there is no minimum-move MV, then LP and QP give the same result; otherwise, they may give different results.

Consider the mixed economic. By invoking LP, the optimization problem is

$$\begin{aligned}
\min_{\varepsilon^{(r_0)}(k),\varepsilon_{eq+}^{(r_0)}(k),\varepsilon_{eq-}^{(r_0)}(k),\delta u_{i,ss}(k),U_i(k)} & \left[\sum_{i \notin \mathcal{I}_{mm} \bigcup \mathcal{I}_t} h_i \delta u_{i,ss}(k) + \sum_{i \in \mathcal{I}_{mm}} h_i U_i(k) \right. \\
& + \sum_{\ell=1}^{d^{(r_0)}} \left(\overline{\varepsilon}_\ell^{(r_0)} \right)^{-1} \varepsilon_\ell^{(r_0)}(k) \\
& \left. + \sum_{\tau=1}^{d_{eq}^{(r_0)}} \left[\left(\overline{\varepsilon}_{eq+,\tau}^{(r_0)} \right)^{-1} \varepsilon_{eq+,\tau}^{(r_0)}(k) + \left(\overline{\varepsilon}_{eq-,\tau}^{(r_0)} \right)^{-1} \varepsilon_{eq-,\tau}^{(r_0)}(k) \right] \right], \\
\text{s.t. } & (4.48),\ i \in \mathcal{I}_{mm};\ (4.44)-(4.45),\ (4.35),\ r = r_0 - 1, \text{ and } \varepsilon \geq 0.
\end{aligned} \tag{4.50}$$

By invoking QP, the optimization problem is

$$\begin{aligned}
\min_{\varepsilon^{(r_0)}(k),\varepsilon_{eq+}^{(r_0)}(k),\varepsilon_{eq-}^{(r_0)}(k),\delta u_{i,ss}(k),U_i(k)} & \left[\left(\sum_{i \notin \mathcal{I}_{mm} \bigcup \mathcal{I}_t} h_i \delta u_{i,ss}(k) - J_{min} \right)^2 \right. \\
& + \sum_{i \in \mathcal{I}_{mm}} h_i^2 U_i(k)^2 + \sum_{\ell=1}^{d^{(r_0)}} (\overline{\varepsilon}_\ell^{(r_0)})^{-2} \varepsilon_\ell^{(r_0)}(k)^2 \\
& \left. + \sum_{\tau=1}^{d_{eq}^{(r_0)}} \left[(\overline{\varepsilon}_{eq+,\tau}^{(r_0)})^{-2} \varepsilon_{eq+,\tau}^{(r_0)}(k)^2 + (\overline{\varepsilon}_{eq-,\tau}^{(r_0)})^{-2} \varepsilon_{eq-,\tau}^{(r_0)}(k)^2 \right] \right], \\
\text{s.t. } & (4.48),\ i \in \mathcal{I}_{mm};\ (4.44)-(4.45),\ (4.35),\ r = r_0 - 1, \text{ and } \varepsilon \geq 0.
\end{aligned} \tag{4.51}$$

In the mixed economic, LP and QP may not give the same result.

After the feasibility and economic stages, $\delta u_{ss}(k)$ has been optimized. Using (4.10) obtains $y_{ss}(k)$. Thus, SSTC is finished.

4.3 Dynamic Calculation Module

Let $\{M, N, P\}$ satisfy $M \leq P \leq N + M$.

At each control interval k, given $Y_N^{ol}(k|k)$ in Section 4.1, we have

$$Y_P^{ol}(k|k) = \begin{bmatrix} y^{ol}(k+1|k) \\ y^{ol}(k+2|k) \\ \vdots \\ y^{ol}(k+P|k) \end{bmatrix}.$$

When $P > N$, it satisfies $y^{\rm ol}(k+j|k) = y^{\rm ol}(k+N|k)$ for all $j > N$. By applying (4.6), changing the superscript fr as ol, it is easy to obtain

$$Y_P(k|k) = Y_P^{\rm ol}(k|k) + \mathscr{S}\,\Delta\tilde{u}(k|k), \tag{4.52}$$

where

$$Y_P(k|k) = \begin{bmatrix} y(k+1|k) \\ y(k+2|k) \\ \vdots \\ y(k+P|k) \end{bmatrix}, \quad \Delta\tilde{u}(k|k) = \begin{bmatrix} \Delta u(k|k) \\ \Delta u(k+1|k) \\ \vdots \\ \Delta u(k+M-1|k) \end{bmatrix},$$

$$\mathscr{S} = \begin{bmatrix} S_1^u & 0 & \cdots & 0 \\ S_2^u & S_1^u & \ddots & \vdots \\ \vdots & \ddots & \ddots & 0 \\ S_M^u & \cdots & S_2^u & S_1^u \\ \vdots & & \vdots & \vdots \\ S_P^u & \cdots & S_{P-M+2}^u & S_{P-M+1}^u \end{bmatrix}.$$

$\mathscr{S}$ is always called the dynamic matrix. Equation (4.52) is the closed-loop prediction equation.

In DC, three objectives should be achieved: (i) the future values of CV tracking $y_{\rm ss}(k)$ as close as possible; (ii) the sharp change of MV being suppressed; (iii) the soft constraint of CV being relaxed as small as possible with respect to feasibility retrieval. Choose the cost function, for the dynamic calculation, as

$$J(k) = \sum_{i=1}^{P} \|y(k+i|k) - y_{\rm ss}(k)\|^2_{Q_i(k)} + \sum_{j=0}^{M-1} \|\Delta u(k+j|k)\|^2_{\Lambda},$$

$$J'(k) = \sum_{i=1}^{P} \|y(k+i|k) - y_{\rm ss}(k)\|^2_{Q_i(k)} + \sum_{j=0}^{M-1} \|\Delta u(k+j|k)\|^2_{\Lambda} + \|\underline{\varepsilon}_{\rm dc}(k)\|^2_{\underline{\Omega}} + \|\overline{\varepsilon}_{\rm dc}(k)\|^2_{\overline{\Omega}},$$

where $\underline{\varepsilon}_{\rm dc}(k)$ and $\overline{\varepsilon}_{\rm dc}(k)$ are the slack variables of CV constraint,

$$Q_i(k) = \mathrm{diag}\{q_{i1}(k)^2, q_{i2}(k)^2, \ldots, q_{i,n_y}(k)^2\},$$

$$q_{ij}(k) = \begin{cases} \underline{q}_j, & \check{y}_j(k+i|k) \le \underline{y}_{j,0} \\ \underline{a}\check{y}_j(k+i|k) + \underline{b}, & \underline{y}_{j,0} \le \check{y}_j(k+i|k) \le \underline{z}_j \\ \check{q}_j, & \underline{z}_j \le \check{y}_j(k+i|k) \le \overline{z}_j \\ \overline{a}\check{y}_j(k+i|k) + \overline{b}, & \overline{z}_j \le \check{y}_j(k+i|k) \le \overline{y}_{j,0} \\ \overline{q}_j, & \check{y}_j(k+i|k) \ge \overline{y}_{j,0} \end{cases},$$

$$\underline{a} = \frac{\underline{q}_j - \check{q}_j}{\underline{y}_{j,0} - \underline{z}_j}, \quad \underline{b} = \frac{\underline{y}_{j,0}\check{q}_j - \underline{z}_j\underline{q}_j}{\underline{y}_{j,0} - \underline{z}_j}, \quad \overline{a} = \frac{\overline{q}_j - \check{q}_j}{\overline{y}_{j,0} - \overline{z}_j}, \quad \overline{b} = \frac{\overline{y}_{j,0}\check{q}_j - \overline{z}_j\overline{q}_j}{\overline{y}_{j,0} - \overline{z}_j},$$

$$\Lambda = \mathrm{diag}\{\lambda_1^2, \lambda_2^2, \ldots, \lambda_{n_u}^2\},$$

$$\underline{\Omega} = \text{diag}\{\underline{\omega}_1^2, \underline{\omega}_2^2, \dots, \underline{\omega}_{n_y}^2\}, \ \overline{\Omega} = \text{diag}\{\overline{\omega}_1^2, \overline{\omega}_2^2, \dots, \overline{\omega}_{n_y}^2\},$$

$$\underline{\omega}_j = (\underline{y}_{j,0} - \underline{y}_{j,0,h})^{-1}\rho, \ \overline{\omega}_j = (\overline{y}_{j,0,h} - \overline{y}_{j,0})^{-1}\rho.$$

$\check{y}_j(k+i|k)$ is determined by $y_j^{\text{ol}}(k+i|k)$ and $y_{j,\text{ss}}(k)$ as in the following:

- if $y_j^{\text{ol}}(k+i|k) > \max\{y_{j,\text{ss}}(k), \overline{y}_{j,0}\}$ or $y_j^{\text{ol}}(k+i|k) < \min\{y_{j,\text{ss}}(k), \underline{y}_{j,0}\}$, then take $\check{y}_j(k+i|k) = y_j^{\text{ol}}(k+i|k)$ in order to drive CV into operating limits as quick as possible;
- if $y_{j,\text{ss}}(k) \geq \max\{y_j^{\text{ol}}(k+i|k), \overline{y}_{j,0}\}$, then take $\check{y}_j(k+i|k) = \overline{z}_j$ in order to slow down the approaching of CV toward upper limit; if $y_{j,\text{ss}}(k) \leq \min\{y_j^{\text{ol}}(k+i|k), \underline{y}_{j,0}\}$, then take $\check{y}_j(k+i|k) = \underline{z}_j$ in order to slow down the approaching of CV toward lower limit;
- otherwise, then take $\check{y}_j(k+i|k) = \frac{1}{2}[y_j^{\text{ol}}(k+i|k) + y_{j,\text{ss}}(k)]$.

ρ, $\{\lambda_1, \lambda_2, \dots, \lambda_{n_u}\}$ and $\{\underline{q}_j, \check{q}_j, \overline{q}_j\}$ are tunable controller parameters (usually, $\underline{q}_j > \check{q}_j$ and $\overline{q}_j > \check{q}_j$). $\underline{z}_j$ and $\overline{z}_j$ are the boundaries of the transition zone. In the three cases (i.e., $\check{y}_j(k+i|k)$ being above CVHL, below CVLL, and between CVLL and CVHL), different weight coefficients are applied. The transition zone is built in order to avoid oscillations between $\underline{q}_j$ and $\check{q}_j$, or between $\check{q}_j$ and $\overline{q}_j$.

Differently from tracking $y_{\text{ss}}(k)$, in order to track $\delta u_{\text{ss}}(k)$ the following constraint is imposed in DC:

$$L\Delta\tilde{u}(k|k) = \delta u_{\text{ss}}(k), \tag{4.53}$$

where $L = [I \ I \ \cdots \ I]$. Moreover, in DC usually the following inequality constraints (on MV rate, MV maginitude, CV magnitude, and slack variable) are considered:

$$|\Delta u(k+j|k)| \leq \Delta\bar{u}, \ \ 0 \leq j \leq M-1, \tag{4.54}$$

$$\underline{u} \leq u(k-1) + \sum_{l=0}^{j} \Delta u(k+l|k) \leq \bar{u}, \ \ 0 \leq j \leq M-1, \tag{4.55}$$

$$\underline{y}_0'(k) \leq y^{\text{ol}}(k+i|k) + \mathcal{S}_i\Delta\tilde{u}(k|k) \leq \overline{y}_0'(k), \ \ 1 \leq i \leq P, \tag{4.56}$$

$$\underline{y}_0'(k) - \underline{\varepsilon}_{\text{dc}}(k) \leq y^{\text{ol}}(k+i|k) + \mathcal{S}_i\Delta\tilde{u}(k|k) \leq \overline{y}_0'(k) + \overline{\varepsilon}_{\text{dc}}(k), \ \ 1 \leq i \leq P, \tag{4.57}$$

$$\underline{\varepsilon}_{\text{dc}}(k) \leq \underline{y}_0'(k) - \underline{y}_{0,h}, \tag{4.58}$$

$$\overline{\varepsilon}_{\text{dc}}(k) \leq \overline{y}_{0,h} - \overline{y}_0'(k), \tag{4.59}$$

where $\mathcal{S}_i$ is the ith row of $\mathcal{S}$,

$$\overline{y}_0'(k) = \max\{\overline{y}_0, y_{\text{ss}}(k)\},$$

$$\underline{y}_0'(k) = \min\{\underline{y}_0, y_{\text{ss}}(k)\}.$$

In summary, at each control interval k, we first solve

$$\min_{\Delta\tilde{u}(k|k)} J(k), \ \text{s.t.}\ (4.53) - (4.56). \tag{4.60}$$

If (4.60) is infeasible, then further solve

$$\min_{\underline{\varepsilon}_{\text{dc}}(k), \overline{\varepsilon}_{\text{dc}}(k), \Delta\tilde{u}(k|k)} J'(k), \ \text{s.t.}\ (4.53) - (4.55),\ (4.57) - (4.59). \tag{4.61}$$

If either (4.60) or (4.61) is feasible, then $\Delta u(k|k)$ is sent to the controlled systems.

4.4 Numerical Example

Apply the heavy oil fractionator model. The continuous-time transfer function matrix around the equilibrium is

$$G^u(s) = \begin{bmatrix} \frac{4.05e^{-27s}}{50s+1} & \frac{1.77e^{-28s}}{60s+1} & \frac{5.88e^{-27s}}{50s+1} \\ \frac{5.39e^{-18s}}{50s+1} & \frac{5.72e^{-14s}}{60s+1} & \frac{6.90e^{-15s}}{40s+1} \\ \frac{4.38e^{-20s}}{33s+1} & \frac{4.42e^{-22s}}{44s+1} & \frac{7.20}{19s+1} \end{bmatrix}, \quad G^f(s) = \begin{bmatrix} \frac{1.20e^{-27s}}{45s+1} & \frac{1.44e^{-27s}}{40s+1} \\ \frac{1.52e^{-15s}}{25s+1} & \frac{1.83e^{-15s}}{20s+1} \\ \frac{1.14}{27s+1} & \frac{1.26}{32s+1} \end{bmatrix}.$$

The sampling time is 4.

Take $N = 100$, and use MATLAB command `tfd2step` to obtain the FSR model shown in (4.4) using the above transfer function matrix (i.e., the step response coefficient matrices S_i^u and S_i^f are obtained, for u and f, respectively). $\underline{u} = u_{\text{eq}} + [-0.5, -0.5, -0.5]^{\text{T}}$, $\bar{u} = u_{\text{eq}} + [0.5, 0.5, 0.5]^{\text{T}}$, $\Delta\bar{u}_i = \delta\bar{u}_{i,\text{ss}} = 0.1$; $\underline{y}_{0,h} = y_{\text{eq}} + [-0.7, -0.7, -0.7]^{\text{T}}$, $\bar{y}_{0,h} = y_{\text{eq}} + [0.7, 0.7, 0.7]^{\text{T}}$, $\underline{y}_0 = y_{\text{eq}} + [-0.5, -0.5, -0.5]^{\text{T}}$, $\bar{y}_0 = y_{\text{eq}} + [0.5, 0.5, 0.5]^{\text{T}}$, $\Delta\bar{y}_{1,\text{ss}} = 0.2$, $\Delta\bar{y}_{2,\text{ss}} = 0.2$, $\Delta\bar{y}_{3,\text{ss}} = 0.3$. Moreover, $\{y_1, y_2, u_3\}$ have ETs and their expected allowable range is 0.5. The parameters of SSTC are shown in Table 4.1; in each priority rank, either there are only equality-type soft constraints, or only inequality-type. It invokes mixed economics.

In the economic optimization, u_2 is the minimum-move MV. Take $h = [-2, 1, 2]$, $J_{\min} = -0.2$. At $k \in [64, 78]$, the disturbance appears with $f_{\text{eq}} + [0.20; 0.10]$; at $k \geq 122$, the disturbance appears with $f_{\text{eq}} + [1; -1]$; at the other intervals the disturbance is f_{eq}. $Y_N^{\text{ol}}(0|0) = [y_{\text{eq}}; \dots; y_{\text{eq}}]$. Take $y_{\text{eq}} = 0$, $u_{\text{eq}} = 0$ and $f_{\text{eq}} = 0$.

Choose $P = 15$, $M = 8$, $\Lambda = \text{diag}\{3, 5, 3\}$, $\bar{z} = y_{\text{eq}} + [0.4; 0.4; 0.4]$, $\underline{z} = y_{\text{eq}} + [-0.4; -0.4; -0.4]$, $\underline{q}_1 = 2.0$, $\check{q}_1 = 0.5$, $\bar{q}_1 = 2.0$; $\underline{q}_2 = 2.0$, $\check{q}_2 = 1.0$, $\bar{q}_2 = 2.0$; $\underline{q}_3 = 2.5$, $\check{q}_3 = 1.0$, $\bar{q}_3 = 4.0$, $\rho = 0.2$. $u(-1) = u_{\text{eq}}$, $y(0) = y_{\text{eq}}$, $\overline{\Delta u}(k) = \Delta u(k|k)$. The true system output is produced using MATLAB Simulink and the above transfer function matrix, which is multiplied by 0.9 to represent the mismatch between the model and the true system. The control results

Table 4.1 Parameters of multi-priority-rank SSTC.

Rank no.	Type of constraint	Variable	Soft constraint	$\bar{\varepsilon}$
1	Inequality	$y_{2,\text{ss}}$	CV lower bound	0.20
1	Inequality	$y_{3,\text{ss}}$	CV upper bound	0.20
2	Equality	$y_{2,\text{ss}}$	$y_{2,\text{eq}} - 0.6$	0.25
3	Inequality	$y_{2,\text{ss}}$	CV upper bound	0.20
3	Inequality	$y_{3,\text{ss}}$	CV lower bound	0.20
4	Equality	$u_{3,\text{ss}}$	$u_{3,\text{eq}} + 0.5$	0.25
4	Equality	$y_{1,\text{ss}}$	$y_{1,\text{eq}} + 0.7$	0.25
5	Inequality	$y_{1,\text{ss}}$	CV lower bound	0.20
5	Inequality	$y_{1,\text{ss}}$	CV upper bound	0.20

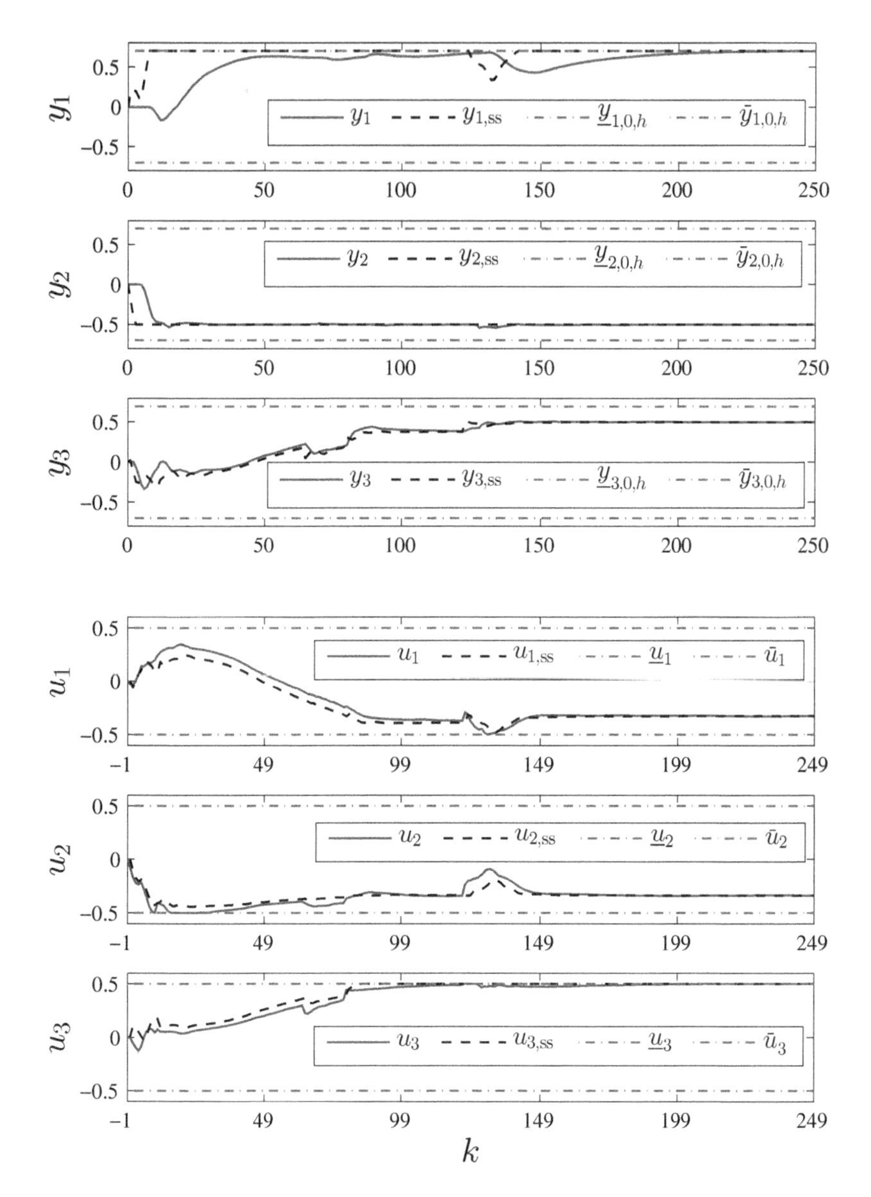

Figure 4.1 The control result

are shown in Figure 4.1, which demonstrates that CVss and MVss given by SSTC can be fully tracked by DC.

Note that in this example, when we take other arbitrary $\{y, u, f\}_{\mathrm{eq}}$ (without satisfying $y_{\mathrm{eq}} = S_N^u u_{\mathrm{eq}} + S_N^f f_{\mathrm{eq}}$), the result is still available. If $\{y, u, f\}_{\mathrm{eq}}$ satisfies $y_{\mathrm{eq}} = S_N^u u_{\mathrm{eq}} + S_N^f f_{\mathrm{eq}}$ and there is no model-plant mismatch, then for the different $\{y, u, f\}_{\mathrm{eq}}$, the simulation results are different in values but same in relative variations (i.e, the shapes of curves). These have been verified by simulation, which are omitted.

In this example, if the model remains unchanged, but the transfer function matrix that produces the true system output is modified as

$$G^u(s) = \begin{bmatrix} \frac{4.05e^{-27s}}{(50s+1)(5000s-1)} & \frac{1.77e^{-28s}}{60s+1} & \frac{5.88e^{-27s}}{50s+1} \\ \frac{5.39e^{-18s}}{50s+1} & \frac{5.72e^{-14s}}{(60s+1)(6000s-1)} & \frac{6.90e^{-15s}}{40s+1} \\ \frac{4.38e^{-20s}}{33s+1} & \frac{4.42e^{-22s}}{44s+1} & \frac{7.20}{(19s+1)(1900s-1)} \end{bmatrix},$$

then the closed-loop system cannot be stabilized; if the model remains unchanged, but the transfer function matrix that produces the true system output is modified as

$$G^u(s) = \begin{bmatrix} \left(\frac{4.05}{50s+1} + \frac{0.001}{s}\right)e^{-27s} & \frac{1.77e^{-28s}}{60s+1} & \frac{5.88e^{-27s}}{50s+1} \\ \frac{5.39e^{-18s}}{50s+1} & \left(\frac{5.72}{60s+1} + \frac{0.001}{s}\right)e^{-14s} & \frac{6.90e^{-15s}}{40s+1} \\ \frac{4.38e^{-20s}}{33s+1} & \frac{4.42e^{-22s}}{44s+1} & \frac{7.20}{19s+1} + \frac{0.001}{s} \end{bmatrix},$$

then the closed-loop system can be stabilized. If the transfer function matrix that produces the true system output remains unchanged, but FSR is changed as

$$S_i^u \leftarrow S_i^u(1 + 0.01i), \quad S_i^f \leftarrow S_i^f(1 + 0.01i),$$

then the closed-loop system can be stabilized. If the transfer function matrix that produces the true system output remains unchanged, but FSR is changed as

$$S_i^u \leftarrow (1 + 0.01e^{0.01i})S_i^u, \quad S_i^f \leftarrow (1 + 0.01e^{0.01i})S_i^f,$$

then the closed-loop system can be stabilized. If the transfer function matrix that produces the true system output remains unchanged, and FSR is changed as

$$S_i^u \leftarrow (1 - 0.01e^{0.01i})S_i^u, \quad S_i^f \leftarrow (1 - 0.01e^{0.01i})S_i^f,$$

then the closed-loop system can be stabilized. The above stabilizable cases need to modify some controller parameters. For a series of simulations, we summarize as follows.

The system cannot be unstable, while the model can be unstable;
The system can be integrals, while the model can have integrals.

5

Two-Layered DMC for Stable and Integral Processes

The notations, if not explained, are the same as in Chapter 4. New notations:

n_y^r:	the number of iCVs;
$\mathbb{N}_y^r$:	$\mathbb{N}_{1y}^r \bigcup \mathbb{N}_{2y}^r \bigcup \mathbb{N}_{3y}^r \bigcup \mathbb{N}_{4y}^r$, the index set of iCVs;
$\mathbb{N}_{1y}^r$:	the index set of type 1 iCVs, where zeroing $y_{\text{Slope,ss}}(k)$ is necessary;
$\mathbb{N}_{2y}^r$:	the index set of type 2 iCVs, where confining $y_{\text{Slope,ss}}(k)$ is necessary and zeroing $y_{\text{Slope,ss}}(k)$ is further expected;
$\mathbb{N}_{3y}^r$:	the index set of type 3 iCVs, where confining $y_{\text{Slope,ss}}(k)$ is necessary;
$\mathbb{N}_{4y}^r$:	the index set of type 4 iCVs, where zeroing $y_{\text{Slope,ss}}(k)$ is expected;
$y_{\text{Slope,ss}}(k)$:	the steady-state change rate of y at time k;
$y_{\text{Slope,ss}}^{\text{ol}}(k)$:	ol steady-state change rate of y predicted at time k;
H_i^u:	impulse-response coefficient matrix for u, $H_i^u = \Delta S_i^u = S_i^u - S_{i-1}^u$;
H_i^f:	impulse-response coefficient matrix for f, $H_i^f = \Delta S_i^f = S_i^f - S_{i-1}^f$.

The two-layered MPC in this chapter consists of three modules, i.e., ol prediction, SSTC, and DC. In ol prediction module, which can be conceptually represented as

$$\begin{aligned} &\left\{y_{\text{ss}}^{\text{ol}}(k), y_{\text{Slope,ss}}^{\text{ol}}(k), \{y^{\text{ol}}(k+n|k),\ n=1,2,\ldots,N\}\right\} \\ &= \Phi_{\text{olp}}\left(y(k), \{y^{\text{ol}}(k+n-1|k-1),\ n=1,2,\ldots,N\}, \Delta u(k-1),\right. \\ &\qquad \left.\Delta f(k), \Delta f(k+1|k), \ldots, \Delta f(k+P'|k)\right), \end{aligned} \tag{5.1}$$

the future values of CV are predicted based on the current CV measurements, assuming that the future MV keeps unchanged. In SSTC, which can be conceptually represented as

$$\begin{aligned} &\left\{\delta u_{\text{ss}}(k), y_{\text{ss}}(k), y_{\text{Slope,ss}}(k)\right\} \\ &= \Phi_{\text{SSTC}}\left(u(k-1), y_{\text{ss}}(k-1), y_{\text{ss}}^{\text{ol}}(k), u_t^{(\text{sm})}(k), y_t^{(\text{sm})}(k), y_{\text{Slope,ss}}^{\text{ol}}(k)\right), \end{aligned} \tag{5.2}$$

some constrained LPs or constrained QPs are solved, which obtain MVss and CVss. In DC, which can be conceptually represented as

$$\begin{aligned} &\Delta u(k|k) \\ &= \Phi_{\text{DC}}\left(u(k-1), \delta u_{\text{ss}}(k), \{y^{\text{ol}}(k+p|k),\ p=1,2,\ldots,P\}, y_{\text{ss}}(k), \{y_{j,\text{Slope,ss}}(k), j\in\mathbb{N}_y^r\}\right), \end{aligned} \tag{5.3}$$

ss from SSTC are tracked by optimizing the future MV moves.

Model Predictive Control, First Edition. Baocang Ding and Yuanqing Yang.

5.1 Open-Loop Prediction Module

When the process is stable, consider the following FSR model (equivalent for all $N' \geq N-1$):

$$\begin{aligned}\nabla y(k) =& \sum_{i=1}^{N'-1} S_i^u \Delta u(k-i) + S_{N'}^u \nabla u(k-N') \\ &+ \sum_{i=1}^{N'-1} S_i^f \Delta f(k-i) + S_{N'}^f \nabla f(k-N'),\end{aligned} \tag{5.4}$$

where the step-response coefficient matrices satisfy $S_{N-1+i}^u = S_{N-1}^u$ and $S_{N-1+i}^f = S_{N-1}^f$ for all $i \geq 0$. Assume that $f(k)$ is measurable at k, and does not change in the future. Let $y^{\mathrm{fr}}(k+p|k)$ be the prediction based on (5.4) by taking $\Delta u(k+i-1|k) = 0$ for $1 \leq i \leq p$. Like Chapter 4, applying (5.4) yields

$$y^{\mathrm{fr}}(k+p|k) - y^{\mathrm{fr}}(k+p|k-1) = S_{p+1}^u \Delta u(k-1) + S_p^f \Delta f(k), \tag{5.5}$$

$$y(k+p|k) - y^{\mathrm{fr}}(k+p|k) = \sum_{i=1}^{p} S_i^u \Delta u(k+p-i|k). \tag{5.6}$$

When there are both sCVs and iCVs, in order to incorporate iCVs, one can change (5.4) to the following incremental FSR model (equivalent for all $N' \geq N$):

$$\begin{aligned}\Delta y(k) =& \sum_{i=1}^{N'-1} \Delta S_i^u \Delta u(k-i) + \Delta S_{N'}^u \nabla u(k-N') \\ &+ \sum_{i=1}^{N'-1} \Delta S_i^f \Delta f(k-i) + \Delta S_{N'}^f \nabla f(k-N') \\ =& \sum_{i=1}^{N'-1} H_i^u \Delta u(k-i) + H_{N'}^u \nabla u(k-N') \\ &+ \sum_{i=1}^{N'-1} H_i^f \Delta f(k-i) + H_{N'}^f \nabla f(k-N'),\end{aligned} \tag{5.7}$$

where the impulse-response coefficient matrices satisfy $H_{N+i}^u = H_N^u$ and $H_{N+i}^f = H_N^f$ for all $i \geq 0$. The operator Δ (before y and S) in (5.7) cannot be removed (without this Δ, (5.7) is not equivalent for all $N' \geq N$). Some predictions based on (5.7) are

$$\begin{aligned}\Delta y(k+p|k) =& \sum_{i=1}^{N'-1} \Delta S_i^u \Delta u(k-i+p|k) + \Delta S_{N'}^u \nabla u(k-N'+p|k) \\ &+ \sum_{i=1}^{N'-1} \Delta S_i^f \Delta f(k-i+p|k) + \Delta S_{N'}^f \nabla f(k-N'+p|k), \\ \Delta y(k+p|k-1) =& \sum_{i=1}^{N'-1} \Delta S_i^u \Delta u(k-i+p|k-1) + \Delta S_{N'}^u \nabla u(k-N'+p|k-1) \\ &+ \sum_{i=1}^{N'-1} \Delta S_i^f \Delta f(k-i+p|k-1) + \Delta S_{N'}^f \nabla f(k-N'+p|k-1).\end{aligned}$$

Let $\Delta y^{\text{fr}}(k+p|k)$ be the prediction based on (5.7) by taking $\Delta u(k+i-1|k)=0$ for $1\leq i\leq p$. Applying these predictions yields

$$\Delta y^{\text{fr}}(k+p|k)-\Delta y^{\text{fr}}(k+p|k-1)=\Delta S_{p+1}^{u}\Delta u(k-1)+\Delta S_{p}^{f}\Delta f(k), \tag{5.8}$$

$$\Delta y(k+p|k)-\Delta y^{\text{fr}}(k+p|k)=\sum_{i=1}^{p}\Delta S_{i}^{u}\Delta u(k+p-i|k). \tag{5.9}$$

By removing the operator Δ from (5.8) and (5.9), we also obtain (5.5) and (5.6).

However, in predicting $y^{\text{fr}}(k+p|k)$, the current measurement $y(k)$ is not applied. In the following, we consider the feedback correction in predicting $y(k+p)$, which yields $y^{\text{ol}}(k+i|k)$. Let us denote

$$Y_{N}^{\text{ol}}(k|k)=\begin{bmatrix} y^{\text{ol}}(k+1|k)\\ y^{\text{ol}}(k+2|k)\\ \vdots\\ y^{\text{ol}}(k+N|k)\end{bmatrix}.$$

At time $k>0$, after $\Delta u(k-1)$ is measured, applying (5.5) yields

$$y^{\text{ol}}(k|k)=y^{\text{ol}}(k|k-1)+S_{1}^{u}\Delta u(k-1).$$

Applying the measured $y(k)$ yields

$$\epsilon(k)=y(k)-y^{\text{ol}}(k|k)=y(k)-y^{\text{ol}}(k|k-1)-S_{1}^{u}\Delta u(k-1). \tag{5.10}$$

$\epsilon(k)$, named prediction error, represents the influence of model-plant mismatch on the prediction of y. Using $\epsilon(k)$ and (5.5), we give the following ol dynamic prediction equation:

$$Y_{N}^{\text{ol}}(k|k)=\mathcal{M}\left\{Y_{N}^{\text{ol}}(k-1|k-1)+\begin{bmatrix} S_{1}^{u}\\ S_{2}^{u}\\ \vdots\\ S_{N}^{u}\end{bmatrix}\Delta u(k-1)\right\}+\begin{bmatrix} S_{1}^{f}\\ S_{2}^{f}\\ \vdots\\ S_{N}^{f}\end{bmatrix}\Delta f(k)+\begin{bmatrix} e(k+1|k)\\ e(k+2|k)\\ \vdots\\ e(k+N|k)\end{bmatrix}, \tag{5.11}$$

where $\mathcal{M}=\begin{bmatrix} 0 & I & 0 & \dots & 0\\ \vdots & \ddots & \ddots & \ddots & \vdots\\ 0 & \dots & 0 & I & 0\\ 0 & \dots & 0 & 0 & I\\ 0 & \dots & 0 & -I & 2I\end{bmatrix}$, with $Y_{N}^{\text{ol}}(0|0)=\begin{bmatrix} y(0)\\ y(0)\\ \vdots\\ y(0)\end{bmatrix}+\begin{bmatrix} S_{1}^{f}\\ S_{2}^{f}\\ \vdots\\ S_{N}^{f}\end{bmatrix}\Delta f(0)$ being the initial value, and

$$e(k+i|k)=(I+i\Sigma)\epsilon(k),\ i\geq 0. \tag{5.12}$$

In (5.12),

$$\Sigma=\text{diag}\{\sigma_{1},\sigma_{2},\dots,\sigma_{n_y}\},$$

where $\sigma_{j}\in[0,1]$ is the rotation factor. If the jth y is stable, then one should take $\sigma_{j}=0$, being consistent with Chapter 4.

If ol prediction of y does not change after $N-1$ control intervals into the future, then ol steady-state prediction is

$$y_{ss}^{ol}(k) = y^{ol}(k+N-1+i|k), \quad i \geq 0. \tag{5.13}$$

Even if (5.13) is not satisfied, we still denote $y_{ss}^{ol}(k) = y^{ol}(k+N|k)$ (as in Chapter 4). Let $y_{\text{Slope,ss}}^{ol}(k)$ be the prediction of the steady-state change rate of y, at time k, by taking $\Delta u(k+i|k) = 0$ for all $i \geq 0$. Here, the change rate is defined as the change in one control interval. Since (5.12) is applied, we obtain

$$y_{\text{Slope,ss}}^{ol}(k) = y^{ol}(k+i|k) - y^{ol}(k+i-1|k), \quad i \geq N, \tag{5.14}$$

or equivalently

$$y^{ol}(k+N-1+i|k) = y^{ol}(k+N-1|k) + i y_{\text{Slope,ss}}^{ol}(k), \quad i \geq 1. \tag{5.15}$$

Remark 5.1 In deducing (5.11), we have applied

$$\begin{aligned} &y^{ol}(k+N+i|k-1) - y^{ol}(k+N-1+i|k-1) \\ &= y^{ol}(k+N-1+i|k-1) - y^{ol}(k+N-2+i|k-1), \quad i \geq 0, \end{aligned} \tag{5.16}$$

with the case $i = 0$. Eq. (5.16) is consistent with (5.14)–(5.15). For any $p > 0$,

$$\begin{aligned} y^{ol}(k+p|k) &= y^{ol}(k+p|k-1) + S_{p+1}^u \Delta u(k-1) + S_p^f \Delta f(k) + e(k+p|k) \\ &= y^{ol}(k+p-1|k-1) + [y^{ol}(k+p|k-1) - y^{ol}(k+p-1|k-1)] \\ &\quad + [S_p^u + (S_{p+1}^u - S_p^u)]\Delta u(k-1) + S_p^f \Delta f(k) + e(k+p|k). \end{aligned}$$

However, only for $p \geq N$, by using (5.16), we have

$$\begin{aligned} y^{ol}(k+p|k) &= y^{ol}(k+p-1|k-1) + [y^{ol}(k+p-1|k-1) - y^{ol}(k+p-2|k-1)] \\ &\quad + [S_p^u + (S_p^u - S_{p-1}^u)]\Delta u(k-1) + S_p^f \Delta f(k) + e(k+p|k) \\ &= -y^{ol}(k+p-2|k-1) + 2y^{ol}(k+p-1|k-1) \\ &\quad + (-S_{p-1}^u + 2S_p^u)\Delta u(k-1) + S_p^f \Delta f(k) + e(k+p|k). \end{aligned}$$

Therefore, by taking $Y_N^{ol}(k|k)$, we can obtain the closed-form equation (5.11).

By applying (5.11) to calculate $y_{\text{Slope,ss}}^{ol}(k) = y^{ol}(k+N|k) - y^{ol}(k+N-1|k)$, we can also obtain the following closed-form equation:

$$y_{\text{Slope,ss}}^{ol}(k) = y_{\text{Slope,ss}}^{ol}(k-1) + H_N^u \Delta u(k-1) + H_N^f \Delta f(k) + \Sigma \epsilon(k). \tag{5.17}$$

As compared with $Y_N^{ol}(k|k)$, $y_{\text{Slope,ss}}^{ol}(k)$ is not an independent signal. Moreover, by applying (5.17), and $y^{ol}(k+N|k)$ in (5.11), we can obtain the following closed-form equation:

$$\begin{aligned} \begin{bmatrix} y_{ss}^{ol}(k) \\ y_{\text{Slope,ss}}^{ol}(k) \end{bmatrix} &= \begin{bmatrix} I & I \\ 0 & I \end{bmatrix} \left\{ \begin{bmatrix} y_{ss}^{ol}(k-1) \\ y_{\text{Slope,ss}}^{ol}(k-1) \end{bmatrix} + \begin{bmatrix} S_N^u \\ H_N^u \end{bmatrix} \Delta u(k-1) \right\} \\ &\quad + \begin{bmatrix} S_N^f \\ H_N^f \end{bmatrix} \Delta f(k) + \begin{bmatrix} I + N\Sigma \\ \Sigma \end{bmatrix} \epsilon(k). \end{aligned}$$

5.2 Steady-State Target Calculation Module

When the process is stable, SSTC optimizes $\delta u_{ss}(k)$ and $y_{ss}(k)$, which satisfy the steady-state CV prediction equation

$$y_{ss}(k) = S_N^u \delta u_{ss}(k) + y_{ss}^{ol}(k), \tag{5.18}$$

where $y_{ss}(k) = y(k + N + M - 1 + i|k)$ for all $i \geq 0$. When there is an iCV, it may not satisfy $y_{ss}(k) = y(k + N + M - 1 + i|k)$ for all $i \geq 0$, but we still denote $y_{ss}(k) = y(k + N + M - 1|k)$. Applying (5.6) and (5.15) yields

$$\begin{aligned}
&y_{ss}(k) \\
&= y(k + N + M - 1|k) \\
&= y^{ol}(k + N + M - 1|k) + \sum_{i=0}^{M-1} S_{N+M-1-i}^u \Delta u(k + i|k) \\
&= y^{ol}(k + N|k) + (M - 1)y_{\text{Slope,ss}}^{ol}(k) + \sum_{i=0}^{M-1} [S_N^u + (M - 1 - i)H_N^u]\Delta u(k + i|k) \\
&= y^{ol}(k + N|k) + (M - 1)y_{\text{Slope,ss}}^{ol}(k) + S_N^u \delta u_{ss}(k) + H_N^u \delta u_b(k) \\
&= y_{ss}^{ol}(k) + (M - 1)y_{\text{Slope,ss}}^{ol}(k) + S_N^u \delta u_{ss}(k) + H_N^u \delta u_b(k),
\end{aligned} \tag{5.19}$$

where (see [Lee and Xiao, 2000])

$$\delta u_b(k) = \sum_{i=0}^{M-2} (M - 1 - i)\Delta u(k + i|k). \tag{5.20}$$

When the process is stable, since $y_{\text{Slope,ss}}^{ol}(k) = 0$ and $H_N^u = 0$, (5.18) is recovered from (5.19). Since $\delta u_{ss}(k)$ and $\delta u_b(k)$ cannot be independently optimized, let us choose

$$H_N^u \delta u_b(k) = \frac{M - 1}{2} \Xi H_N^u \delta u_{ss}(k), \tag{5.21}$$

where $\Xi = \text{diag}\{\xi_1, \xi_2, \dots, \xi_{n_y}\}$, with ξ_j being the tuning parameters. By applying (5.21), it is shown that (5.19) becomes

$$y_{ss}(k) = y_{ss}^{ol}(k) + (M - 1)y_{\text{Slope,ss}}^{ol}(k) + K^u \delta u_{ss}(k), \tag{5.22}$$

where $K^u = S_N^u + \frac{M-1}{2} \Xi H_N^u$.

Similarly to (5.22), we can obtain

$$y(k + N + M|k) = y^{ol}(k + N|k) + My_{\text{Slope,ss}}^{ol}(k) + (K^u + H_N^u)\delta u_{ss}(k). \tag{5.23}$$

When u and f become unchanged, the change rate of y will become constant after $N - 1$ control intervals in the future. Hence, the prediction of the steady-state change rate of y is

$$y_{\text{Slope,ss}}(k) = y(k + M + i|k) - y(k + M - 1 + i|k), \quad i \geq N. \tag{5.24}$$

By applying (5.22) and (5.23) on (5.24), it yields

$$y_{\text{Slope,ss}}(k) = y_{\text{Slope,ss}}^{ol}(k) + H_N^u \delta u_{ss}(k). \tag{5.25}$$

Applying (5.22), and (5.24)–(5.25), yields

$$
\begin{aligned}
& y(k+N+M+L-1|k) \\
&= y_{ss}(k) + Ly_{\text{Slope,ss}}(k) \\
&- y_{ss}^{ol}(k) + (M+L-1)y_{\text{Slope,ss}}^{ol}(k) + (K^u + LH_N^u)\delta u_{ss}(k),
\end{aligned} \tag{5.26}
$$

where L is a nonnegative integer. When the process is stable, $y_{\text{Slope,ss}}(k) = 0$ and $y(k+N+M+L-1|k) = y(k+N+M-1|k) = y_{ss}(k)$.

SSTC receives ET from RTO, process operator or control engineer. For iCV, we can utilize ET to enhance safety. In SSTC, each constraint in LP and QP is either hard or soft. The hard constraints cannot be violated. For the feasibility of SSTC, any soft constraint can be relaxed.

5.2.1 Hard and Soft Constraints

Similarly to Chapter 4, each MVss has a hard constraint, i.e.,

$$
\underline{u}''(k) \leq \delta u_{ss}(k) \leq \bar{u}''(k), \tag{5.27}
$$

where

$$
\begin{aligned}
\underline{u}''(k) &= \min\{\max\{\underline{u} - u(k-1), -\delta\bar{u}'_{ss}\}, \bar{u} - u(k-1)\}, \\
\bar{u}''(k) &= \max\{\min\{\bar{u} - u(k-1), \delta\bar{u}'_{ss}\}, \underline{u} - u(k-1)\}, \\
\delta\bar{u}'_{ss} &= \min\{M\Delta\bar{u}, \delta\bar{u}_{ss}\}.
\end{aligned}
$$

There is the soft constraints for $u_{ss}(k)$, i.e.,

$$
\underline{u}_{i,ss}(k) \leq \delta u_{i,ss}(k) \leq \bar{u}_{i,ss}(k), \quad \delta u_{i,ss}(k) = u_{i,t}(k) - u_i(k-1), \quad i \in \mathcal{I}_t, \tag{5.28}
$$

where

$$
\begin{aligned}
\underline{u}_{i,ss}(k) &= \min\{\max\{u_{i,t}(k) - \frac{1}{2}u_{i,\text{ss,range}} - u_i(k-1), \underline{u}_i''(k)\}, \bar{u}_i''(k)\}, \\
\bar{u}_{i,ss}(k) &= \max\{\min\{u_{i,t}(k) + \frac{1}{2}u_{i,\text{ss,range}} - u_i(k-1), \bar{u}_i''(k)\}, \underline{u}_i''(k)\}.
\end{aligned}
$$

Each CVss has hard magnitude constraints (engineering limits), i.e.,

$$
\underline{y}_{0,h} \leq y_{ss}(k) \leq \bar{y}_{0,h}, \tag{5.29}
$$

$$
\underline{y}_{j,0,h} \leq y_j(k+N+M+L-1|k) \leq \bar{y}_{j,0,h}, \ j \in \mathbb{N}_{2y}^r \bigcup \mathbb{N}_{3y}^r. \tag{5.30}
$$

For $j \in \mathbb{N}_{2y}^r \bigcup \mathbb{N}_{3y}^r$, (5.29)–(5.30) introduce a hard constraint on the steady-state change rate $y_{j,\text{Slope,ss}}(k)$; when L becomes larger, this constraint becomes tighter. An additional constraint is imposed, i.e.,

$$
|\delta y_{ss}(k)| \leq \delta\bar{y}_{ss}, \tag{5.31}
$$

where $\delta y_{ss}(k) = y_{ss}(k) - y_{ss}^{ol}(k) - (\mathrm{M}-1)y_{\text{Slope,ss}}^{ol}(k)$. If (5.31) conflicts with (5.29), then (5.29) has to be satisfied. Hence, by applying (5.22), it is shown that (5.29) and (5.31) are combined to yield the hard constraint

$$
\underline{y}_h(k) \leq K^u\delta u_{ss}(k) \leq \bar{y}_h(k), \tag{5.32}
$$

where

$$\underline{y}_h(k) = \min\{\max\{-\delta\bar{y}_{ss}, \underline{y}_{0,h} - \heartsuit\}, \bar{y}_{0,h} - \heartsuit\},$$
$$\bar{y}_h(k) = \max\{\min\{\delta\bar{y}_{ss}, \bar{y}_{0,h} - \heartsuit\}, \underline{y}_{0,h} - \heartsuit\},$$
$$\heartsuit = y_{ss}^{ol}(k) + (M-1)y_{\text{Slope,ss}}^{ol}(k).$$

By applying (5.26), (5.30) becomes

$$\underline{y}_{j,h}^{L}(k) \leq (K_j^u + LH_{j,N}^u)\delta u_{ss}(k) \leq \bar{y}_{j,h}^{L}(k),\ \ j \in \mathbb{N}_{2y}^r \bigcup \mathbb{N}_{3y}^r, \tag{5.33}$$

where

$$\underline{y}_{j,h}^{L}(k) = \underline{y}_{j,0,h} - y_{j,ss}^{ol}(k) - (M+L-1)y_{j,\text{Slope,ss}}^{ol}(k),$$
$$\bar{y}_{j,h}^{L}(k) = \bar{y}_{j,0,h} - y_{j,ss}^{ol}(k) - (M+L-1)y_{j,\text{Slope,ss}}^{ol}(k).$$

For type 1 iCV, an additional hard constraint is

$$H_{j,N}^u \delta u_{ss}(k) = -y_{j,\text{Slope,ss}}^{ol}(k),\ \ j \in \mathbb{N}_{1y}^r, \tag{5.34}$$

which is obtained by taking $y_{j,\text{Slope,ss}}(k) = 0$ in (5.25).

Each CVss is expected to satisfy (operating limits)

$$\underline{y}_0 \leq y_{ss}(k) \leq \bar{y}_0, \tag{5.35}$$

$$\underline{y}_{j,0} \leq y_j(k+N+M+L-1|k) \leq \bar{y}_{j,0},\ \ j \in \mathbb{N}_{2y}^r \bigcup \mathbb{N}_{3y}^r. \tag{5.36}$$

For $j \in \mathbb{N}_{2y}^r \bigcup \mathbb{N}_{3y}^r$, (5.35) and (5.36) introduce a soft constraint on the steady-state change rate $y_{j,\text{Slope,ss}}(k)$; when L becomes larger, this constraint becomes tighter. Since (5.32) has to be satisfied, (5.35) is revised as the soft constraints

$$K^u \delta u_{ss}(k) \leq \bar{y}(k), \tag{5.37}$$

$$K^u \delta u_{ss}(k) \geq \underline{y}(k), \tag{5.38}$$

where

$$\underline{y}(k) = \min\{\max\{\underline{y}_0 - y_{ss}^{ol}(k) - (M-1)y_{\text{Slope,ss}}^{ol}(k), \underline{y}_h(k)\}, \bar{y}_h(k)\},$$
$$\bar{y}(k) = \max\{\min\{\bar{y}_0 - y_{ss}^{ol}(k) - (M-1)y_{\text{Slope,ss}}^{ol}(k), \bar{y}_h(k)\}, \underline{y}_h(k)\}.$$

By applying (5.26), (5.36) becomes

$$(K^u + LH_{j,N}^u)\delta u_{ss}(k) \leq \bar{y}_j^L(k),\ \ j \in \mathbb{N}_{2y}^r \bigcup \mathbb{N}_{3y}^r, \tag{5.39}$$

$$(K^u + LH_{j,N}^u)\delta u_{ss}(k) \geq \underline{y}_j^L(k),\ \ j \in \mathbb{N}_{2y}^r \bigcup \mathbb{N}_{3y}^r, \tag{5.40}$$

where

$$\underline{y}_j^L(k) = \min\{\max\{\underline{y}_{j,0} - y_{j,ss}^{ol}(k) - (M+L-1)y_{j,\text{Slope,ss}}^{ol}(k), \underline{y}_{j,h(k)}^L\}, \bar{y}_{j,h}^L(k)\},$$
$$\bar{y}_j^L(k) = \max\{\min\{\bar{y}_{j,0} - y_{j,ss}^{ol}(k) - (M+L-1)y_{j,\text{Slope,ss}}^{ol}(k), \bar{y}_{j,h}^L(k)\}, \underline{y}_{j,h}^L(k)\}.$$

For type 2 and type 4 iCVs, an additional soft constraint is

$$H_{j,N}^u \delta u_{ss}(k) = -y_{j,\text{Slope,ss}}^{ol}(k),\ \ j \in \mathbb{N}_{2y}^r \bigcup \mathbb{N}_{4y}^r, \tag{5.41}$$

which is obtained by taking $y_{j,\text{Slope,ss}}(k) = 0$ in (5.25).

If $j \in \mathcal{O}_t \subseteq \mathbb{N}_y$, then $y_{j,\text{ss}}(k)$ is expected to satisfy

$$y_{j,t}(k) - \frac{1}{2}y_{j,\text{ss,range}} \leq y_{j,\text{ss}}(k) \leq y_{j,t}(k) + \frac{1}{2}y_{j,\text{ss,range}}, \quad k \geq 0. \tag{5.42}$$

Since (5.32) has to be satisfied, (5.42) is revised as the soft constraint

$$\underline{y}_{j,\text{ss}}(k) \leq K_j^u \delta u_{\text{ss}}(k) \leq \bar{y}_{j,\text{ss}}(k), \quad j \in \mathcal{O}_t, \tag{5.43}$$

where

$$\underline{y}_{j,\text{ss}}(k) = \min\{\max\{y_{j,t}(k) - \frac{1}{2}y_{j,\text{ss,range}} - \heartsuit, \underline{y}_{j,h}(k)\}, \bar{y}_{j,h}(k)\}, \tag{5.44}$$

$$\bar{y}_{j,\text{ss}}(k) = \max\{\min\{y_{j,t}(k) + \frac{1}{2}y_{j,\text{ss,range}} - \heartsuit, \bar{y}_{j,h}(k)\}, \underline{y}_{j,h}(k)\},$$

$$\heartsuit = y_{\text{ss}}^{\text{ol}}(k) + (M-1)y_{\text{Slope,ss}}^{\text{ol}}(k). \tag{5.45}$$

By adding CVET to (5.43), it becomes

$$\underline{y}_{j,\text{ss}}(k) \leq K_j^u \delta u_{\text{ss}}(k) \leq \bar{y}_{j,\text{ss}}(k),$$
$$K_j^u \delta u_{\text{ss}}(k) = y_{j,t}(k) - y_{j,\text{ss}}^{\text{ol}}(k) - (M-1)y_{j,\text{Slope,ss}}^{\text{ol}}(k), \quad j \in \mathcal{O}_t. \tag{5.46}$$

5.2.2 Priority Rank of Soft Constraints

The soft constraint set {(5.28), (5.37)–(5.40), (5.41), (5.46)} need to be grouped into a number of (say, r_0) subsets, each corresponding to a priority rank. The soft constraints are handled in the order of priority rank. The soft constraints with lower value of r have higher priority. In rank r ($r = 1, 2, \ldots, r_0$), the following soft constraints are included:

- the upper bounds for some CVs indexed in $\tilde{\mathcal{O}}_u^{(r)} \subseteq \mathbb{N}_y$, i.e., (5.37) and (5.39) for $j \in \tilde{\mathcal{O}}_u^{(r)}$. These constraints can be relaxed as

$$K_j^u \delta u_{\text{ss}}(k) \leq \bar{y}_j(k) + \varepsilon_{\bar{y},j}, \quad j \in \tilde{\mathcal{O}}_u^{(r)},$$
$$(K_j^u + LH_{j,N}^u)\delta u_{\text{ss}}(k) \leq \bar{y}_j^L(k) + \varepsilon_{\bar{y},j}^L, \quad j \in \tilde{\mathcal{O}}_u^{(r)} \bigcap (\mathbb{N}_{2y}^r \bigcup \mathbb{N}_{3y}^r), \tag{5.47}$$

  where $\varepsilon_{\bar{y},j}$ and $\varepsilon_{\bar{y},j}^L$ are the nonnegative slack variable;
- the lower bounds for some CVs indexed in $\tilde{\mathcal{O}}_l^{(r)} \subseteq \mathbb{N}_y$, i.e., (5.38) and (5.40) for $j \in \tilde{\mathcal{O}}_l^{(r)}$. These constraints can be relaxed as

$$K_j^u \delta u_{\text{ss}}(k) \geq \underline{y}_j(k) - \varepsilon_{\underline{y},j}, \quad j \in \tilde{\mathcal{O}}_l^{(r)},$$
$$(K_j^u + LH_{j,N}^u)\delta u_{\text{ss}}(k) \geq \underline{y}_j^L(k) - \varepsilon_{\underline{y},j}^L, \quad j \in \tilde{\mathcal{O}}_l^{(r)} \bigcap (\mathbb{N}_{2y}^r \bigcup \mathbb{N}_{3y}^r), \tag{5.48}$$

  where $\varepsilon_{\underline{y},j}$ and $\varepsilon_{\underline{y},j}^L$ are the nonnegative slack variable;
- the expected value and upper/lower bounds for some CVs indexed in $\tilde{\mathbb{O}}^{(r)} \subseteq \mathcal{O}_t$, i.e., (5.46) for $j \in \tilde{\mathbb{O}}^{(r)}$. These constraints can be relaxed as

$$K_j^u \delta u_{\text{ss}}(k) \geq \underline{y}_{j,\text{ss}}(k) - \varepsilon_{\underline{y},j,\text{ss}}, \quad K_j^u \delta u_{\text{ss}}(k) \leq \bar{y}_{j,\text{ss}}(k) + \varepsilon_{\bar{y},j,\text{ss}},$$
$$K_j^u \delta u_{\text{ss}}(k) = y_{j,t}(k) - y_{j,\text{ss}}^{\text{ol}}(k) - (M-1)y_{j,\text{Slope,ss}}^{\text{ol}}(k) + \varepsilon_{y,j,t,+} - \varepsilon_{y,j,t,-},$$
$$j \in \tilde{\mathbb{O}}^{(r)}, \tag{5.49}$$

  where $\varepsilon_{\underline{y},j,\text{ss}}$, $\varepsilon_{\bar{y},j,\text{ss}}$, $\varepsilon_{y,j,t,+}$, $\varepsilon_{y,j,t,-}$ are the nonnegative slack variables;

- the expected value and upper/lower bounds for some MVs indexed in $\tilde{\mathbb{I}}^{(r)} \subseteq \mathcal{I}_t$, i.e., (5.28) for $i \in \tilde{\mathbb{I}}^{(r)}$. These constraints can be relaxed as

$$\delta u_{i,\mathrm{ss}}(k) \geq \underline{u}_{i,\mathrm{ss}}(k) - \varepsilon_{\underline{u},i,\mathrm{ss}}, \quad \delta u_{i,\mathrm{ss}}(k) \leq \bar{u}_{i,\mathrm{ss}}(k) + \varepsilon_{\bar{u},i,\mathrm{ss}},$$
$$\delta u_{i,\mathrm{ss}}(k) = u_{i,t}(k) - u_i(k-1) + \varepsilon_{u,i,t,+} - \varepsilon_{u,i,t,-}, \quad i \in \tilde{\mathbb{I}}^{(r)}, \tag{5.50}$$

where $\varepsilon_{\underline{u},i,\mathrm{ss}}$, $\varepsilon_{\bar{u},i,\mathrm{ss}}$, $\varepsilon_{u,i,t,+}$, $\varepsilon_{u,i,t,-}$ are the nonnegative slack variables;
- the expected rate balance equation for some CVs indexed in $\tilde{\mathcal{O}}^{(r)} \subseteq \mathbb{N}^r_{2y} \bigcup \mathbb{N}^r_{4y}$, i.e., (5.41) for $j \in \tilde{\mathcal{O}}^{(r)}$. This constraint can be relaxed as

$$H^u_{j,N} \delta u_{\mathrm{ss}}(k) = -y^{\mathrm{ol}}_{j,\mathrm{Slope},\mathrm{ss}}(k) + \varepsilon_{j,\mathrm{Slope},+} - \varepsilon_{j,\mathrm{Slope},-}, \quad j \in \tilde{\mathcal{O}}^{(r)}, \tag{5.51}$$

where $\varepsilon_{j,\mathrm{Slope},+}$ and $\varepsilon_{j,\mathrm{Slope},-}$ are the nonnegative slack variables.

The soft constraints in rank r (i.e., (5.47)–(5.51)) are expressed as

$$\begin{cases} \tilde{C}^{(r)} \delta u_{\mathrm{ss}}(k) \leq \tilde{c}^{(r)}(k) + \varepsilon^{(r)}(k) \\ \tilde{C}^{(r)}_{\mathrm{eq}} \delta u_{\mathrm{ss}}(k) = \tilde{c}^{(r)}_{\mathrm{eq}}(k) + \varepsilon^{(r)}_{\mathrm{eq+}}(k) - \varepsilon^{(r)}_{\mathrm{eq-}}(k) \end{cases}, \tag{5.52}$$

where $\varepsilon^{(r)}(k)$, $\varepsilon^{(r)}_{\mathrm{eq+}}(k)$ and $\varepsilon^{(r)}_{\mathrm{eq-}}(k)$ are nonnegative slack variables, appropriately composed of ε in (5.47)–(5.51).

Let us denote $\mathcal{O}^{(r)}_u = \bigcup_{k=1}^r \tilde{\mathcal{O}}^{(k)}_u$, $\mathcal{O}^{(r)}_l = \bigcup_{k=1}^r \tilde{\mathcal{O}}^{(k)}_l$, $\mathbb{O}^{(r)} = \bigcup_{k=1}^r \tilde{\mathbb{O}}^{(k)}$, $\mathbb{I}^{(r)} = \bigcup_{k=1}^r \tilde{\mathbb{I}}^{(k)}$, $\mathcal{O}^{(r)} = \bigcup_{k=1}^r \tilde{\mathcal{O}}^{(k)}$.

5.2.3 Feasibility Stage

In each rank, the hard constraints {(5.27), (5.32), (5.33), (5.34)} have to be satisfied. In the rank r, the soft-to-hard constraints for all ranks $l < r$ (with $\varepsilon \geq 0$ fixed) are taken as hard, then (5.52) is handled. Handling (5.52) means finding/fixing/minimizing ε in the rank r so that (5.52) does not conflict in itself or with the hard constraints. Denote the soft-to-hard constraints, from ranks 1 to r, as

$$\begin{cases} C^{(r)} \delta u_{\mathrm{ss}}(k) \leq c^{(r)}(k) \\ C^{(r)}_{\mathrm{eq}} \delta u_{\mathrm{ss}}(k) = c^{(r)}_{\mathrm{eq}}(k) \end{cases}. \tag{5.53}$$

For the rank $r+1$, (5.53) is the hard constraint. By applying (5.47)–(5.51), it is easy to show that (5.53) includes

$$K^u_j \delta u_{\mathrm{ss}}(k) \leq \bar{y}'_j(k), \quad j \in \mathcal{O}^{(r)}_u,$$
$$(K^u_j + LH^u_{j,N}) \delta u_{\mathrm{ss}}(k) \leq \bar{y}_j^{'L}(k), \quad j \in \mathcal{O}^{(r)}_u \bigcap (\mathbb{N}^r_{2y} \bigcup \mathbb{N}^r_{3y}), \tag{5.54}$$

$$K^u_j \delta u_{\mathrm{ss}}(k) \geq \underline{y}'_j(k), \quad j \in \mathcal{O}^{(r)}_l,$$
$$(K^u_j + LH^u_{j,N}) \delta u_{\mathrm{ss}}(k) \geq \underline{y}_j^{'L}(k), \quad j \in \mathcal{O}^{(r)}_l \bigcap (\mathbb{N}^r_{2y} \bigcup \mathbb{N}^r_{3y}), \tag{5.55}$$

$$K^u_j \delta u_{\mathrm{ss}}(k) = y_{j,\mathrm{ss}}(k) - y^{\mathrm{ol}}_{j,\mathrm{ss}}(k) - (M-1) y^{\mathrm{ol}}_{j,\mathrm{Slope},\mathrm{ss}}(k), \quad j \in \mathbb{O}^{(r)}, \tag{5.56}$$

$$\delta u_{i,\mathrm{ss}}(k) = u_{i,\mathrm{ss}}(k) - u_i(k-1), \quad i \in \mathbb{I}^{(r)}, \tag{5.57}$$

$$H^u_{j,N} \delta u_{\mathrm{ss}}(k) = y'_{j,\mathrm{Slope},\mathrm{ss}}(k), \quad j \in \mathcal{O}^{(r)}, \tag{5.58}$$

where $\bar{y}'_j(k) = \bar{y}_j(k) + \varepsilon_{\bar{y},j}$, $\bar{y}_j'^L(k) = \bar{y}_j^L(k) + \varepsilon^L_{\bar{y},j}$, $\underline{y}'_j(k) = \underline{y}_j(k) - \varepsilon_{\underline{y},j}$, $\underline{y}_j'^L(k) = \underline{y}_j^L(k) - \varepsilon^L_{\underline{y},j}$, $y_{j,\mathrm{ss}}(k) = y_{j,t}(k) + \varepsilon_{y,j,t,+} - \varepsilon_{y,j,t,-}$, $u_{i,\mathrm{ss}}(k) = u_{i,t}(k) + \varepsilon_{u,i,t,+} - \varepsilon_{u,i,t,-}$ and $y'_{j,\mathrm{Slope,ss}}(k) = -y^{\mathrm{ol}}_{j,\mathrm{Slope,ss}}(k) + \varepsilon_{j,\mathrm{Slope},+} - \varepsilon_{j,\mathrm{Slope},-}$.

Lemma 5.1 With (5.54)–(5.57) satisfied, the hard constraints in {(5.27), (5.32), (5.33)} can be simplified as

$$\underline{u}''_i(k) \leq \delta u_{i,\mathrm{ss}}(k) \leq \bar{u}''_i(k), \; i \in \mathbb{N}_u \backslash \mathbb{I}^{(r)}, \tag{5.59}$$

$$K^u_j \delta u_{\mathrm{ss}}(k) \leq \bar{y}_{j,h}(k), \; j \in \mathbb{N}_y \backslash \mathbb{O}^{(r)} \backslash \mathcal{O}^{(r)}_u, \tag{5.60}$$

$$- K^u_j \delta u_{\mathrm{ss}}(k) \leq -\underline{y}_{j,h}(k), \; j \in \mathbb{N}_y \backslash \mathbb{O}^{(r)} \backslash \mathcal{O}^{(r)}_l, \tag{5.61}$$

$$(K^u_j + LH^u_{j,N})\delta u_{\mathrm{ss}}(k) \leq \bar{y}^L_{j,h}(k), \; j \in \mathbb{N}^r_{2y} \bigcup \mathbb{N}^r_{3y} \backslash \mathbb{O}^{(r)} \backslash \mathcal{O}^{(r)}_u, \tag{5.62}$$

$$- (K^u_j + LH^u_{j,N})\delta u_{\mathrm{ss}}(k) \leq -\underline{y}^L_{j,h}(k), \; j \in \mathbb{N}^r_{2y} \bigcup \mathbb{N}^r_{3y} \backslash \mathbb{O}^{(r)} \backslash \mathcal{O}^{(r)}_l. \tag{5.63}$$

For the rank $r+1$, by summarizing (5.52), (5.53) and Lemma 5.1, it is shown that the hard/soft constraints are composed of {(5.34), (5.59)–(5.63)} and

$$\begin{cases} \begin{bmatrix} C^{(r)} \\ \tilde{C}^{(r+1)} \end{bmatrix} \delta u_{\mathrm{ss}}(k) \leq \begin{bmatrix} c^{(r)}(k) \\ \tilde{c}^{(r+1)}(k) + \varepsilon^{(r+1)}(k) \end{bmatrix} \\ \begin{bmatrix} C^{(r)}_{\mathrm{eq}} \\ \tilde{C}^{(r+1)}_{\mathrm{eq}} \end{bmatrix} \delta u_{\mathrm{ss}}(k) = \begin{bmatrix} c^{(r)}_{\mathrm{eq}}(k) \\ \tilde{c}^{(r+1)}_{\mathrm{eq}}(k) + \varepsilon^{(r+1)}_{\mathrm{eq+}}(k) - \varepsilon^{(r+1)}_{\mathrm{eq-}}(k) \end{bmatrix} \end{cases}. \tag{5.64}$$

In the rank $r+1$, either LP or QP is invoked, so as to find ε as small as possible. In each rank, if multiple ε are to be optimized, then they are equally important. For each ε, an equal concern error $\bar{\varepsilon}$ is applied, i.e.,

$$\begin{aligned} \bar{\varepsilon}_{\bar{y},j} &= \bar{y}_{j,h}(k) - \bar{y}_j(k), & \bar{\varepsilon}_{\underline{y},j} &= \underline{y}_j(k) - \underline{y}_{j,h}(k), \\ \bar{\varepsilon}^L_{\bar{y},j} &= \bar{y}^L_{j,h}(k) - \bar{y}^L_j(k), & \bar{\varepsilon}^L_{\underline{y},j} &= \underline{y}^L_j(k) - \underline{y}^L_{j,h}(k), \\ \bar{\varepsilon}_{\bar{y},j,\mathrm{ss}} &= \bar{y}_{j,h}(k) - \bar{y}_{j,\mathrm{ss}}(k), & \bar{\varepsilon}_{\underline{y},j,\mathrm{ss}} &= \underline{y}_{j,\mathrm{ss}}(k) - \underline{y}_{j,h}(k), \\ \bar{\varepsilon}_{y,j,t,+} &= \frac{1}{2}[\bar{y}_{j,h}(k) - \underline{y}_{j,h}(k)], & \bar{\varepsilon}_{y,j,t,-} &= \frac{1}{2}[\bar{y}_{j,h}(k) - \underline{y}_{j,h}(k)], \\ \bar{\varepsilon}_{\bar{u},i,\mathrm{ss}} &= \bar{u}''_i(k) - \bar{u}_{i,\mathrm{ss}}(k), & \bar{\varepsilon}_{\underline{u},i,\mathrm{ss}} &= \underline{u}_{i,\mathrm{ss}}(k) - \underline{u}''_i(k), \\ \bar{\varepsilon}_{u,i,t,+} &= \frac{1}{2}[\bar{u}''_i(k) - \underline{u}''_i(k)], & \bar{\varepsilon}_{u,i,t,-} &= \frac{1}{2}[\bar{u}''_i(k) - \underline{u}''_i(k)], \\ \bar{\varepsilon}_{j,\mathrm{Slope},+} &= \frac{1}{L}[\bar{y}_{j,h}(k) - \underline{y}_{j,h}(k)], & \bar{\varepsilon}_{j,\mathrm{Slope},-} &= \frac{1}{L}[\bar{y}_{j,h}(k) - \underline{y}_{j,h}(k)]. \end{aligned}$$

If a $\bar{\varepsilon} \leq 0$, then its corresponding soft constraint cannot be relaxed; in this case, we can remove the corresponding soft constraint, or simply take this $\bar{\varepsilon} = 1$ (it is equivalent to taking any value for this $\bar{\varepsilon}$).

By invoking LP, the optimization problem is

$$\min_{\epsilon^{(r+1)}(k),\epsilon^{(r+1)}_{\text{eq+}}(k),\epsilon^{(r+1)}_{\text{eq-}}(k),\delta u_{\text{ss}}(k)} \left[\sum_{\ell=1}^{d^{(r+1)}} \left(\overline{\epsilon}^{(r+1)}_{\ell}\right)^{-1} \epsilon^{(r+1)}_{\ell}(k) + \sum_{\tau=1}^{d^{(r+1)}_{\text{eq}}} \left[\left(\overline{\epsilon}^{(r+1)}_{\text{eq+},\tau}\right)^{-1} \epsilon^{(r+1)}_{\text{eq+},\tau}(k) + \left(\overline{\epsilon}^{(r+1)}_{\text{eq-},\tau}\right)^{-1} \epsilon^{(r+1)}_{\text{eq-},\tau}(k) \right] \right],$$

$$\text{s.t. } (5.34),\ (5.59)-(5.63),\ (5.64) \text{ and } \epsilon \geq 0, \tag{5.65}$$

where the subscript ℓ corresponds to the ℓ-th element of $\epsilon^{(r+1)}(k)$, and $d^{(r+1)}$ is the dimension of $\epsilon^{(r+1)}$; the subscript τ corresponds to the τth element of $\epsilon^{(r+1)}_{\text{eq}}(k)$, and $d^{(r+1)}_{\text{eq}}$ is the dimension of $\epsilon^{(r+1)}_{\text{eq}}(k)$.

By invoking QP, the optimization problem is

$$\min_{\epsilon^{(r+1)}(k),\epsilon^{(r+1)}_{\text{eq+}}(k),\epsilon^{(r+1)}_{\text{eq-}}(k),\delta u_{\text{ss}}(k)} \left[\sum_{\ell=1}^{d^{(r+1)}} (\overline{\epsilon}^{(r+1)}_{\ell})^{-2} \epsilon^{(r+1)}_{\ell}(k)^2 + \sum_{\tau=1}^{d^{(r+1)}_{\text{eq}}} \left[(\overline{\epsilon}^{(r+1)}_{\text{eq+},\tau})^{-2} \epsilon^{(r+1)}_{\text{eq+},\tau}(k)^2 + (\overline{\epsilon}^{(r+1)}_{\text{eq-},\tau})^{-2} \epsilon^{(r+1)}_{\text{eq-},\tau}(k)^2 \right] \right],$$

$$\text{s.t. } (5.34),\ (5.59)-(5.63),\ (5.64), \text{ and } \epsilon \geq 0. \tag{5.66}$$

After the optimization for the rank $r+1$, the constraint (5.64) is expressed as (5.53) with r being changed as $r+1$.

5.2.4 Economic Stage

Since only a part of CVs/MVs have their ETs, ss of the other CVs/MVs are evaluated by an economic cost. This economic cost evaluates the benefit/loss when changing $u_{\text{ss}}(k)$ and $y_{\text{ss}}(k)$, so the cost can be expressed in $\delta u_{\text{ss}}(k)$. Denote the index set of minimum-move MV as $\mathcal{I}_{\text{mm}}$. Minimizing $|\delta u_{i,\text{ss}}(k)|$ is equivalent to minimizing $U_i(k)$ satisfying

$$-U_i(k) \leq \delta u_{i,\text{ss}}(k) \leq U_i(k). \tag{5.67}$$

The minimum-move penalty term is also added to the economic cost.

Lemma 5.2 After feasibility stage optimizations from rank 1 to rank r_0, all the hard (including soft-to-hard) constraints can be expressed as

$$\underline{u}''_i(k) \leq \delta u_{i,\text{ss}}(k) \leq \bar{u}''_i(k),\ i \in \mathbb{N}_u \backslash \mathcal{I}_t, \tag{5.68}$$

$$\underline{y}'_j(k) \leq K^u_j \delta u_{\text{ss}}(k) \leq \overline{y}'_j(k),\ j \in \mathbb{N}_y \backslash \mathcal{O}_t, \tag{5.69}$$

$$\underline{y}'^{L}_j(k) \leq (K^u_j + LH^u_{j,N})\delta u_{\text{ss}}(k) \leq \overline{y}'^{L}_j(k),\ j \in \mathbb{N}^r_{2y} \bigcup \mathbb{N}^r_{3y} \backslash \mathcal{O}_t, \tag{5.70}$$

$$K^u_j \delta u_{\text{ss}}(k) = y_{j,\text{ss}}(k) - y^{\text{ol}}_{j,\text{ss}}(k) - (M-1)y^{\text{ol}}_{j,\text{Slope,ss}}(k),\ j \in \mathcal{O}_t, \tag{5.71}$$

$$\delta u_{i,\text{ss}}(k) = u_{i,\text{ss}}(k) - u_i(k-1),\ i \in \mathcal{I}_t, \tag{5.72}$$

$$H^u_{j,N}\delta u_{ss}(k) = -y^{ol}_{j,\mathrm{Slope},ss}(k),\ j \in \mathbb{N}^r_{1y}, \tag{5.73}$$

$$H^u_{j,N}\delta u_{ss}(k) = y'_{j,\mathrm{Slope},ss}(k),\ j \in \mathbb{N}^r_{2y} \bigcup \mathbb{N}^r_{4y}. \tag{5.74}$$

Consider the pure economic. By invoking LP, the optimization problem is

$$\min_{\delta u_{i,ss}(k),U_i(k)} \left[\sum_{i \notin \mathcal{I}_{mm} \bigcup \mathcal{I}_t} h_i \delta u_{i,ss}(k) + i\mathcal{I}_{mm} \in \sum h_i U_i(k) \right],$$
$$\text{s.t. } (5.67),\ i \in \mathcal{I}_{mm};\ (5.68) - (5.74),$$

where h_i is the weight coefficient. By invoking QP, the optimization problem is

$$\min_{\delta u_{i,ss}(k),U_i(k)} \left[\left(\sum_{i \notin \mathcal{I}_{mm} \bigcup \mathcal{I}_t} h_i \delta u_{i,ss}(k) - J_{\min} \right)^2 + \sum_{i \in \mathcal{I}_{mm}} h_i^2 U_i(k)^2 \right]$$
$$\text{s.t. } (5.67),\ i \in \mathcal{I}_{mm};\ (5.68) - (5.74),$$

where $J_{\min}$ is equal to, or smaller than, the minimum of $\sum_{i \notin \mathcal{I}_{mm} \bigcup \mathcal{I}_t} h_i \delta u_{i,ss}(k)$. One can utilize the upper/lower bounds of $\delta u_{i,ss}(k)$, and the signs of h_i, to determine a $J_{\min}$.

Consider the mixed economic. By invoking LP, the optimization problem is

$$\min_{\varepsilon^{(r_0)}(k),\varepsilon^{(r_0)}_{eq+}(k),\varepsilon^{(r_0)}_{eq-}(k),\delta u_{i,ss}(k),U_i(k)} \left[\sum_{i \notin \mathcal{I}_{mm} \bigcup \mathcal{I}_t} h_i \delta u_{i,ss}(k) + \sum_{i \in \mathcal{I}_{mm}} h_i U_i(k) \right.$$
$$+ \sum_{\ell=1}^{d^{(r_0)}} \left(\overline{\varepsilon}^{(r_0)}_\ell\right)^{-1} \varepsilon^{(r_0)}_\ell(k)$$
$$\left. + \sum_{\tau=1}^{d^{(r_0)}_{eq}} \left[\left(\overline{\varepsilon}^{(r_0)}_{eq+,\tau}\right)^{-1} \varepsilon^{(r_0)}_{eq+,\tau}(k) + \left(\overline{\varepsilon}^{(r_0)}_{eq-,\tau}\right)^{-1} \varepsilon^{(r_0)}_{eq-,\tau}(k) \right] \right]$$
$$\text{s.t. } (5.67),\ i \in \mathcal{I}_{mm};\ (5.34),\ (5.59) - (5.63),\ (5.64),\ r = r_0 - 1,\ \text{and } \varepsilon \geq 0. \tag{5.75}$$

By invoking QP, the optimization problem is

$$\min_{\varepsilon^{(r_0)}(k),\varepsilon^{(r_0)}_{eq+}(k),\varepsilon^{(r_0)}_{eq-}(k),\delta u_{i,ss}(k),U_i(k)} \left[\left(\sum_{i \notin \mathcal{I}_{mm} \bigcup \mathcal{I}_t} h_i \delta u_{i,ss}(k) - J_{\min} \right)^2 \right.$$
$$+ \sum_{i \in \mathcal{I}_{mm}} h_i^2 U_i(k)^2 + \sum_{\ell=1}^{d^{(r_0)}} (\overline{\varepsilon}^{(r_0)}_\ell)^{-2} \varepsilon^{(r_0)}_\ell(k)^2$$
$$\left. + \sum_{\tau=1}^{d^{(r_0)}_{eq}} \left[(\overline{\varepsilon}^{(r_0)}_{eq+,\tau})^{-2} \varepsilon^{(r_0)}_{eq+,\tau}(k)^2 + (\overline{\varepsilon}^{(r_0)}_{eq-,\tau})^{-2} \varepsilon^{(r_0)}_{eq-,\tau}(k)^2 \right] \right]$$
$$\text{s.t. } (5.67),\ i \in \mathcal{I}_{mm};\ (5.34),\ (5.59) - (5.63),\ (5.64),\ r = r_0 - 1,\ \text{and } \varepsilon \geq 0. \tag{5.76}$$

In the mixed economic, LP and QP may not give the same result.

After the feasibility and economic stages, $\delta u_{ss}(k)$ has been optimized. Using (5.22) obtains $y_{ss}(k)$. Using (5.25), $y_{\mathrm{Slope},ss}(k)$ is obtained. Hence, SSTC is finished.

5.3 Dynamic Calculation Module

Let $\{M, N, P\}$ satisfy $M \leq P \leq N + M$.

At each control interval k, given $Y_N^{\text{ol}}(k|k)$, we have

$$Y_P^{\text{ol}}(k|k) = \begin{bmatrix} y^{\text{ol}}(k+1|k) \\ y^{\text{ol}}(k+2|k) \\ \vdots \\ y^{\text{ol}}(k+P|k) \end{bmatrix}.$$

By applying (5.6), it is easy to obtain

$$Y_P(k|k) = Y_P^{\text{ol}}(k|k) + \mathcal{S}\,\Delta\tilde{u}(k|k), \tag{5.77}$$

where

$$Y_P(k|k) = \begin{bmatrix} y(k+1|k) \\ y(k+2|k) \\ \vdots \\ y(k+P|k) \end{bmatrix}, \quad \Delta\tilde{u}(k|k) = \begin{bmatrix} \Delta u(k|k) \\ \Delta u(k+1|k) \\ \vdots \\ \Delta u(k+M-1|k) \end{bmatrix},$$

$$\mathcal{S} = \begin{bmatrix} S_1^u & 0 & \cdots & 0 \\ S_2^u & S_1^u & \ddots & \vdots \\ \vdots & \ddots & \ddots & 0 \\ S_M^u & \cdots & S_2^u & S_1^u \\ \vdots & & \vdots & \vdots \\ S_P^u & \cdots & S_{P-M+2}^u & S_{P-M+1}^u \end{bmatrix}.$$

$\mathcal{S}$ is called the dynamic matrix. Equation (5.77) is the closed-loop prediction equation.

In DC, three objectives should be achieved: (i) the future values of CV tracking $y_{\text{ss}}(k+i|k)$ as close as possible; (ii) the sharp change of MV being suppressed; (iii) the soft constraint of CV being relaxed as small as possible with respect to feasibility retrieval. Choose the cost function, for DC, as

$$J(k) = \sum_{i=1}^{P} \|y(k+i|k) - y_{\text{ss}}(k+i|k)\|_{Q_i(k)}^2 + \sum_{j=0}^{M-1} \|\Delta u(k+j|k)\|_{\Lambda}^2,$$

$$J'(k) = \sum_{i=1}^{P} \|y(k+i|k) - y_{\text{ss}}(k+i|k)\|_{Q_i(k)}^2 + \sum_{j=0}^{M-1} \|\Delta u(k+j|k)\|_{\Lambda}^2 + \|\underline{\varepsilon}_{\text{dc}}(k)\|_{\underline{\Omega}}^2 + \|\overline{\varepsilon}_{\text{dc}}(k)\|_{\overline{\Omega}}^2,$$

where $y_{\text{ss}}(k+i|k) = y_{\text{ss}}(k) + iy_{\text{Slope,ss}}(k)$; $\underline{\varepsilon}_{\text{dc}}(k)$ and $\overline{\varepsilon}_{\text{dc}}(k)$ are the slack variables of CV constraint;

$$Q_i(k) = \text{diag}\{q_{i1}(k)^2, q_{i2}(k)^2, \ldots, q_{i,n_y}(k)^2\},$$

$$q_{ij}(k) = \begin{cases} \underline{q}_j, & \check{y}_j(k+i|k) \le \underline{y}_{j,0} \\ \underline{a}\check{y}_j(k+i|k) + \underline{b}, & \underline{y}_{j,0} \le \check{y}_j(k+i|k) \le \underline{z}_j \\ \check{q}_j, & \underline{z}_j \le \check{y}_j(k+i|k) \le \overline{z}_j \\ \bar{a}\check{y}_j(k+i|k) + \bar{b}, & \overline{z}_j \le \check{y}_j(k+i|k) \le \overline{y}_{j,0} \\ \overline{q}_j, & \check{y}_j(k+i|k) \ge \overline{y}_{j,0} \end{cases},$$

$$\underline{a} = \frac{\underline{q}_j - \check{q}_j}{\underline{y}_{j,0} - \underline{z}_j}, \quad \underline{b} = \frac{\underline{y}_{j,0}\check{q}_j - \underline{z}_j\underline{q}_j}{\underline{y}_{j,0} - \underline{z}_j}, \quad \bar{a} = \frac{\overline{q}_j - \check{q}_j}{\overline{y}_{j,0} - \overline{z}_j}, \quad \bar{b} = \frac{\overline{y}_{j,0}\check{q}_j - \overline{z}_j\overline{q}_j}{\overline{y}_{j,0} - \overline{z}_j},$$

$$\Lambda = \text{diag}\{\lambda_1^2, \lambda_2^2, \dots, \lambda_{n_u}^2\},$$

$$\underline{\Omega} = \text{diag}\{\underline{\omega}_1^2, \underline{\omega}_2^2, \dots, \underline{\omega}_{n_y}^2\}, \quad \overline{\Omega} = \text{diag}\{\overline{\omega}_1^2, \overline{\omega}_2^2, \dots, \overline{\omega}_{n_y}^2\},$$

$$\underline{\omega}_j = (\underline{y}_{j,0} - \underline{y}_{j,0,h})^{-1}\rho, \quad \overline{\omega}_j = (\overline{y}_{j,0,h} - \overline{y}_{j,0})^{-1}\rho.$$

$\check{y}_j(k+i|k)$ is determined by $y_j^{\text{ol}}(k+i|k)$ and $y_{j,\text{ss}}(k+i|k)$ as in the following:

- if $y_j^{\text{ol}}(k+i|k) > \max\{y_{j,\text{ss}}(k+i|k), \overline{y}_{j,0}\}$ or $y_j^{\text{ol}}(k+i|k) < \min\{y_{j,\text{ss}}(k+i|k), \underline{y}_{j,0}\}$, then take $\check{y}_j(k+i|k) = y_j^{\text{ol}}(k+i|k)$ in order to drive CV into operating limits as quick as possible;
- if $y_{j,\text{ss}}(k+i|k) \ge \max\{y_j^{\text{ol}}(k+i|k), \overline{y}_{j,0}\}$, then take $\check{y}_j(k+i|k) = \overline{z}_j$ in order to slow down the approaching of CV toward HL; if $y_{j,\text{ss}}(k+i|k) \le \min\{y_j^{\text{ol}}(k+i|k), \underline{y}_{j,0}\}$, then take $\check{y}_j(k+i|k) = \underline{z}_j$ in order to slow down the approaching of CV toward LL;
- otherwise, then take $\check{y}_j(k+i|k) = \frac{1}{2}[y_j^{\text{ol}}(k+i|k) + y_{j,\text{ss}}(k+i|k)]$.

Differently from tracking $y_{\text{ss}}(k)$, in order to track $\delta u_{\text{ss}}(k)$ the following constraint is imposed in DC:

$$L_{\text{ss}}\Delta\tilde{u}(k|k) = \delta u_{\text{ss}}(k), \quad H^{u,r}L_b\Delta\tilde{u}(k|k) = \frac{M-1}{2}\Xi H^{u,r}\delta u_{\text{ss}}(k), \tag{5.78}$$

where $L_{\text{ss}} = [I \; I \; \cdots \; I]$ and $L_b = [(M-1)I \; (M-2)I \; \cdots \; 0]$. Moreover, in DC usually the following inequality constraints (on MV rate, MV maginitude, CV magnitude, and slack variable) are considered:

$$|\Delta u(k+j|k)| \le \Delta\bar{u}, \quad 0 \le j \le M-1, \tag{5.79}$$

$$\underline{u} \le u(k-1) + \sum_{l=0}^{j} \Delta u(k+l|k) \le \bar{u}, \quad 0 \le j \le M-1, \tag{5.80}$$

$$\underline{y}_0'(k) \le y^{\text{ol}}(k+i|k) + \mathcal{S}_i\Delta\tilde{u}(k|k) \le \overline{y}_0'(k), \quad 1 \le i \le P, \tag{5.81}$$

$$\underline{y}_0'(k) - \underline{\varepsilon}_{\text{dc}}(k) \le y^{\text{ol}}(k+i|k) + \mathcal{S}_i\Delta\tilde{u}(k|k) \le \overline{y}_0'(k) + \overline{\varepsilon}_{\text{dc}}(k), \quad 1 \le i \le P, \tag{5.82}$$

$$\underline{\varepsilon}_{\text{dc}}(k) \le \underline{y}_0'(k) - \underline{y}_{0,h}, \tag{5.83}$$

$$\overline{\varepsilon}_{\text{dc}}(k) \le \overline{y}_{0,h} - \overline{y}_0'(k), \tag{5.84}$$

where $\mathcal{S}_i$ is the ith row of $\mathcal{S}$,

$$\overline{y}_0'(k) = \min\{\max\{\overline{y}_0, y_{\text{ss}}(k), y_{\text{ss}}(k) + Py_{\text{Slope,ss}}(k)\}, \overline{y}_{0,h}\},$$

$$\underline{y}_0'(k) = \max\{\min\{\underline{y}_0, y_{\text{ss}}(k), y_{\text{ss}}(k) + Py_{\text{Slope,ss}}(k)\}, \underline{y}_{0,h}\}.$$

In summary, at each control interval k, we first solve

$$\min_{\Delta\tilde{u}(k|k)} J(k), \text{ s.t. } (5.78)-(5.81). \tag{5.85}$$

If (5.85) is infeasible, then further solve

$$\min_{\underline{\varepsilon}_{\mathrm{dc}}(k),\overline{\varepsilon}_{\mathrm{dc}}(k),\Delta\tilde{u}(k|k)} J'(k), \text{ s.t. } (5.78)-(5.80),\ (5.82)-(5.84). \tag{5.86}$$

If either (5.85) or (5.86) is feasible, then $\Delta u(k|k)$ is sent to the controlled system.

5.4 Numerical Example

In order to obtain the mathematical model of the integral process, we modify the mathematical model of the heavy oil fractionator provided in Chapter 4. Here, y_2 is changed as iCV, and the related continuous-time transfer function matrix is described as

$$G^u(s) = \begin{bmatrix} \frac{4.05e^{-27s}}{50s+1} & \frac{1.77e^{-28s}}{60s+1} & \frac{5.88e^{-27s}}{50s+1} \\ \frac{0.0539e^{-18s}}{s(50s+1)} & \frac{0.0572e^{-14s}}{s(60s+1)} & \frac{0.069e^{-15s}}{s(40s+1)} \\ \frac{4.38e^{-20s}}{33s+1} & \frac{4.42e^{-22s}}{44s+1} & \frac{7.20}{19s+1} \end{bmatrix},$$

$$G^f(s) = \begin{bmatrix} \frac{1.20e^{-27s}}{45s+1} & \frac{1.44e^{-27s}}{40s+1} \\ \frac{0.0152e^{-15s}}{s(25s+1)} & \frac{0.0183e^{-15s}}{s(20s+1)} \\ \frac{1.14}{27s+1} & \frac{1.26}{32s+1} \end{bmatrix},$$

where MV includes u_1, u_2 and u_3; CV includes y_1, y_2 and y_3; DV includes f_1 and f_2. Take the sampling time as 1, $N = 100$, and use MATLAB command `tfd2step` to obtain the FSR model using the above transfer function matrix (i.e., obtain the step response coefficient matrix $S_i^{u,s}$, $S_i^{u,r}$, $S_i^{f,s}$, and $S_i^{f,r}$ for u and f, respectively). Take $y_{\mathrm{eq}} = 0$, $u_{\mathrm{eq}} = 0$ and $f_{\mathrm{eq}} = 0$.

In ol prediction part, the rotation factor is $\sigma_2 = 0.1$. Take $\underline{u} = u_{\mathrm{eq}} + [-0.5; -0.5; -0.5]$, $\bar{u} = u_{\mathrm{eq}} + [0.5; 0.5; 0.5]$, $\Delta\bar{u}_i = \delta\bar{u}_{i,\mathrm{ss}} = 0.2$; $\underline{y}_{0,h} = y_{\mathrm{eq}} + [-0.7; -0.7; -0.7]$, $\bar{y}_{0,h} = y_{\mathrm{eq}} + [0.7; 0.7; 0.7]$, $\underline{y}_0 = y_{\mathrm{eq}} + [-0.5; -0.5; -0.5]$, $\bar{y}_0 = y_{\mathrm{eq}} + [0.5; 0.5; 0.5]$, $\Delta\bar{y}_{1,\mathrm{ss}} = 0.2$, $\Delta\bar{y}_{2,\mathrm{ss}} = 0.3$, $\Delta\bar{y}_{3,\mathrm{ss}} = 0.2$. y_1, y_2, u_3 have ETs, and their expected allowable range is 0.5.

In SSTC, for the parameters of soft constraints, see Table 5.1 for types 1 and 3 iCV, and Table 5.2 for types 2 and 4 iCV; it applies the mixed economic. In the economic stage, u_2 is the minimum-move MV, and take $h = [1, 2, 2]$, $J_{\min} = -0.2$. At $k \geq 99$, there exists the disturbance with $f_{\mathrm{eq}} + [0.20; 0.10]$; at $k \geq 122$, the disturbance appears with $f_{\mathrm{eq}} + [0.20; 0.10]$; at the other intervals the disturbance is f_{eq}. $Y_N^{s,\mathrm{ol}}(0|0) = \left[y_{\mathrm{eq}}^s; \dots; y_{\mathrm{eq}}^s\right]$, $Y_N^{r,\mathrm{ol}}(0|0) = \left[y_{\mathrm{eq}}^r; \dots; y_{\mathrm{eq}}^r\right]$.

The feasibility stage of SSTC applies LP, and economic stage, QP. After obtaining ss, use DC to obtain MV dynamic moves. Choose the parameters for DC as $P = 15$, $M = 8$, $\Lambda = \mathrm{diag}\{7, 9, 7\}$, $\bar{z} = y_{\mathrm{eq}} + [0.4; 0.4; 0.4]$, $\underline{z} = y_{\mathrm{eq}} + [-0.4; -0.4; -0.4]$, $\underline{q}_1 = 2.0$, $\check{q}_1 = 1.0$, $\bar{q}_1 = 2.0$; $\underline{q}_2 = 1.0$, $\check{q}_2 = 0.8$, $\bar{q}_2 = 1.0$; $\underline{q}_3 = 2.5$, $\check{q}_3 = 1.0$, $\bar{q}_3 = 4.0$. For determining $\underline{\Omega}$ and $\overline{\Omega}$, take $\rho = 1$. The initial value $u(-1) = u_{\mathrm{eq}}$, $y(0) = y_{\mathrm{eq}}$, and $u(k) = u(k|k)$. The true system output is produced using MATLAB Simulink and the above transfer function matrix, which is multiplied by 0.9 to represent the mismatch between the model and the true system.

Next, we discuss four types of iCV.

Table 5.1 Parameters of multi-priority-rank SSTC (types 1 and 3 iCV).

Rank no.	Type of constraint	Variable	Soft constraint	$\overline{\varepsilon}$
1	Inequality	$y_{2,ss}$	CV lower bound	0.20
1	Inequality	$y_{3,ss}$	CV upper bound	0.20
2	Equality	$y_{2,ss}$	$y_{2,eq} - 0.3$	0.25
3	Inequality	$y_{2,ss}$	CV upper bound	0.20
3	Inequality	$y_{3,ss}$	CV lower bound	0.20
4	Equality	$u_{3,ss}$	$u_{3,eq} + 0.5$	0.25
4	Equality	$y_{1,ss}$	$y_{1,eq} + 0.7$	0.25
5	Inequality	$y_{1,ss}$	CV upper, lower bound	0.20

Table 5.2 Parameters of multi-priority-rank SSTC (types 2 and 4 iCV).

Rank no.	Type of constraint	Variable	Soft constraint	$\overline{\varepsilon}$
1	Inequality	$y_{2,ss}$	CV lower bound	0.20
1	Inequality	$y_{3,ss}$	CV upper bound	0.20
2	Equality	$u_{3,ss}$	$u_{3,eq} + 0.2$	0.25
2	Equality	$y_{1,ss}$	$y_{1,eq} + 0.7$	0.25
3	Inequality	$y_{2,ss}$	CV upper bound	0.20
3	Inequality	$y_{3,ss}$	CV lower bound	0.20
4	Equality	$y_{2,ss}$	rate balance equation	1.40
5	Inequality	$y_{1,ss}$	CV upper, lower bound	0.20

(1) type 1 iCV

Take $\xi_2 = 1.25$. The control result is shown in Figure 5.1. The CVss and MVss, given by SSTC, can be tracked by DC with offset-free.

(2) type 2 iCV

Take $\xi_2 = 1.25$ and $L = 200$. The control result is shown in Figure 5.2. During the simulation, sometimes (5.41) is satisfied, and sometimes (5.41) is not satisfied (i.e., it is relaxed).

(4) type 3 iCV

Take $\xi_2 = 1.25$ and $L = 200$. The control result is shown in Figure 5.3, and the change of y_2 is smooth. The system is finally stabilized. CVss and MVss given by SSTC can be tracked by DC with offset-free.

(5) type 4 iCV

Take $\xi_2 = 1.25$. We modify the mathematical model of the heavy oil fractionator provided in Chapter 4. Here, y_2 is changed as the pseudo iCV, and the related continuous-time

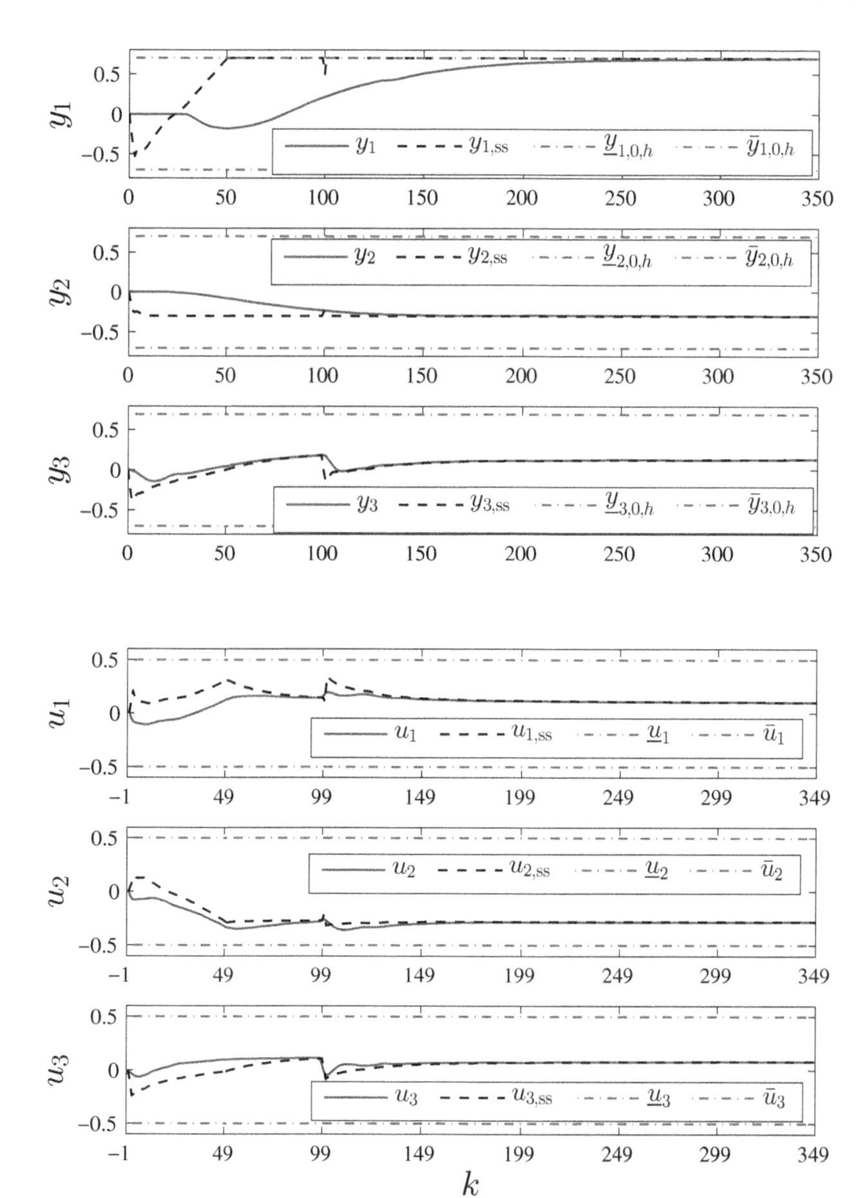

Figure 5.1 The control result of type 1 iCV.

transfer function matrix is described as

$$G^u(s) = \begin{bmatrix} \frac{4.05e^{-27s}}{50s+1} & \frac{1.77e^{-28s}}{60s+1} & \frac{5.88e^{-27s}}{50s+1} \\ \frac{5.39e^{-18s}}{250s+1} & \frac{5.72e^{-14s}}{320s+1} & \frac{6.9e^{-15s}}{200s+1} \\ \frac{4.38e^{-20s}}{33s+1} & \frac{4.42e^{-22s}}{44s+1} & \frac{7.20}{19s+1} \end{bmatrix},$$

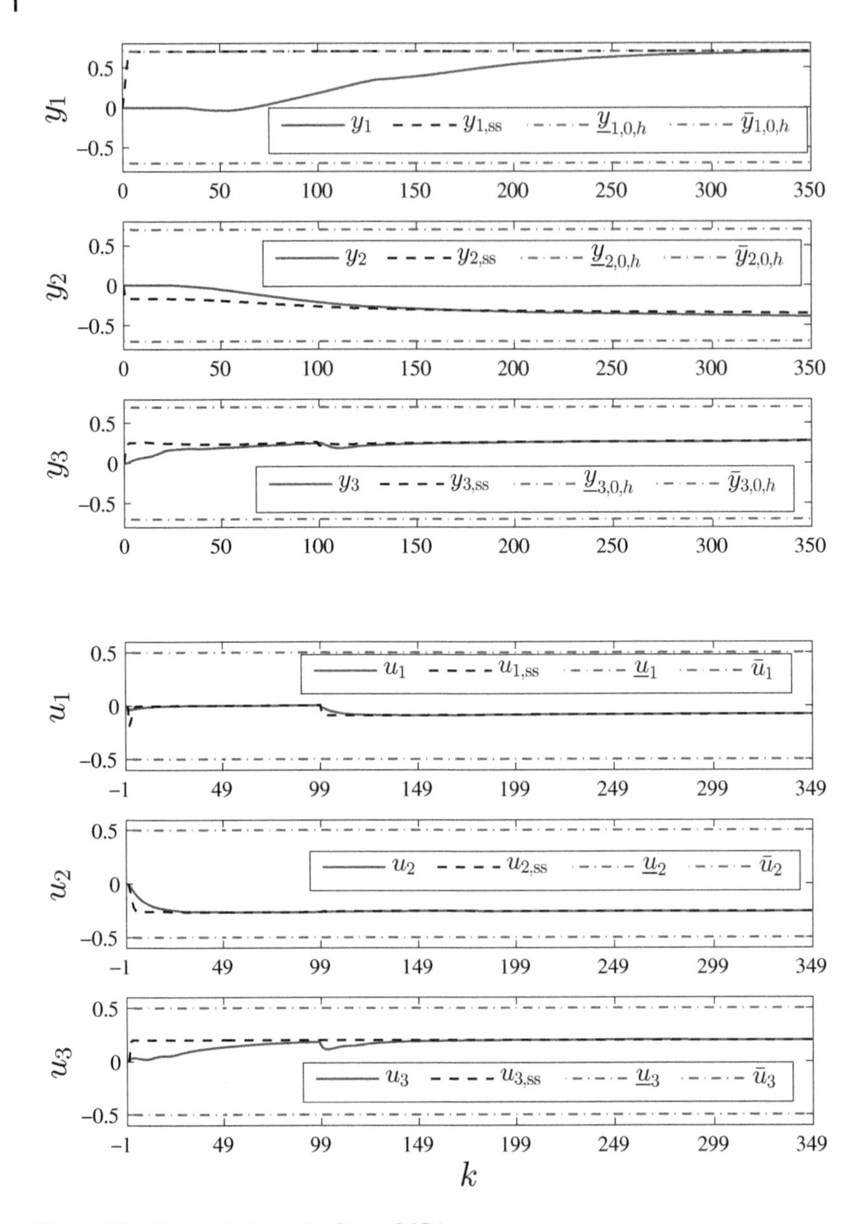

Figure 5.2 The control result of type 2 iCV.

$$G^f(s) = \begin{bmatrix} \frac{1.20e^{-27s}}{45s+1} & \frac{1.44e^{-27s}}{40s+1} \\ \frac{1.52e^{-15s}}{90s+1} & \frac{1.83e^{-15s}}{80s+1} \\ \frac{1.14}{27s+1} & \frac{1.26}{32s+1} \end{bmatrix}.$$

The step response curve of the aforementioned model is shown in Figure 5.4. From the figure, it can be seen that TTSS of y_2 is significantly larger (about $3 \sim 4$ times as large as the other CVs). Take the sampling time as 4, TTSS as 400, and model horizon as $N = 100$. Use

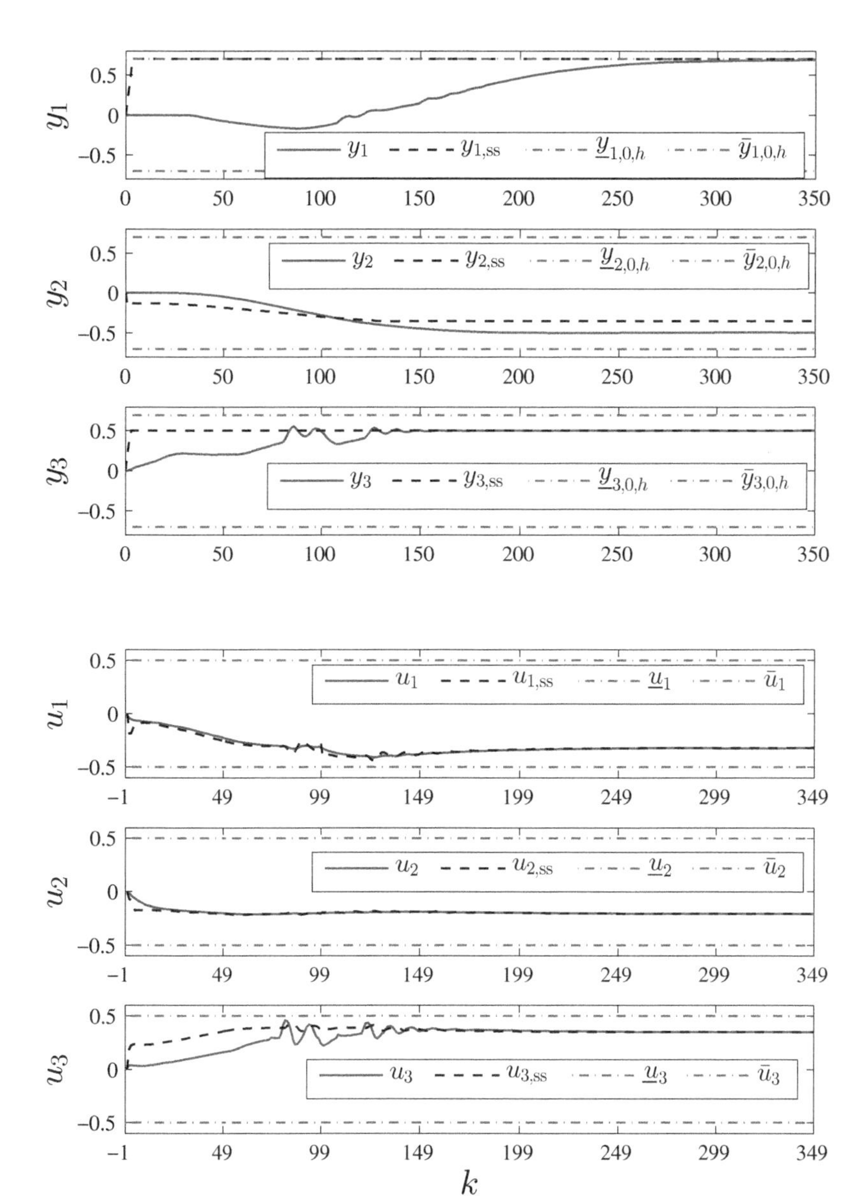

Figure 5.3 The control result of type 3 iCV.

MATLAB command `tfd2step` to obtain FSR model using the above transfer function matrix. Take $y_{\rm eq} = 0$, $u_{\rm eq} = 0$ and $f_{\rm eq} = 0$.

In ol prediction part, the rotation factor is $\sigma_2 = 0.1$. Different from types 1–3 iCV, reset parameters as $\underline{u} = u_{\rm eq} + [-0.5; -0.5; -0.5]$, $\bar{u} = u_{\rm eq} + [0.5; 0.5; 0.5]$, $\Delta\bar{u}_i = \delta\bar{u}_{i,{\rm ss}} = 0.3$; $\underline{y}_{0,h} = y_{\rm eq} + [-0.7; -0.7; -0.7]$, $\bar{y}_{0,h} = y_{\rm eq} + [0.7; 0.7; 0.7]$, $\underline{y}_0 = y_{\rm eq} + [-0.5; -0.5; -0.5]$, $\bar{y}_0 = y_{\rm eq} + [0.5; 0.5; 0.5]$, $\Delta\bar{y}_{1,{\rm ss}} = \Delta\bar{y}_{2,{\rm ss}} = \Delta\bar{y}_{3,{\rm ss}} = 0.3$. y_1, y_2, u_3 have ETs and their expected allowable range is 0.5. In the economic stage, u_2 is the minimum-move MV, and we take $h = [1, 2, 2]$, $J_{\rm min} = -0.3$. At $k \geq 99$, there exists the disturbance with $f_{\rm eq} + [0.20; 0.10]$;

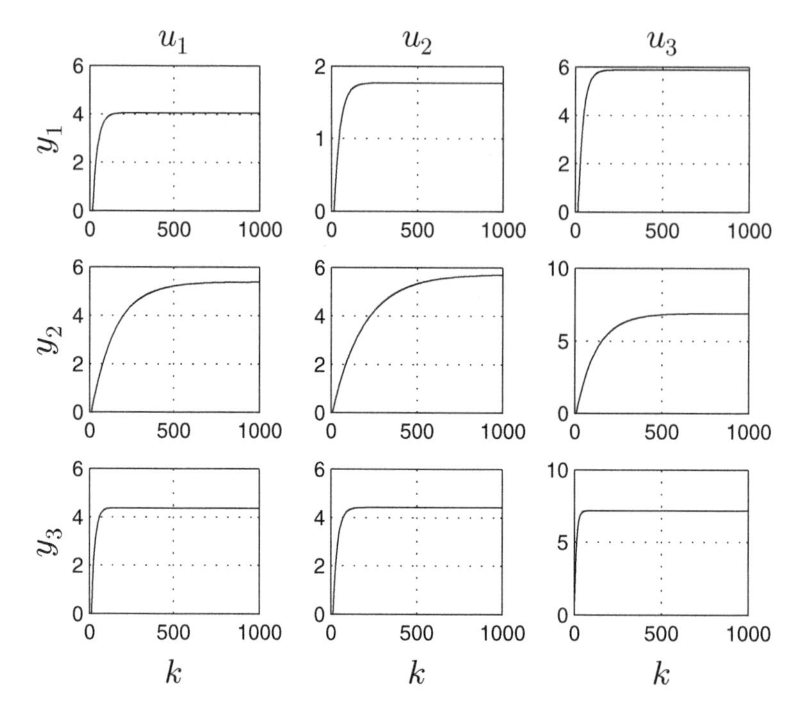

Figure 5.4 The step response curve including the pseudo iCV.

at $k \geq 122$, the disturbance appears with $f_{\text{eq}} + [0.20; 0.10]$; at the other intervals the disturbance is f_{eq}.

The feasibility stage of SSTC applies LP, and economic stage, QP. After obtaining ss, use DC to obtain MV dynamic moves. Choose $P = 15$, $M = 8$, $\Lambda = \text{diag}\{7, 9, 7\}$, $\overline{z} = y_{\text{eq}} + [0.4; 0.4; 0.4]$, $\underline{z} = y_{\text{eq}} + [-0.4; -0.4; -0.4]$, $\underline{q}_1 = 0.6$, $\check{q}_1 = 0.3$, $\overline{q}_1 = 0.6$; $\underline{q}_2 = 0.6$, $\check{q}_2 = 0.3$, $\overline{q}_2 = 0.6$; $\underline{q}_3 = 0.6$, $\check{q}_3 = 0.3$, $\overline{q}_3 = 0.6$. For determining $\underline{\Omega}$ and $\overline{\Omega}$, take $\rho = 1$. $u(-1) = u_{\text{eq}}$, $y(0) = y_{\text{eq}}$, and $u(k) = u(k|k)$. The true system output is produced using MATLAB Simulink and the above transfer function matrix, which is multiplied by 0.9 to represent the mismatch between the model and the true system.

The control result is shown in Figure 5.5. CVss and MVss given by SSTC can be tracked by DC with offset-free.

Note that in this example, when we take other suitable $\{y, u, f\}_{\text{eq}}$, the result is still available; it is easily misunderstood here that $y^s_{\text{eq}} = S^{u,s}_N u_{\text{eq}} + S^{f,s}_N f_{\text{eq}}$ and $y^r_{\text{Slope,eq}} = H^{u,r} u_{\text{eq}} + H^{f,r} f_{\text{eq}}$ should be satisfied, where $y^r_{\text{Slope,eq}} = 0$ is the integral steady-state slope at the equilibrium, and $H^{f,r} = S^{f,r}_{N+1} - S^{f,r}_N$.

Consider the following transfer function matrix description:

$$G^u(s) = \begin{bmatrix} \frac{4.05e^{-27s}}{50s+1} & \frac{1.77e^{-28s}}{60s+1} & \frac{5.88e^{-27s}}{50s+1} \\ \frac{0.0539e^{-18s}}{s(50s+1)} & \frac{0.0572e^{-14s}}{s(60s+1)} & \frac{0.069e^{-15s}}{s(40s+1)} \\ \frac{0.0438e^{-20s}}{s(33s+1)} & \frac{0.0442e^{-22s}}{s(44s+1)} & \frac{0.0720}{s(19s+1)} \end{bmatrix},$$

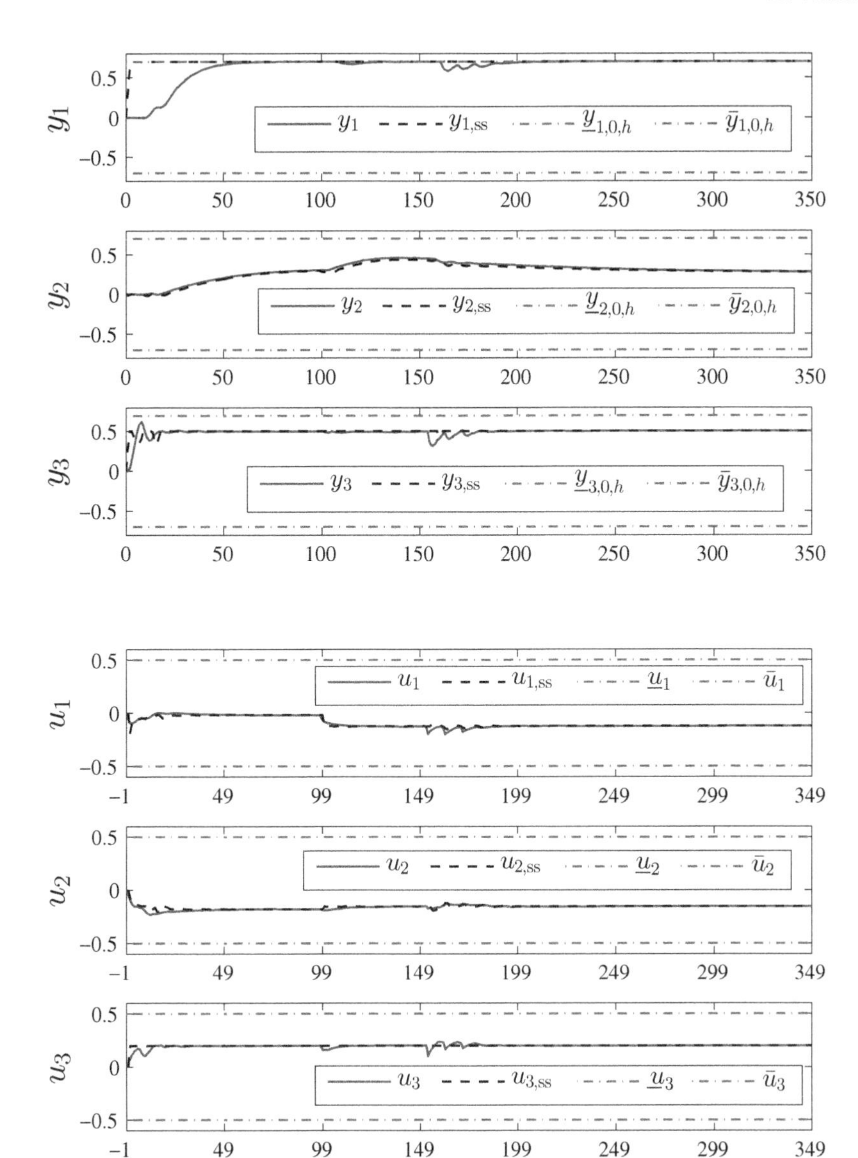

Figure 5.5 The control result of type 4 iCV.

$$G^f(s) = \begin{bmatrix} \frac{1.20e^{-27s}}{45s+1} & \frac{1.44e^{-27s}}{40s+1} \\ \frac{0.0152e^{-15s}}{s(25s+1)} & \frac{0.0183e^{-15s}}{s(20s+1)} \\ \frac{0.0114}{s(27s+1)} & \frac{0.0126}{s(32s+1)} \end{bmatrix}.$$

If we treat all three CVs as type 2, then the closed-loop system can be stabilized. Further, if the model remains unchanged, and the transfer function matrix that produces the true system output is modified as

$$G^u(s) = \begin{bmatrix} \frac{4.05e^{-27s}}{(50s+1)(5000s-1)} & \frac{1.77e^{-28s}}{60s+1} & \frac{5.88e^{-27s}}{50s+1} \\ \frac{0.0539e^{-18s}}{s(50s+1)} & \frac{0.0572e^{-14s}}{s(60s+1)(6000s-1)} & \frac{0.069e^{-15s}}{s(40s+1)} \\ \frac{0.0438e^{-20s}}{s(33s+1)} & \frac{0.0442e^{-22s}}{s(44s+1)} & \frac{0.0720}{s(19s+1)(1900s-1)} \end{bmatrix},$$

then the closed-loop system fails to be stabilized. If the transfer function matrix that produces the true system output remains unchanged, and FSR is changed as

$$S_i^u \leftarrow (1 + 0.01e^{0.01i})S_i^u, \; S_i^f \leftarrow (1 + 0.01e^{0.01i})S_i^f,$$

then the closed-loop system can be stabilized; if the transfer function matrix that produces the true system output remains unchanged, and FSR is changed as

$$S_i^u \leftarrow (1 - 0.001e^{0.01i})S_i^u, \; S_i^f \leftarrow (1 - 0.001e^{0.01i})S_i^f,$$

then the closed-loop system can be stabilized. The above stabilizable cases need to modify some controller parameters. For a series of simulations, we summarize it as follows.

The system cannot be unstable, while the model can be inaccurate.

6

Two-Layered DMC for State-Space Model

The notations, if not explained, are the same as in Chapter 4. New notations:

N: prediction horizon;
N_c: control horizon.

The two-layered MPC consists of three modules, i.e., ol prediction module, SSTC, and DC. In ol prediction module, which can be conceptually represented as

$$\begin{aligned}&\Big\{u_{\rm ss}^{\rm ol}(k), y_{\rm ss}^{\rm ol}(k), \{y^{\rm ol}(k+n|k),\ n=1,2,\dots,N\},\\&\quad\{\hat{x}^{\rm ol}(k+n_c|k),\ n_c=1,2,\dots,N_c-1\}, \hat{x}_{k|k}, \hat{d}_{k|k}, \hat{p}_{k|k}\Big\}\\&\quad=\Phi_{\rm olp}\Big(y(k), u(k-1), \triangledown y(k-1), \triangledown y(k),\\&\qquad\triangledown v(k-1), \triangledown f(k-1), \triangledown f(k), \hat{x}_{k-1|k-1}, \hat{d}_{k-1|k-1}, \hat{p}_{k-1|k-1}\Big),\end{aligned}\tag{6.1}$$

the future values of CV are predicted based on the current CV measurements, assuming that the future MV keeps unchanged. In SSTC, which can be conceptually represented as

$$\{\delta v_{\rm ss}(k), y_{\rm ss}(k)\}=\Phi_{\rm SSTC}\Big(u(k-1), u_{\rm ss}^{\rm ol}(k), y_{\rm ss}^{\rm ol}(k), u_t^{\rm (sm)}(k), y_t^{\rm (sm)}(k)\Big),\tag{6.2}$$

some constrained LPs or constrained QPs are solved, which obtains MVss and CVss. In DC, which can be conceptually represented as

$$\begin{aligned}&\Delta v(k|k)\\&\quad=\Phi_{\rm DC}\Big(u(k-1), \delta v_{\rm ss}(k), \{y^{\rm ol}(k+n|k),\ n=1,2,\dots,N\}, y_{\rm ss}(k),\\&\qquad\{\hat{x}^{\rm ol}(k+n_c|k),\ n_c=0,1,\dots,N_c-1\}\Big),\end{aligned}\tag{6.3}$$

ss from SSTC are tracked by optimizing the future MV moves.

6.1 Artificial Disturbance Model

Consider the linear time-invariant discrete-time model

$$\begin{cases}x_{k+1}=Ax_k+B\triangledown u_k+F\triangledown f_k\\ \triangledown y_k=Cx_k\end{cases}.\tag{6.4}$$

Model Predictive Control, First Edition. Baocang Ding and Yuanqing Yang.

Assume that (A, B) is stabilizable (i.e., there is an appropriate K such that $A + BK$ has all its eigenvalues being stable), and (C, A) is detectable (i.e., there is an appropriate L such that $A + LC$ has all its eigenvalues being stable). Note that if (A, B, C) is a stable or integral process (like in Chapters 4 and 5), then we can apply DMC to (6.4); the following approach can be applied to the unstable (A, B, C).

6.1.1 Basic Model

If we utilize (6.4) directly to design MPC, then there are a number of mature methods. However, if there are un-modeled dynamics, then it may fail to achieve offset-free. In MPC techniques, artificial disturbance has been introduced into (6.4) to tackle this issue. Let us augment the model (see [Pannocchia and Rawlings, 2003])

$$\begin{cases} \tilde{x}_{k+1} = \tilde{A}\tilde{x}_k + \tilde{B}\nabla u_k + \tilde{F}\nabla f_k \\ \nabla y_k = \tilde{C}\tilde{x}_k \end{cases}, \tag{6.5}$$

where

$$\begin{aligned} &\tilde{x}_k = \begin{bmatrix} x_k \\ s_k \end{bmatrix}, \ \tilde{A} = \begin{bmatrix} A & G_\sigma \\ 0 & I \end{bmatrix}, \\ &\tilde{B} = \begin{bmatrix} B \\ 0 \end{bmatrix}, \ \tilde{F} = \begin{bmatrix} F \\ 0 \end{bmatrix}, \ \tilde{C} = \begin{bmatrix} C & G_s \end{bmatrix}, \end{aligned} \tag{6.6}$$

and $s \in \mathbb{R}^{n_s}$ is the introduced artificial disturbance. In $\{(6.5), (6.6)\}$, $(\tilde{A}, \tilde{B})$ is not stabilizable.

Theorem 6.1 ***(see [Pannocchia and Rawlings, 2003])*** In (6.5) and (6.6), $(\tilde{C}, \tilde{A})$ is detectable, if and only if (C, A) is detectable and

$$\text{rank} \begin{bmatrix} I - A & -G_\sigma \\ C & G_s \end{bmatrix} = n_x + n_s. \tag{6.7}$$

A necessary condition for (6.7) is $n_s \le n_y$. Let us, specially, take (see [Muske and Badgwell, 2002])

$$\begin{aligned} &s = \begin{bmatrix} d \\ p \end{bmatrix}, \ G_\sigma = \begin{bmatrix} G_d & 0 \end{bmatrix}, \ G_s = \begin{bmatrix} 0 & G_p \end{bmatrix}, \\ &d \in \mathbb{R}^{n_d}, \ p \in \mathbb{R}^{n_p}, \ n_s = n_d + n_p, \end{aligned}$$

then (6.6) becomes

$$\begin{aligned} &\tilde{x}_k = \begin{bmatrix} x_k \\ d_k \\ p_k \end{bmatrix}, \ \tilde{A} = \begin{bmatrix} A & G_d & 0 \\ 0 & I & 0 \\ 0 & 0 & I \end{bmatrix}, \\ &\tilde{B} = \begin{bmatrix} B \\ 0 \\ 0 \end{bmatrix}, \ \tilde{F} = \begin{bmatrix} F \\ 0 \\ 0 \end{bmatrix}, \ \tilde{C} = \begin{bmatrix} C & 0 & G_p \end{bmatrix}. \end{aligned} \tag{6.8}$$

Theorem 6.2 ***(see [Muske and Badgwell, 2002])*** In $\{(6.5), (6.8)\}$, $(\tilde{C}, \tilde{A})$ is detectable if and only if (C, A) is detectable and

$$\text{rank}\begin{bmatrix} I - A & -G_d & 0 \\ C & 0 & G_p \end{bmatrix} = n_x + n_d + n_p. \tag{6.9}$$

Theorem 6.3 ***(see [Muske and Badgwell, 2002)]*** For all detectable (C, A), by applying $\{(6.5), (6.8)\}$ where $n_p + n_d = n_y$, there exist detectable $(\tilde{C}, \tilde{A})$.

Hence, taking $\{(6.5), (6.8)\}$, we can asymptotically estimate $\tilde{x}$ for appropriate $\{n_d, n_p, G_d, G_p\}$ and $\tilde{F}\nabla f_k$.

6.1.2 Controlled Variable as Additional State

Applying (6.4), we can obtain

$$\begin{cases} \Delta x_{k+1} = A\Delta x_k + B\Delta u_k + F\Delta f_k \\ \Delta y_k = C\Delta x_k \end{cases}. \tag{6.10}$$

Applying (6.10), we obtain $\Delta y_{k+1} = C\Delta x_{k+1} = CA\Delta x_k + CB\Delta u_k + CF\Delta f_k$. Since $y_{k+1} = y_k + \Delta y_{k+1}$, we can obtain the augmented state-space model (see [González et al., 2008])

$$\begin{cases} \tilde{x}_{k+1} = \tilde{A}\tilde{x}_k + \tilde{B}\Delta u_k + \tilde{F}\Delta f_k \\ y_k = \tilde{C}\tilde{x}_k \end{cases}, \tag{6.11}$$

where

$$\tilde{x}_k = \begin{bmatrix} \Delta x_k \\ y_k \end{bmatrix}, \ \tilde{A} = \begin{bmatrix} A & 0 \\ CA & I \end{bmatrix},$$
$$\tilde{B} = \begin{bmatrix} B \\ CB \end{bmatrix}, \ \tilde{F} = \begin{bmatrix} F \\ CF \end{bmatrix}, \ \tilde{C} = \begin{bmatrix} 0 & I \end{bmatrix}. \tag{6.12}$$

Since (C, A) is detectable, by applying Hautus's Theorem (see [Sontag, 1990]), it is shown that $(\tilde{C}, \tilde{A})$ is detectable. Hence, we can asymptotically estimate $\tilde{x}$ for appropriate $\tilde{F}\Delta f_k$. However, since (A, B) is stabilizable, $(\tilde{A}, \tilde{B})$ is stabilizable if and only if

$$\text{rank}\begin{bmatrix} I - A & B \\ -CA & CB \end{bmatrix} = n_x + n_y. \tag{6.13}$$

Hence, if (6.13) holds, then we can find a stabilizing dynamic feedback law $\Delta u = K\Delta x + K_y y$ for (6.11)–(6.12) and appropriate $\tilde{F}\Delta f_k$. A necessary condition for (6.13) is $n_y \leq n_u$. An equivalent model to (6.11) and (6.12) is to take

$$\tilde{x}_k = \begin{bmatrix} \Delta x_k \\ y_{k-1} \end{bmatrix}, \ \tilde{A} = \begin{bmatrix} A & 0 \\ C & I \end{bmatrix},$$
$$\tilde{B} = \begin{bmatrix} B \\ 0 \end{bmatrix}, \ \tilde{F} = \begin{bmatrix} F \\ 0 \end{bmatrix}, \ \tilde{C} = \begin{bmatrix} C & I \end{bmatrix}. \tag{6.14}$$

For (6.11) and (6.12), let us augment the model as $\Delta x_{k+1} = A\Delta x_k + B\Delta u_k + F\Delta f_k + G_d\Delta d_k$ and $\Delta y_{k+1} = C\Delta x_{k+1} + G_p\Delta p_k = CA\Delta x_k + CB\Delta u_k + CF\Delta f_k + CG_d\Delta d_k + G_p\Delta p_k$, i.e., give the augmented state-space model

$$\begin{cases} \tilde{x}_{k+1} = \tilde{A}\tilde{x}_k + \tilde{B}\Delta u_k + \tilde{F}\Delta f_k \\ y_k = \tilde{C}\tilde{x}_k \end{cases}, \tag{6.15}$$

where

$$\tilde{x}_k = \begin{bmatrix} \Delta x_k \\ y_k \\ \Delta s_k \end{bmatrix},\ \tilde{A} = \begin{bmatrix} A & 0 & G_{x,s} \\ CA & I & G_{y,s} \\ 0 & 0 & I \end{bmatrix}$$
$$\tilde{B} = \begin{bmatrix} B \\ CB \\ 0 \end{bmatrix},\ \tilde{F} = \begin{bmatrix} F \\ CF \\ 0 \end{bmatrix},\ \tilde{C} = \begin{bmatrix} 0 & I & 0 \end{bmatrix}, \tag{6.16}$$

and

$$\Delta s = \begin{bmatrix} \Delta d \\ \Delta p \end{bmatrix},\ \begin{bmatrix} G_{x,s} \\ G_{y,s} \end{bmatrix} = \begin{bmatrix} G_d & 0 \\ CG_d & G_p \end{bmatrix}.$$

For (6.15) and (6.16), $(\tilde{C}, \tilde{A})$ is detectable if and only if

$$\text{rank} \begin{bmatrix} I - A & 0 & -G_{x,s} \\ -CA & 0 & -G_{y,s} \\ 0 & I & 0 \end{bmatrix} = n_x + n_y + n_s. \tag{6.17}$$

A necessary condition for (6.17) is $n_s \leq n_y$.

6.1.3 Manipulated Variable as Additional State

Since $\nabla u_k = \nabla u_{k-1} + \Delta u_k$, the following augmented form of (6.4) is obtained (see [González et al., 2008]):

$$\begin{cases} \tilde{x}_{k+1} = \tilde{A}\tilde{x}_k + \tilde{B}\Delta u_k + \tilde{F}\nabla f_k \\ \nabla y_k = \tilde{C}\tilde{x}_k \end{cases}, \tag{6.18}$$

where

$$\tilde{x}_k = \begin{bmatrix} x_k \\ \nabla u_{k-1} \end{bmatrix},\ \tilde{A} = \begin{bmatrix} A & B \\ 0 & I \end{bmatrix},$$
$$\tilde{B} = \begin{bmatrix} B \\ I \end{bmatrix},\ \tilde{F} = \begin{bmatrix} F \\ 0 \end{bmatrix},\ \tilde{C} = \begin{bmatrix} C & 0 \end{bmatrix}. \tag{6.19}$$

Since (A, B) is stabilizable, by applying Hautus's Theorem (see [Sontag, 1990]), it is shown that $(\tilde{A}, \tilde{B})$ is stabilizable. Hence, we can find a stabilizing augmented feedback law $\Delta u_k = Kx_k + K_u\nabla u_{k-1}$ for (6.18) and (6.19) and appropriate $\tilde{F}\nabla f_k$. However, since (C, A) is detectable, $(\tilde{C}, \tilde{A})$ is detectable if and only if

$$\text{rank} \begin{bmatrix} I - A & -B \\ C & 0 \end{bmatrix} = n_x + n_u. \tag{6.20}$$

Hence, if (6.20) holds, we can asymptotically estimate $\tilde{x}$ for appropriate $\tilde{F}\triangledown f_k$. A necessary condition for (6.20) is $n_u \leq n_y$.

Since $\triangledown u_k = \triangledown u_{k-1} + \Delta u_k$ is a fact, we will not introduce artificial disturbance in this equation. Let us introduce d and p into (6.18) and (6.19), and give the augmented state-space model

$$\begin{cases} \tilde{x}_{k+1} = \tilde{A}\tilde{x}_k + \tilde{B}\Delta u_k + \tilde{F}\triangledown f_k \\ \triangledown y_k = \tilde{C}\tilde{x}_k \end{cases}, \tag{6.21}$$

where

$$\tilde{x}_k = \begin{bmatrix} x_k \\ \triangledown u_{k-1} \\ d_k \\ p_k \end{bmatrix}, \quad \tilde{A} = \begin{bmatrix} A & B & G_d & 0 \\ 0 & I & 0 & 0 \\ 0 & 0 & I & 0 \\ 0 & 0 & 0 & I \end{bmatrix}$$
$$\tilde{B} = \begin{bmatrix} B \\ I \\ 0 \\ 0 \end{bmatrix}, \quad \tilde{F} = \begin{bmatrix} F \\ 0 \\ 0 \\ 0 \end{bmatrix}, \quad \tilde{C} = \begin{bmatrix} C & 0 & 0 & G_p \end{bmatrix}. \tag{6.22}$$

For (6.21) and (6.22), $(\tilde{C}, \tilde{A})$ is detectable if and only if

$$\text{rank} \begin{bmatrix} I - A & -B & -G_d & 0 \\ C & 0 & 0 & G_p \end{bmatrix} = n_x + n_u + n_d + n_p. \tag{6.23}$$

A necessary condition for (6.23) is $n_u + n_d + n_p \leq n_y$. We can put $\triangledown u_{k-1} = \triangledown u_{k-1}$ into the output equation in (6.21) in order to enhance the detectability of $(\tilde{C}, \tilde{A})$.

If we take both MVs and CVs as the additional states, then we should consider both y and u as the outputs of the state-space equation; otherwise, the resultant $(\tilde{C}, \tilde{A})$ would be undetectable. For example, we can give the state-space model

$$\begin{cases} \tilde{x}_{k+1} = \tilde{A}\tilde{x}_k + \tilde{B}\Delta u_k + \tilde{F}\Delta f_k \\ \tilde{y}_k = \tilde{C}\tilde{x}_k \end{cases}, \tag{6.24}$$

where

$$\tilde{x}_k = \begin{bmatrix} \Delta x_k \\ y_k \\ u_{k-1} \end{bmatrix}, \quad \tilde{y}_k = \begin{bmatrix} y_k \\ u_{k-1} \end{bmatrix}, \quad \tilde{A} = \begin{bmatrix} A & 0 & 0 \\ CA & I & 0 \\ 0 & 0 & I \end{bmatrix}$$
$$\tilde{B} = \begin{bmatrix} B \\ CB \\ I \end{bmatrix}, \quad \tilde{F} = \begin{bmatrix} F \\ CF \\ 0 \end{bmatrix}, \quad \tilde{C} = \begin{bmatrix} 0 & I & 0 \\ 0 & 0 & I \end{bmatrix}. \tag{6.25}$$

Since (C, A) is detectable, by applying Hautus's Theorem (see [Sontag, 1990]), it is shown that $(\tilde{C}, \tilde{A})$ is detectable. Let us introduce Δs into (6.24)–(6.25), and renew the augmented state-space model (6.24) with

$$\tilde{x}_k = \begin{bmatrix} \Delta x_k \\ y_k \\ u_{k-1} \\ \Delta s_k \end{bmatrix}, \ \tilde{A} = \begin{bmatrix} A & 0 & 0 & G_{x,s} \\ CA & I & 0 & G_{y,s} \\ 0 & 0 & I & 0 \\ 0 & 0 & 0 & I \end{bmatrix}$$

$$\tilde{B} = \begin{bmatrix} B \\ CB \\ I \\ 0 \end{bmatrix}, \ \tilde{F} = \begin{bmatrix} F \\ CF \\ 0 \\ 0 \end{bmatrix}, \ \tilde{C} = \begin{bmatrix} 0 & I & 0 & 0 \\ 0 & 0 & I & 0 \end{bmatrix}. \tag{6.26}$$

For {(6.24), (6.26)}, $(\tilde{C}, \tilde{A})$ is detectable if and only if

$$\text{rank}\begin{bmatrix} I-A & 0 & 0 & -G_{x,s} \\ -CA & 0 & 0 & -G_{y,s} \\ 0 & I & 0 & 0 \\ 0 & 0 & I & 0 \end{bmatrix} = n_x + n_y + n_u + n_s. \tag{6.27}$$

A necessary condition for (6.27) is $n_s \leq n_y$. One can especially, take

$$\Delta s = \begin{bmatrix} \Delta d \\ \Delta p \end{bmatrix}, \ \begin{bmatrix} G_{x,s} \\ G_{y,s} \end{bmatrix} = \begin{bmatrix} G_d & 0 \\ CG_d & G_p \end{bmatrix}.$$

6.1.4 Kalman Filter

Let us write {(6.5), (6.11), (6.15), (6.18), (6.21), (6.24)} uniformly as

$$\begin{cases} \tilde{x}_{k+1} = \tilde{A}\tilde{x}_k + \tilde{B}\tilde{u}_k + \tilde{F}\tilde{f}_k \\ \tilde{y}_k = \tilde{C}\tilde{x}_k \end{cases}, \tag{6.28}$$

where $\tilde{u}_k$ is ∇u_k, u_k or Δu_k, $\tilde{f}_k$ is ∇f_k or Δf_k. Solve the algebraic Riccati equation

$$\Sigma = \tilde{A}\Sigma\tilde{A}^{\mathrm{T}} - (\tilde{A}\Sigma\tilde{C}^{\mathrm{T}} + R_{12})(\tilde{C}\Sigma\tilde{C}^{\mathrm{T}} + R_2)^{-1}(\tilde{A}\Sigma\tilde{C}^{\mathrm{T}} + R_{12})^{\mathrm{T}} + R_1$$

to obtain Σ where R_1, R_{12}, and R_2, satisfying $\begin{bmatrix} R_1 & R_{12} \\ R_{12}^{\mathrm{T}} & R_2 \end{bmatrix} \geq 0$, are tuning parameters. For $k > 0$, adopt ssKF for the augmented model (6.28), i.e.,

$$\begin{aligned} \hat{\tilde{x}}_{(k|k-1)|k} &= \tilde{A}\hat{\tilde{x}}_{k-1|k-1} + \tilde{B}\tilde{u}_{k-1} + \tilde{F}\tilde{f}_{k-1} + R_{12}R_2^{-1}(\tilde{y}_{k-1} - \tilde{C}\hat{\tilde{x}}_{k-1|k-1}), \\ \hat{\tilde{x}}_{k|k} &= \hat{\tilde{x}}_{(k|k-1)|k} + L(\tilde{y}_k - \tilde{C}\hat{\tilde{x}}_{(k|k-1)|k}), \end{aligned} \tag{6.29}$$

with pre-given $\hat{\tilde{x}}_{0|0}$, where the estimator gain matrix is

$$L = \Sigma\tilde{C}^{\mathrm{T}}(\tilde{C}\Sigma\tilde{C}^{\mathrm{T}} + R_2)^{-1}. \tag{6.30}$$

In (6.29), we use the notation $\hat{\tilde{x}}_{(k|k-1)|k}$ instead of $\hat{\tilde{x}}_{k|k-1}$, since it is unnecessary that $u(k-1) = u(k-1|k-1)$.

Consider ssKF for the augmented model {(6.5), (6.8)}, let us further augment (6.28) as

$$\begin{cases} \tilde{x}_{k+1} = \tilde{A}\tilde{x}_k + \tilde{B}u_k + \tilde{F}f_k - \tilde{B}u_{\mathrm{eq}} - \tilde{F}f_{\mathrm{eq}} + \xi_k \\ y_k = \tilde{C}\tilde{x}_k + y_{\mathrm{eq}} + \zeta_k \end{cases}, \tag{6.31}$$

where the inaccuracies of equilibrium are amended by $\{d, p\}$ and the noises for measurements $\{u, f, y\}$ are included in $\{\xi, \zeta\}$. In the sequel, we often use the notion $\triangledown$, for simplicity. Hence, for $k > 0$, the special ssKF for $\{(6.5), (6.8)\}$ is

$$\begin{aligned}
\begin{bmatrix} \hat{x}_{(k|k-1)|k} \\ \hat{d}_{k|k-1} \\ \hat{p}_{k|k-1} \end{bmatrix} &= \tilde{A} \begin{bmatrix} \hat{x}_{k-1|k-1} \\ \hat{d}_{k-1|k-1} \\ \hat{p}_{k-1|k-1} \end{bmatrix} + R_{12}R_2^{-1} \left(\triangledown y_{k-1} - \tilde{C} \begin{bmatrix} \hat{x}_{k-1|k-1} \\ \hat{d}_{k-1|k-1} \\ \hat{p}_{k-1|k-1} \end{bmatrix} \right) \\
&\quad + \tilde{B} \triangledown u_{k-1} + \tilde{F} \triangledown f_{k-1}, \\
\begin{bmatrix} \hat{x}_{k|k} \\ \hat{d}_{k|k} \\ \hat{p}_{k|k} \end{bmatrix} &= \begin{bmatrix} \hat{x}_{(k|k-1)|k} \\ \hat{d}_{k|k-1} \\ \hat{p}_{k|k-1} \end{bmatrix} + \begin{bmatrix} L_x \\ L_d \\ L_p \end{bmatrix} \left(\triangledown y_k - \tilde{C} \begin{bmatrix} \hat{x}_{(k|k-1)|k} \\ \hat{d}_{k|k-1} \\ \hat{p}_{k|k-1} \end{bmatrix} \right),
\end{aligned} \tag{6.32}$$

with pre-given $\begin{bmatrix} \hat{x}_{0|0} \\ \hat{d}_{0|0} \\ \hat{p}_{0|0} \end{bmatrix}$, where

$$R_{12} = \begin{bmatrix} R_{12}^x \\ R_{12}^d \\ R_{12}^p \end{bmatrix}, \quad L = \begin{bmatrix} L_x \\ L_d \\ L_p \end{bmatrix}. \tag{6.33}$$

In (6.32), we use the notation $\{\hat{d}_{k|k-1}, \hat{p}_{k|k-1}\}$ instead of $\{\hat{d}_{(k|k-1)|k}, \hat{p}_{(k|k-1)|k}\}$ due to the special structure of $\tilde{B}$.

Let us adopt the estimator-state feedback law to pre-stabilize the model, so the actual control move is calculated as

$$\triangledown u(k+i|k) = K\hat{x}(k+i|k) + v(k+i|k), \quad i \geq 0, \quad k \geq 0, \tag{6.34}$$

where K is the feedback gain matrix such that $A_c = A + BK$ is asymptotically stable, and v is the perturbation item. At time $k-1$, (6.34) becomes

$$\begin{aligned}
\triangledown u(k-1+i|k-1) &= K\hat{x}(k-1+i|k-1) + v(k-1+i|k-1), \\
& i \geq 0, \quad k \geq 1.
\end{aligned} \tag{6.35}$$

Define

$$\begin{aligned}
\Delta\hat{x}(k|k) &= \hat{x}(k|k) - \hat{x}(k-1|k-1), \quad k \geq 1, \\
\Delta\hat{x}(k+i+1|k) &= \hat{x}(k+i+1|k) - \hat{x}(k+i|k), \quad i \geq 0, \quad k \geq 0.
\end{aligned}$$

Subtracting (6.35) from (6.34) yields

$$\Delta u(k+i|k) = K\Delta\hat{x}(k+i|k) + \Delta v(k+i|k), \quad i \geq 0, \quad k \geq 1. \tag{6.36}$$

Equation (6.36) cannot be utilized for $k = 0$ and $i = 0$. However, if we pseudo-take $K\hat{x}(-1|-1) = \triangledown u(k-1)$ and $v(-1) = 0$ (so that $\triangledown u(k-1) = K\hat{x}(-1|-1) + v(-1)$, although $\hat{x}(-1|-1)$ and $v(-1)$ do not exist), then

$$\Delta u(k+i|k) = K\Delta\hat{x}(k+i|k) + \Delta v(k+i|k), \quad i \geq 0, \quad k \geq 0. \tag{6.37}$$

Assume that f keeps unchanged in the future. Based on the $i \geq 1$ step ahead Kalman predictor, the future predictions are

$$\hat{x}(k+1|k) = A\hat{x}(k|k) + B\nabla u(k|k) + F\nabla f(k) + G_d\hat{d}(k|k) + R_{12}^x R_2^{-1}[\nabla y(k) - C\hat{x}(k|k) - G_p\hat{p}(k|k)], \tag{6.38}$$

$$\hat{d}(k+1|k) = \hat{d}(k|k) + R_{12}^d R_2^{-1}[\nabla y(k) - C\hat{x}(k|k) - G_p\hat{p}(k|k)], \tag{6.39}$$

$$\hat{p}(k+1|k) = \hat{p}(k|k) + R_{12}^p R_2^{-1}[\nabla y(k) - C\hat{x}(k|k) - G_p\hat{p}(k|k)], \tag{6.40}$$

$$\hat{x}(k+i+1|k) = A\hat{x}(k+i|k) + B\nabla u(k+i|k) + F\nabla f(k) + G_d\hat{d}(k+1|k) \tag{6.41}$$

$$= A\hat{x}(k|k) + A_c\sum_{j=1}^{i}\Delta\hat{x}(k+j|k) + B\nabla u(k|k) + B\sum_{j=1}^{i}\Delta v(k+j|k) + F\nabla f(k) + G_d\hat{d}(k+1|k), \quad i \geq 1, \tag{6.42}$$

$$\nabla y(k+i|k) = C\hat{x}(k+i|k) + G_p\hat{p}(k+i|k), i \geq 1, \tag{6.43}$$

$$\hat{d}(k+i|k) = \hat{d}(k+1|k), \quad i \geq 1, \tag{6.44}$$

$$\hat{p}(k+i|k) = \hat{p}(k+1|k), \quad i \geq 1. \tag{6.45}$$

Subtracting both sides of (6.38) by $\hat{x}(k|k)$, yields

$$\Delta\hat{x}(k+1|k) = (A-I)\hat{x}(k|k) + B\nabla u(k|k) + F\nabla f(k) + G_d\hat{d}(k|k) + R_{12}^x R_2^{-1}[\nabla y(k) - C\hat{x}(k|k) - G_p\hat{p}(k|k)]. \tag{6.46}$$

For $i = 1$, subtracting (6.38) from (6.42), yields

$$\Delta\hat{x}(k+2|k) = A_c\Delta\hat{x}(k+1|k) + B\Delta v(k+1|k) + G_d\Delta\hat{d}(k+1|k) - R_{12}^x R_2^{-1}[\Delta y(k) - C\Delta\hat{x}(k|k) - G_p\Delta\hat{p}(k|k)]. \tag{6.47}$$

For $i > 1$, the incremental form of (6.42) is

$$\Delta\hat{x}(k+i+1|k) = A_c\Delta\hat{x}(k+i|k) + B\Delta v(k+i|k), \quad i \geq 2. \tag{6.48}$$

Denote

$$\nabla y(k|k) = C\hat{x}(k|k) + G_p\hat{p}(k|k). \tag{6.49}$$

Based on (6.43) and (6.49), the predictions of the future CV are

$$y(k+i|k) = y(k|k) + G_p\Delta\hat{p}(k+1|k) + \sum_{j=1}^{i}C\Delta\hat{x}(k+j|k), \quad i \geq 1. \tag{6.50}$$

Moreover, the predictions of the future MV are

$$u(k+i|k) = u(k-1) + \sum_{j=0}^{i}[K\Delta\hat{x}(k+j|k) + \Delta v(k+j|k)], \quad i \geq 0. \tag{6.51}$$

Due to the use of $\{\nabla y, \nabla u, \nabla f\}$ in ssKF, it cannot remove the equilibrium in calculating $y(k+i|k)$ and $u(k+i|k)$.

6.2 Open-Loop Prediction Module

For the feedback control law and estimator, we have several choices, e.g.,

(a) feedback law $\triangledown u = Kx$, and estimator for {(6.5), (6.8)} satisfying (6.9);
(b) feedback law $\Delta u = K\Delta x$, and estimator for {(6.5), (6.8)} satisfying (6.9);
(c) feedback law $\Delta u_k = K\Delta x_k + K_y y_k$, and estimator for (6.15) and (6.16) satisfying (6.17);
(d) feedback law $\Delta u_k = Kx_k + K_u \triangledown u_{k-1}$, and estimator for (6.21) and (6.22) satisfying (6.23);
(e) feedback law $\Delta u_k = K\Delta x_k + K_y y_k + K_u u_{k-1}$, and estimator for {(6.24), (6.26)} satisfying (6.27).

Choosing (c) and (e) can avoid the use of equilibrium in the two-layered MPC.

In the following two-layered MPC, we select (b). We apply {(6.5), (6.8)} to obtain the 1-step ahead estimation of $[x; d; p]$, where that of $[d; p]$ will be the real-time steady-state estimation, and that of x, the initial value of the dynamic ol predictions. Thus, the 1-step ahead estimation of $[d; p]$ affects the initial value of the dynamic ol predictions of x and, consequently, the steady-state/dynamic predictions of the state/CV; this shows why introducing $[d; p]$ improves the control performance. Since $[d; p]$ in {(6.5), (6.8)} is not stabilizable, we cannot utilize $[d; p]$ for feedback. Hence, in both SSTC and DC, the model and estimation of $[d; p]$ are not involved.

Define $\Delta u^{\text{ol}}(k+i|k) = K\Delta\hat{x}^{\text{ol}}(k+i|k)\,(i \geq 0)$ as the open-loop control move. The so-called open-loop prediction is estimated under the open-loop control moves. According to {(6.46), (6.47), (6.48)}, we obtain

$$\begin{aligned}\Delta\hat{x}^{\text{ol}}(k+1|k) &= (A-I)\hat{x}^{\text{ol}}(k|k) + B\triangledown u(k-1) + BK\Delta\hat{x}^{\text{ol}}(k|k) + F\triangledown f(k)\\ &\quad + G_d\hat{d}(k|k) + R_{12}^{x}R_2^{-1}[\triangledown y(k) - C\hat{x}(k|k) - G_p\hat{p}(k|k)],\end{aligned} \tag{6.52}$$

$$\begin{aligned}\Delta\hat{x}^{\text{ol}}(k+2|k) &= A_c\Delta\hat{x}^{\text{ol}}(k+1|k) + G_d\Delta\hat{d}(k+1|k)\\ &\quad - R_{12}^{x}R_2^{-1}[\Delta y(k) - C\Delta\hat{x}(k|k) - G_p\Delta\hat{p}(k|k)],\end{aligned} \tag{6.53}$$

$$\Delta\hat{x}^{\text{ol}}(k+i+1|k) = A_c\Delta\hat{x}^{\text{ol}}(k+i|k), \quad i \geq 2, \tag{6.54}$$

$$\triangledown y^{\text{ol}}(k+i|k) = C\hat{x}^{\text{ol}}(k+i|k) + G_p\hat{p}(k+1|k), i \geq 1, \tag{6.55}$$

where the initial prediction is $\hat{x}^{\text{ol}}(k|k) = \hat{x}(k|k)$. Based on (6.49) and (6.52)–(6.55), ol predictions of the future CV and MV are

$$y^{\text{ol}}(k+i|k) = y(k|k) + G_p\Delta\hat{p}(k+1|k) + \sum_{j=1}^{i} C\Delta\hat{x}^{\text{ol}}(k+j|k), \quad i \geq 1, \tag{6.56}$$

$$u^{\text{ol}}(k+i|k) = u(k-1) + \sum_{j=0}^{i} K\Delta\hat{x}^{\text{ol}}(k+j|k), \quad i \geq 0. \tag{6.57}$$

According to (6.52)–(6.57), since A_c is asymptotically stable, we obtain the steady-state ol prediction model,

$$\delta\hat{x}_{\text{ss}}^{\text{ol}}(k) = \sum_{j=1}^{\infty} \Delta\hat{x}^{\text{ol}}(k+j|k) = (I - A_c)^{-1}\mathcal{T}_1(k), \tag{6.58}$$

$$u_{ss}^{ol}(k) = u(k-1) + K\Delta\hat{x}^{ol}(k|k) + K(I-A_c)^{-1}\mathcal{T}_1(k), \tag{6.59}$$

$$y_{ss}^{ol}(k) = y(k|k) + G_p\Delta\hat{p}(k+1|k) + C(I-A_c)^{-1}\mathcal{T}_1(k), \tag{6.60}$$

where $\mathcal{T}_1(k) = (A-I)\hat{x}^{ol}(k|k) + B\nabla u(k-1) + BK\Delta\hat{x}^{ol}(k|k) + F\nabla f(k) + G_d\hat{d}(k+1|k)$. If the two-layered MPC is closed-loop stable and $\lim_{k\to\infty}\Delta f(k) = 0$, then it is clear that $\lim_{k\to\infty}\mathcal{T}_1(k) = 0$.

6.3 Steady-State Target Calculation Module

According to (6.46)–(6.51), we obtain the steady-state closed-loop prediction model, i.e.,

$$\delta\hat{x}_{ss}(k) = \sum_{j=1}^{\infty}\Delta\hat{x}(k+j|k) = (I-A_c)^{-1}[B\delta v_{ss}(k) + \mathcal{T}_1(k)], \tag{6.61}$$

$$u_{ss}(k) = u(k-1) + K\Delta\hat{x}^{ol}(k|k) + K_c\delta v_{ss}(k) + K(I-A_c)^{-1}\mathcal{T}_1(k), \tag{6.62}$$

$$y_{ss}(k) = y(k|k) + G_p\Delta\hat{p}(k+1|k) + G_c\delta v_{ss}(k) + C(I-A_c)^{-1}\mathcal{T}_1(k), \tag{6.63}$$

where $K_c = K(I-A_c)^{-1}B + I$ and $G_c = C(I-A_c)^{-1}B$ are steady-state gain matrices. Subtracting (6.59) from (6.62) yields

$$u_{ss}(k) = K_c\delta v_{ss}(k) + u_{ss}^{ol}(k). \tag{6.64}$$

Since $u_{ss}(k) = u_{ss}^{ol}(k) + \delta u_{ss}(k)$, it is shown that $\delta u_{ss}(k) = K_c\delta v_{ss}(k)$. Subtracting (6.60) from (6.63), yields

$$y_{ss}(k) = G_c\delta v_{ss}(k) + y_{ss}^{ol}(k). \tag{6.65}$$

Noting that $y_{ss}(k) = y_{ss}^{ol}(k) + \delta y_{ss}(k)$, it is shown that $\delta y_{ss}(k) = G_c\delta v_{ss}(k)$.

6.3.1 Constraints on Steady-State Perturbation Increment

Similarly to Chapter 4, the following magnitude and rate constraints will be considered for SSTC:

$$\underline{u} \le u_{ss}(k) \le \bar{u}, \qquad |\delta u_{ss}(k)| \le N_c\Delta\bar{u}, \qquad |\delta u_{ss}(k)| \le \delta\bar{u}_{ss}, \tag{6.66}$$

$$\underline{y}_{0,h} \le y_{ss}(k) \le \bar{y}_{0,h}, \qquad |\delta y_{ss}(k)| \le \delta\bar{y}_{ss}, \tag{6.67}$$

$$\underline{y}_0 \le y_{ss}(k) \le \bar{y}_0. \tag{6.68}$$

According to (6.64), the hard constraints in (6.66) are represented as

$$\underline{u}''(k) \le K_c\delta v_{ss}(k) \le \bar{u}''(k), \tag{6.69}$$

where

$$\underline{u}''(k) = \min\{\max\{\underline{u} - u(k-1), -\delta\bar{u}_{ss}'\}, \bar{u} - u(k-1)\},$$
$$\bar{u}''(k) = \max\{\min\{\bar{u} - u(k-1), \delta\bar{u}_{ss}'\}, \underline{u} - u(k-1)\},$$
$$\delta\bar{u}_{ss}' = \min\{N_c\Delta\bar{u}, \delta\bar{u}_{ss}\}.$$

For ET of MV, there are also soft inequality constraints

$$\underline{u}_{i,\mathrm{ss}}(k) \leq K_{c,i}\delta v_{\mathrm{ss}}(k) \leq \bar{u}_{i,\mathrm{ss}}(k), \quad i \in \mathcal{I}_t, \tag{6.70}$$

where $K_{c,i}$ is the ith row of K_c, and

$$\underline{u}_{i,\mathrm{ss}}(k) = \min\{\max\{u_{i,t}(k) - \frac{1}{2}u_{i,\mathrm{ss,range}} - u^{\mathrm{ol}}_{i,\mathrm{ss}}(k), \underline{u}'_i(k)\}, \bar{u}'_i(k)\}, \quad i \in \mathcal{I}_t,$$

$$\bar{u}_{i,\mathrm{ss}}(k) = \max\{\min\{u_{i,t}(k) + \frac{1}{2}u_{i,\mathrm{ss,range}} - u^{\mathrm{ol}}_{i,\mathrm{ss}}(k), \bar{u}'_i(k)\}, \underline{u}'_i(k)\}, \quad i \in \mathcal{I}_t.$$

By adding the equality constraints, the soft constraints for MVET become

$$\underline{u}_{i,\mathrm{ss}}(k) \leq K_{c,i}\delta v_{\mathrm{ss}}(k) \leq \bar{u}_{i,\mathrm{ss}}(k), \; K_{c,i}\delta v_{\mathrm{ss}}(k) = u_{i,t}(k) - u^{\mathrm{ol}}_{i,\mathrm{ss}}(k), \quad i \in \mathcal{I}_t. \tag{6.71}$$

According to (6.65), the hard constraints in (6.67) become

$$\underline{y}_h(k) \leq G_c\delta v_{\mathrm{ss}}(k) \leq \bar{y}_h(k), \tag{6.72}$$

where

$$\underline{y}_h(k) = \min\{\max\{\underline{y}_{0,h} - y^{\mathrm{ol}}_{\mathrm{ss}}(k), -\delta\bar{y}_{\mathrm{ss}}\}, \bar{y}_{0,h} - y^{\mathrm{ol}}_{\mathrm{ss}}(k)\},$$

$$\bar{y}_h(k) = \max\{\min\{\bar{y}_{0,h} - y^{\mathrm{ol}}_{\mathrm{ss}}(k), \delta\bar{y}_{\mathrm{ss}}\}, \underline{y}_{0,h} - y^{\mathrm{ol}}_{\mathrm{ss}}(k)\}.$$

The soft constraint (6.68) becomes

$$G_c\delta v_{\mathrm{ss}}(k) \leq \bar{y}(k), \tag{6.73}$$

$$G_c\delta v_{\mathrm{ss}}(k) \geq \underline{y}(k), \tag{6.74}$$

where

$$\bar{y}(k) = \max\{\min\{\bar{y}_0 - y^{\mathrm{ol}}_{\mathrm{ss}}(k), \bar{y}_h(k)\}, \underline{y}_h(k)\},$$

$$\underline{y}(k) = \min\{\max\{\underline{y}_0 - y^{\mathrm{ol}}_{\mathrm{ss}}(k), \underline{y}_h(k)\}, \bar{y}_h(k)\}.$$

For CVET, there are also soft inequality constraints

$$\underline{y}_{j,\mathrm{ss}}(k) \leq G_{c,j}\delta v_{\mathrm{ss}}(k) \leq \bar{y}_{j,\mathrm{ss}}(k), \; j \in \mathcal{J}_t, \tag{6.75}$$

where $G_{c,j}$ is the jth row of G_c, and

$$\underline{y}_{j,\mathrm{ss}}(k) = \min\{\max\{y_{j,t}(k) - \frac{1}{2}y_{j,\mathrm{ss,range}} - y^{\mathrm{ol}}_{j,\mathrm{ss}}(k), \underline{y}_{j,h}(k)\}, \bar{y}_{j,h}(k)\},$$

$$\bar{y}_{j,\mathrm{ss}}(k) = \max\{\min\{y_{j,t}(k) + \frac{1}{2}y_{j,\mathrm{ss,range}} - y^{\mathrm{ol}}_{j,\mathrm{ss}}(k), \bar{y}_{j,h}(k)\}, \underline{y}_{j,h}(k)\}.$$

By adding the equality constraints, the soft constraints for CVET become

$$\underline{y}_{j,\mathrm{ss}}(k) \leq G_{c,j}\delta v_{\mathrm{ss}}(k) \leq \bar{y}_{j,\mathrm{ss}}(k), \; G_{c,j}\delta v_{\mathrm{ss}}(k) = y_{j,t}(k) - y^{\mathrm{ol}}_{j,\mathrm{ss}}(k), \; j \in \mathcal{J}_t. \tag{6.76}$$

For handling the soft constraints {(6.71), (6.73), (6.74), (6.76)}, we need to partition them into several priority ranks; the details are the same as in Chapter 4. In the following, the feasibility and economic stages of SSTC module will be explained.

6.3.2 Feasibility Stage

In each priority rank, there are hard constraints, i.e.,

$$\begin{bmatrix} K_c \\ -K_c \\ G_c \\ -G_c \end{bmatrix} \delta v_{ss}(k) \leq \begin{bmatrix} \bar{u}''(k) \\ -\underline{u}''(k) \\ \bar{y}_h(k) \\ -\underline{y}_h(k) \end{bmatrix}, \tag{6.77}$$

which are composed of (6.69) and (6.72). In rank r, the soft constraints in this rank may be relaxed so as to be consistent with all the hard constraints. Let us denote the soft-to-hard constraints from rank 1 to rank r as

$$\begin{cases} C^{(r)}\delta v_{ss}(k) \leq c^{(r)}(k) \\ C_{eq}^{(r)}\delta v_{ss}(k) = c_{eq}^{(r)}(k) \end{cases}. \tag{6.78}$$

The constraint (6.78) is hard for the rank $r+1$. It is easy to know that (6.78) is composed of the following constraints:

$$G_{c,j}\delta v_{ss}(k) \leq \bar{y}'_j(k), \quad j \in \mathcal{O}_u^{(r)}, \tag{6.79}$$

$$-G_{c,j}\delta v_{ss}(k) \leq -\underline{y}'_j(k), \quad j \in \mathcal{O}_l^{(r)}, \tag{6.80}$$

$$K_{c,i}\delta v_{ss}(k) = u_{i,ss}(k) - u_{i,ss}^{ol}(k), \quad i \in \mathbb{I}^{(r)}, \tag{6.81}$$

$$G_{c,j}\delta v_{ss}(k) = y_{j,ss}(k) - y_{j,ss}^{ol}(k), \quad j \in \mathbb{O}^{(r)}, \tag{6.82}$$

where $\{\mathcal{O}_u^{(r)}, \mathcal{O}_l^{(r)}\} \in \mathbb{N}_y$, $\mathbb{I}^{(r)} \subseteq \mathcal{I}_t$, $\mathbb{O}^{(r)} \subseteq \mathcal{J}_t$; $\bar{y}'_j(k) \geq \bar{y}_j(k)$, $\underline{y}'_j(k) \leq \underline{y}_j(k)$. The constrains (6.79)–(6.82) are the relaxed collections of {(6.73), (6.74), (6.71), (6.76)}, respectively.

Lemma 6.1 Under (6.79)–(6.82), the constraint (6.77) can be simplified as

$$\begin{cases} \underline{u}''_i(k) \leq K_{c,i}\delta v_{ss}(k) \leq \bar{u}''_i(k), \quad i \in \mathbb{N}_y \backslash \mathbb{I}^{(r)} \\ \underline{y}_{j,h}(k) \leq G_{c,j}\delta v_{ss}(k) \leq \bar{y}_{j,h}(k), \ j \in \mathbb{N}_y \backslash \mathbb{O}^{(r)} \end{cases}. \tag{6.83}$$

Hence, in the rank $r+1$, the constraints are (6.83) and

$$\begin{cases} \begin{bmatrix} C^{(r)} \\ \tilde{C}^{(r+1)} \end{bmatrix} \delta v_{ss}(k) \leq \begin{bmatrix} c^{(r)}(k) \\ \tilde{c}^{(r+1)}(k) + \varepsilon^{(r+1)}(k) \end{bmatrix} \\ \begin{bmatrix} C_{eq}^{(r)} \\ \tilde{C}_{eq}^{(r+1)} \end{bmatrix} \delta v_{ss}(k) = \begin{bmatrix} c_{eq}^{(r)}(k) \\ \tilde{c}_{eq}^{(r+1)}(k) + \varepsilon_{eq+}^{(r+1)}(k) - \varepsilon_{eq-}^{(r+1)}(k) \end{bmatrix} \end{cases}, \tag{6.84}$$

where $\varepsilon^{(r+1)}(k)$, $\varepsilon_{eq+}^{(r+1)}(k)$ and $\varepsilon_{eq-}^{(r+1)}(k)$ are nonnegative slack variables. The optimization in the rank $r+1$ tries to find the smallest ε.

For the rank $r+1$, either LP or QP is applied. If, in one rank, there are multiple equality/inequality soft constraints, then we take them as equally important. For each scalar slack variable ε, denote its equal concern error as $\overline{\varepsilon}$ which are calculated the same as in Chapter 4. By invoking LP, the optimization problem is

$$\min_{\varepsilon_{\text{eq}+}^{(r+1)}(k),\varepsilon_{\text{eq}-}^{(r+1)}(k),\varepsilon^{(r+1)}(k),\delta v_{\text{ss}}(k)} \left[\sum_{\ell=1}^{d^{(r+1)}} \left(\overline{\varepsilon}_{\ell}^{(r+1)}\right)^{-1} \varepsilon_{\ell}^{(r+1)}(k) + \sum_{\tau=1}^{d_{\text{eq}}^{(r+1)}} \left[\left(\overline{\varepsilon}_{\text{eq}+,\tau}^{(r+1)}\right)^{-1} \varepsilon_{\text{eq}+,\tau}^{(r+1)}(k) + \left(\overline{\varepsilon}_{\text{eq}-,\tau}^{(r+1)}\right)^{-1} \varepsilon_{\text{eq}-,\tau}^{(r+1)}(k) \right] \right],$$

$$\text{s.t. } (6.83)-(6.84), \text{ and } \varepsilon \geq 0,$$

where the subscript τ denotes the τth element of $\varepsilon_{\text{eq}}^{(r+1)}(k)$, and $d_{\text{eq}}^{(r+1)}$ is the dimension of $\varepsilon_{\text{eq}}^{(r+1)}(k)$; the subscript ℓ denote the ℓth element of $\varepsilon^{(r+1)}(k)$, and $d^{(r+1)}$ is the dimension of $\varepsilon^{(r+1)}$. By invoking QP, the optimization problem is

$$\min_{\varepsilon_{\text{eq}+}^{(r+1)}(k),\varepsilon_{\text{eq}-}^{(r+1)}(k),\varepsilon^{(r+1)}(k),\delta v_{\text{ss}}(k)} \left[\sum_{\ell=1}^{d^{(r+1)}} \left(\overline{\varepsilon}_{\ell}^{(r+1)}\right)^{-2} \varepsilon_{\ell}^{(r+1)}(k)^2 + \sum_{\tau=1}^{d_{\text{eq}}^{(r+1)}} \left[\left(\overline{\varepsilon}_{\text{eq}+,\tau}^{(r+1)}\right)^{-2} \varepsilon_{\text{eq}+,\tau}^{(r+1)}(k)^2 + \left(\overline{\varepsilon}_{\text{eq}-,\tau}^{(r+1)}\right)^{-2} \varepsilon_{\text{eq}-,\tau}^{(r+1)}(k)^2 \right] \right],$$

$$\text{s.t. } (6.83)-(6.84), \text{ and } \varepsilon \geq 0.$$

When the rank $r+1$ is completed, (6.84) is expressed as (6.78) with r being replaced by $r+1$.

6.3.3 Economic Stage Without Soft Constraint

Lemma 6.2 After the whole feasibility stage, all the hard and soft constraints are combined as

$$\underline{u}_i''(k) \leq K_{c,i}\delta v_{\text{ss}}(k) \leq \bar{u}_i''(k), \quad i \in \mathbb{N}_u \backslash \mathcal{I}_t, \tag{6.85}$$

$$\underline{y}_j'(k) \leq G_{c,j}\delta v_{\text{ss}}(k) \leq \overline{y}_j'(k), \quad j \in \mathbb{N}_y \backslash \mathcal{J}_t, \tag{6.86}$$

$$K_{c,i}\delta v_{\text{ss}}(k) = u_{i,\text{ss}}(k) - u_{i,\text{ss}}^{\text{ol}}(k), \quad i \in \mathcal{I}_t, \tag{6.87}$$

$$G_{c,j}\delta v_{\text{ss}}(k) = y_{j,\text{ss}}(k) - y_{j,\text{ss}}^{\text{ol}}(k), \quad j \in \mathcal{J}_t. \tag{6.88}$$

The so-called economic stage of SSTC is to find $\delta v_{\text{ss}}(k)$ satisfying (6.85)–(6.88). Denote the index set of the minimum-move MV as $\mathcal{I}_{\text{mm}}$. Minimizing $|\delta u_{i,\text{ss}}(k)|$ is equivalent to minimize $U_i(k)$ satisfying

$$-U_i(k) \leq K_{c,i}\delta v_{\text{ss}}(k) \leq U_i(k). \tag{6.89}$$

By invoking LP, the optimization is

$$\min_{\delta v_{\text{ss}}(k),U_i(k)} J = \sum_{i\in\mathcal{I}_{\text{mm}}} h_i U_i(k) + \sum_{i\notin\mathcal{I}_{\text{mm}}\bigcup\mathcal{I}_t} h_i K_{c,i}\delta v_{\text{ss}}(k),$$

$$\text{s.t. } (6.85)-(6.88);\ (6.89), i \in \mathcal{I}_{\text{mm}},$$

where h_i is the cost weight. By invoking QP, the optimization is

$$\min_{\delta v_{ss}(k),U_i(k)} J = \sum_{i\in\mathcal{I}_{mm}} h_i^2 U_i(k)^2 + \left(\sum_{i\notin\mathcal{I}_{mm}\bigcup\mathcal{I}_t} h_i K_{c,i}\delta v_{ss}(k) - J_{\min}\right)^2,$$

$$\text{s.t. } (6.85)-(6.88);\ (6.89), i\in\mathcal{I}_{mm},$$

where $J_{\min}$ is not larger than the minimum of $\sum_{i\notin\mathcal{I}_{mm}\bigcup\mathcal{I}_t} h_i K_{c,i}\delta v_{ss}(k)$.

6.4 Dynamic Calculation Module

Subtracting (6.52)–(6.54) from (6.46)–(6.48), yields

$$\begin{aligned}&\Delta\hat{x}(k+i+1|k) - \Delta\hat{x}^{ol}(k+i+1|k)\\ &\quad = A_c[\Delta\hat{x}(k+i|k) - \Delta\hat{x}^{ol}(k+i|k)] + B\Delta v(k+i|k),\ \ i=0,\cdots,N-1.\end{aligned} \tag{6.90}$$

Iterating via (6.90), yields

$$\begin{aligned}&\Delta\hat{x}(k+i|k) = \Delta\hat{x}^{ol}(k+i|k) + \sum_{l=0}^{\min\{i-1,N_c-1\}} A_c^{i-1-l}B\Delta v(k+l|k),\\ &\quad i=1,\dots,N.\end{aligned} \tag{6.91}$$

Applying (6.50), (6.91), and (6.56), yields

$$\begin{aligned}&y(k+i|k) = y^{ol}(k+i|k) + \sum_{j=0}^{\min\{i-1,N_c-1\}} \left(\sum_{l=0}^{i-1-j} CA_c^l B\right)\Delta v(k+j|k),\\ &\quad i=1,\dots,N.\end{aligned} \tag{6.92}$$

In DC, let us choose the cost functions as

$$J(k) = \sum_{i=1}^{N} \|y(k+i|k) - y_{ss}(k)\|^2_{Q(k)} + \sum_{j=0}^{N_c-1} \|\Delta v(k+j|k)\|^2_{\Lambda},$$

$$J'(k) = \sum_{i=1}^{N} \|y(k+i|k) - y_{ss}(k)\|^2_{Q(k)} + \sum_{j=0}^{N_c-1} \|\Delta v(k+j|k)\|^2_{\Lambda} + \|\underline{\varepsilon}_{dc}(k)\|^2_{\underline{\Omega}} + \|\overline{\varepsilon}_{dc}(k)\|^2_{\overline{\Omega}},$$

where $\underline{\varepsilon}_{dc}(k)$ and $\overline{\varepsilon}_{dc}(k)$ are CV slack variables. The weight matrices $\{Q(k),\Lambda,\underline{\Omega},\overline{\Omega}\}$ are the same as in Chapter 4.

In DC, usually, the following constraints (i.e., MV rate constraint, MV magnitude constraint, CV magnitude constraint, constraint on slack variable) are considered:

$$|\Delta u(k+i|k)| \le \Delta\bar{u},\ \ 0\le i\le N_c-1, \tag{6.93}$$

$$\underline{u} \le u(k+i|k) \le \bar{u},\ \ 0\le i\le N_c-1, \tag{6.94}$$

$$\underline{y}'_0(k) \le y(k+i|k) \le \overline{y}'_0(k),\ \ 1\le i\le N, \tag{6.95}$$

$$\underline{y}'_0(k) - \underline{\varepsilon}_{dc}(k) \le y(k+i|k) \le \overline{y}'_0(k) + \overline{\varepsilon}_{dc}(k),\ \ 1\le i\le N, \tag{6.96}$$

$$\underline{\varepsilon}_{\rm dc}(k) \le \underline{y}_0'(k) - \underline{y}_{0,h}, \tag{6.97}$$

$$\overline{\varepsilon}_{\rm dc}(k) \le \overline{y}_{0,h} - \overline{y}_0'(k), \tag{6.98}$$

where

$$\overline{y}_0'(k) = \max\{\overline{y}_0, y_{\rm ss}(k)\},$$
$$\underline{y}_0'(k) = \min\{\underline{y}_0, y_{\rm ss}(k)\}.$$

We need to unify the above constraints (6.93)–(6.96) via $\Delta v(k+l-1|k)$ $(l = 1, \dots, N_c)$. Substituting (6.91) into (6.37), yields

$$\begin{aligned}\Delta u(k+i|k) &= K\Delta(\hat{x}^{\rm ol}(k+i|k)) + K\sum_{l=0}^{i-1} A_c^{i-1-l}B\Delta v(k+l|k) + \Delta v(k+i|k)\\ &= K\Delta\hat{x}^{\rm ol}(k+i|k) + K\sum_{l=0}^{i-1} A_c^{i-1-l}B\Delta v(k+l|k) + \Delta v(k+i|k),\\ &\quad i = 0, \dots, N_c - 1.\end{aligned} \tag{6.99}$$

Applying (6.99), yields

$$\begin{aligned}&u(k+i|k)\\ &= u(k-1) + \sum_{j=0}^{i} \Delta u(k+j|k)\\ &= u(k-1) + \sum_{j=0}^{i} K\Delta\hat{x}^{\rm ol}(k+i|k)\\ &\quad + \sum_{j=0}^{i}\left(K\sum_{l=0}^{j-1} A_c^{j-1-l}B\Delta v(k+l|k) + \Delta v(k+j|k)\right)\\ &= u(k-1) + \sum_{j=0}^{i} K\Delta\hat{x}^{\rm ol}(k+i|k) + \sum_{j=0}^{i-1}\left(I + \sum_{l=0}^{i-1-j} KA_c^l B\right)\Delta v(k+j|k)\\ &\quad + \Delta v(k+i|k),\ i = 0, \dots, N_c - 1.\end{aligned} \tag{6.100}$$

Define

$$\Delta\tilde{v}(k|k) = \begin{bmatrix} \Delta v(k|k) \\ \Delta v(k+1|k) \\ \vdots \\ \Delta v(k+N_c-1|k) \end{bmatrix}.$$

Differently from tracking $y_{\rm ss}(k)$, the tracking of $\delta v_{\rm ss}(k)$ is achieved by imposing

$$L\Delta\tilde{v}(k|k) = \delta v_{\rm ss}(k), \tag{6.101}$$

where $L = [I\ I\ \cdots\ I]$.

In summary, at each k, we first solve

$$\min_{\Delta\tilde{v}(k|k)} J(k), \text{ s.t. } (6.93) - (6.95), (6.101). \tag{6.102}$$

If (6.102) is infeasible, further solve

$$\min_{\underline{\varepsilon}_{dc}(k),\overline{\varepsilon}_{dc}(k),\Delta\tilde{v}(k|k)} J'(k), \text{ s.t. } (6.93)-(6.94), (6.96)-(6.98), (6.101). \tag{6.103}$$

In solving the aforementioned two optimizations, we should substitute (6.99), (6.100) and (6.92) into (6.93)–(6.96). With the optimal $\Delta\tilde{v}(k|k)$ being obtained, $\Delta u(k|k) = K\Delta\hat{x}(k|k) + \Delta v(k|k)$ will be sent to the actual controlled system. It may happen that $u(k) \neq u(k|k)$.

6.5 Numerical Example

We apply the heavy oil fractionator model in Chapter 4 and the continuous-time transfer function matrix around the equilibrium is

$$G^u(s) = \begin{bmatrix} \frac{4.05e^{-27s}}{50s+1} & \frac{1.77e^{-28s}}{60s+1} & \frac{5.88e^{-27s}}{50s+1} \\ \frac{5.39e^{-18s}}{50s+1} & \frac{5.72e^{-14s}}{60s+1} & \frac{6.90e^{-15s}}{40s+1} \\ \frac{4.38e^{-20s}}{33s+1} & \frac{4.42e^{-22s}}{44s+1} & \frac{7.20}{19s+1} \end{bmatrix},$$

$$G^f(s) = \begin{bmatrix} \frac{1.20e^{-27s}}{45s+1} & \frac{1.44e^{-27s}}{40s+1} \\ \frac{1.52e^{-15s}}{25s+1} & \frac{1.83e^{-15s}}{20s+1} \\ \frac{1.14}{27s+1} & \frac{1.26}{32s+1} \end{bmatrix}.$$

The sampling time is 4. The state-space model is obtained by the subspace identification method, $n_x = 20$.

$A = [0.9413, -0.0805, 0.1018, -0.0170, 0.0324, 0.0612, -0.0247, 0.0375, -0.0379, 0.0102, 0.0309, 0.0113, 0.0158, 0.0003, -0.0119, 0.0058, -0.0056, 0.0080, -0.0030, 0.0026; 0.1503, 0.8091, 0.0659, -0.2503, -0.1564, -0.0628, -0.0226, -0.0081, -0.0026, -0.0097, 0.0393, -0.0757, 0.0429, 0.0320, 0.0112, 0.0176, 0.0371, -0.0114, 0.0096, -0.0109; 0.0520, -0.1057, 0.7855, 0.1981, -0.1129, -0.2358, 0.0250, -0.1078, 0.0764, -0.0369, -0.0898, -0.0302, -0.0273, 0.0154, 0.0506, -0.0194, -0.0005, -0.0058, 0.0247, -0.0026; -0.0471, 0.3606, -0.0534, 0.6205, 0.2867, -0.0380, -0.2017, -0.0442, -0.0773, -0.1115, 0.1527, -0.0817, 0.0967, 0.0786, 0.0063, -0.0042, 0.0084, 0.0510, 0.0460, -0.0074; -0.0579, 0.0966, 0.1623, -0.5332, 0.4457, 0.1168, -0.3324, -0.2801, 0.1323, 0.0172, -0.1092, 0.1109, -0.0954, 0.0217, 0.1199, 0.0085, -0.0182, 0.0280, 0.0435, -0.0058; -0.0373, 0.0226, 0.2258, 0.1498, 0.0047, 0.6707, 0.2950, -0.2837, 0.2660, -0.1047, 0.1105, -0.1505, -0.0366, 0.0277, -0.0321, 0.0111, 0.1078, -0.0151, -0.0292, -0.0304; 0.0259, -0.0686, 0.0269, 0.1424, 0.6051, -0.1391, -0.0522, 0.1634, 0.1535, 0.3781, -0.0450, -0.1500, 0.0375, 0.0056, 0.0834, 0.1300, 0.1409, -0.1052, -0.0412, -0.0422; -0.0195, 0.0171, 0.0630, -0.0015, -0.1116, -0.0198, -0.3297, 0.5687, 0.6149, -0.2482, 0.0366, -0.0075, -0.0637, -0.0897, -0.0749, -0.0287, 0.0259, 0.0215, -0.0478, -0.0259; -0.0259, 0.0181, 0.0707, 0.0185, -0.1176, -0.3721, 0.0205, -0.2676, 0.2021, 0.3555, 0.5390, 0.2699, -0.1441, -0.0476, -0.1918, 0.0679, 0.0483, 0.0203, -0.0639, 0.0135; 0.0113, -0.0534, 0.0032, 0.0933, -0.0413, 0.0244, -0.5971, -0.3129,$

−0.2012, −0.2042, −0.0840, −0.1406, −0.2123, −0.2076, −0.3389, −0.0713, 0.0230, −0.0915, −0.1303, 0.0314; −0.0054, 0.0030, 0.0402, −0.0020, 0.0735, −0.0230, 0.0893, 0.2804, −0.2739, −0.2502, 0.1374, 0.0281, −0.7205, 0.2645, 0.1312, 0.1272, 0.1339, −0.0387, 0.0615, 0.0100; 0.0041, 0.0084, −0.0333, −0.0177, −0.1159, 0.1163, −0.0777, 0.0944, 0.1122, 0.5309, −0.0645, −0.5524, −0.2588, 0.1979, −0.1790, −0.2134, −0.1565, 0.0789, 0.0944, 0.0836; −0.0242, 0.0288, 0.0736, −0.0379, −0.0346, −0.0786, 0.0768, 0.0181, −0.2791, 0.1148, 0.0557, −0.3531, −0.2099, −0.5218, 0.2236, 0.0039, 0.0537, 0.0718, −0.1162, −0.1388; 0.0210, −0.0325, −0.0617, −0.0060, 0.0331, 0.1098, −0.0789, −0.0239, 0.1457, −0.0760, 0.5376, −0.0083, −0.0978, −0.2300, 0.4239, −0.2563, −0.2717, −0.0865, 0.0631, 0.0815; −0.0084, 0.0002, 0.0434, 0.0139, −0.0342, 0.0207, −0.1019, 0.0180, −0.1123, 0.1306, 0.0164, 0.2429, −0.0508, 0.4771, 0.0659, −0.5591, 0.1086, 0.0026, −0.2335, −0.1433; −0.0016, 0.0090, 0.0135, 0.0079, 0.0102, 0.0763, −0.0007, 0.1374, −0.1004, 0.1222, −0.0430, 0.1848, 0.0814, −0.3805, −0.0104, −0.5011, 0.5073, −0.0048, 0.2820, 0.0496; −0.0100, −0.0050, 0.0303, −0.0048, −0.0437, −0.0929, −0.0510, 0.0053, −0.0411, −0.0722, −0.0268, −0.2352, 0.2440, 0.1242, 0.4474, −0.1256, 0.0449, −0.1369, −0.3878, 0.2340; 0.0034, −0.0039, −0.0119, 0.0192, −0.0147, −0.0073, −0.0250, −0.0471, 0.0587, 0.0027, −0.0764, 0.0384, −0.0659, −0.0208, 0.1610, 0.0685, 0.1737, 0.9047, −0.1543, 0.1420; −0.0004, −0.0008, −0.0076, −0.0118, −0.0374, 0.0014, −0.0197, 0.0062, 0.0450, 0.0166, 0.0194, −0.0327, −0.0079, 0.0157, −0.0163, 0.2114, 0.5075, −0.2252, −0.0932, 0.4154; −0.0083, 0.0053, 0.0312, 0.0018, −0.0117, −0.0494, 0.0131, 0.0594, −0.0049, −0.0191, −0.0843, 0.0074, 0.0224, −0.0721, −0.0575, −0.0320, −0.1756, 0.0132, 0.1292, 0.0789];

B = [0.1144, 0.0915, 0.2635; −0.0652, −0.0499, −0.0631; 0.1182, 0.1006, −0.1273; 0.0851, 0.0590, 0.0617; 0.0468, 0.0322, −0.0035; −0.0179, −0.0100, 0.0075; −0.0469, −0.0154, −0.0258; 0.0002, −0.0359, 0.0479; 0.0071, −0.0541, 0.0429; −0.0243, −0.0102, −0.0136; 0.0419, −0.0549, 0.0200; 0.0298, −0.0093, 0.0033; −0.0435, 0.0431, 0.0100; 0.0189, 0.0049, −0.0116; −0.0207, 0.0332, 0.0070; 0.0113, −0.0214, 0.0000; 0.0036, −0.0264, 0.0113; −0.0176, 0.0195, 0.0070; 0.0000, 0.0092, 0.0086; −0.0113, 0.0058, 0.0207];

C = [−0.2130, 1.3911, 0.1415, 2.1119, −1.8960, 1.6202, −1.2679, 0.1790, −0.0992, 1.2321, −0.6145, 0.9597, −0.3086, −0.1634, 0.6258, 0.4651, −0.0828, −0.1210, 0.0082, −0.0177; 0.8230, 3.5952, 0.1546, 1.1836, 1.7018, −0.0610, 1.4942, 0.2360, 0.2840, 0.0205, −0.8418, 0.7809, −0.7187, −0.6172, −0.4146, −0.4073, −0.4772, −0.0425, −0.2900, 0.1887; 3.2990, −0.2109, −3.7836, 1.0539, −0.8301, −1.3328, 0.2028, −1.3189, 1.1641, −0.2938, −0.7396, −0.2430, −0.4887, 0.1007, 0.5026, −0.0950, 0.2100, −0.0921, 0.1363, −0.1056];

F = [0.0558, 0.0589; −0.0157, −0.0169; −0.0057, 0.0076; 0.0150, 0.0183; −0.0032, −0.0119; −0.0026, −0.0163; −0.0043, −0.0060; 0.0245, 0.0515; 0.0030, 0.0035; −0.0028, −0.0066; −0.0084, −0.0186; −0.0081, −0.0133; 0.0207, 0.0391; −0.0204, −0.0380; −0.0027, −0.0125; 0.0012, 0.0046; 0.0002, −0.0014; −0.0128, −0.0344; −0.0066, −0.0072; −0.0115, −0.0202].

Take $y_{\rm eq} = 0$, $u_{\rm eq} = 0$ and $f_{\rm eq} = 0$. I_n denotes the n-order identity matrix.

Take $\underline{u} = u_{\rm eq} + [-0.5; -0.5; -0.5]$, $\bar{u} = u_{\rm eq} + [0.5; 0.5; 0.5]$, $\Delta\bar{u}_i = \delta\bar{u}_{i,\rm ss} = 0.1$; $\underline{y}_{0,h} = y_{\rm eq} + [-0.7; -0.7; -0.7]$, $\bar{y}_{0,h} = y_{\rm eq} + [0.7; 0.7; 0.7]$, $\underline{y}_0 = y_{\rm eq} + [-0.5; -0.5; -0.5]$, $\bar{y}_0 = y_{\rm eq} + [0.5; 0.5; 0.5]$, $\delta\bar{y}_{1,\rm ss} = 0.2$, $\delta\bar{y}_{2,\rm ss} = 0.2$, $\delta\bar{y}_{3,\rm ss} = 0.3$, where $y_{1,\rm ss}, y_{2,\rm ss}, u_{3,\rm ss}$ have ETs and their expected allowable range is 0.5. Take $R_1 = I_{23}$, and all elements of R_{12} are 0.1,

Table 6.1 Parameters of multi-priority-rank SSTC.

Rank no.	Type of constraint	Variable	Soft constraint	$\bar{\varepsilon}$
1	Inequality	$y_{2,\text{ss}}$	CV lower bound	0.20
1	Inequality	$y_{3,\text{ss}}$	CV upper bound	0.20
2	Equality	$y_{2,\text{ss}}$	$y_{2,\text{eq}} - 0.3$	0.25
3	Inequality	$y_{2,\text{ss}}$	CV upper bound	0.20
3	Inequality	$y_{3,\text{ss}}$	CV lower bound	0.20
4	Equality	$u_{3,\text{ss}}$	$u_{3,\text{eq}} + 0.2$	0.25
4	Equality	$y_{1,\text{ss}}$	$y_{1,\text{eq}} + 0.7$	0.25
5	Inequality	$y_{1,\text{ss}}$	CV upper, lower bound	0.20

$R_2 = 3I_3$, and L is obtained by (6.33). Take $Q_{\text{LQR}} = I_{20}$, $R_{\text{LQR}} = I_3$, and solve discrete-time Riccati equation

$$P_{\text{LQR}} = Q_{\text{LQR}} + A^{\text{T}}P_{\text{LQR}}A - A^{\text{T}}P_{\text{LQR}}B(R_{\text{LQR}} + B^{\text{T}}P_{\text{LQR}}B)^{-1}B^{\text{T}}P_{\text{LQR}}A$$

and calculate $K = -(R_{\text{LQR}} + B^{\text{T}}P_{\text{LQR}}B)^{-1}B^{\text{T}}P_{\text{LQR}}A$. $G_d = [I_2, 0]^{\text{T}}$ and $G_p = [1, 0, 0]^{\text{T}}$.

The parameters of the feasibility stage of SSTC are shown in Table 6.1. Take $h = [-2, -1, 2]$. There is no minimum-move MV. Choose $J_{\min} = -0.4$. At $k \in [62, 76]$, the disturbance appears with $f_{\text{eq}} + [0.2; -0.1]$; at $k \geq 120$, the disturbance appears with $f_{\text{eq}} + [1; -1]$; at the other intervals the disturbance is f_{eq}. $Y_N^{\text{ol}}(0|0) = [y_{\text{eq}}; \dots; y_{\text{eq}}]$. $\hat{x}(0|0) = 0$, $u(-1) = u_{\text{eq}}$, $y(0) = y_{\text{eq}}$.

The parameters of DC are $N = 15$, $N_c = 8$, $\Lambda = \text{diag}\{3, 5, 3\}$, $\bar{z} = y_{\text{eq}} + [0.4; 0.4; 0.4]$, $\underline{z} = y_{\text{eq}} + [-0.4; -0.4; -0.4]$, $\underline{q}_1 = 2.0$, $\check{q}_1 = 0.5$, $\bar{q}_1 = 2.0$; $\underline{q}_2 = 2.0$, $\check{q}_2 = 1.0$, $\bar{q}_2 = 2.0$; $\underline{q}_3 = 2.5$, $\check{q}_3 = 1.0$, $\bar{q}_3 = 4.0$, $\rho = 0.2$. $u(k) = u(k|k)$. The real controlled plant is $A_r = 0.9A$, $B_r = 0.9B$, $C_r = 0.9C$, $F_r = 0.9F$. The feasibility stage of SSTC applies LP, and economic stage, QP. DC result is shown in Figure 6.1.

Note that in this example, when we take any other suitable equilibrium (without satisfying $(I - A)x_{\text{eq}} = Bu_{\text{eq}} + Ff_{\text{eq}}$ and $y_{\text{eq}} = Cx_{\text{eq}}$), the result is still available. If there is no model-plant mismatch, and $(I - A)x_{\text{eq}} = Bu_{\text{eq}} + Ff_{\text{eq}}$ and $y_{\text{eq}} = Cx_{\text{eq}}$ are satisfied, then any move of the equilibrium affects the value of each variable but does not affect their relative difference (curve shape). These results have been verified by simulation, which is omitted here.

Moreover, in this example, if A is replaced by $1.15A$ (i.e., the model is unstable in ol, having eigenvalues outside the unit circle), and the true plant is consistent with the model, then the system can still be stabilized after appropriately tuning the controller parameters. Through the simulations, we summarize as follows.

The system can be unstable and the model can be inaccurate.

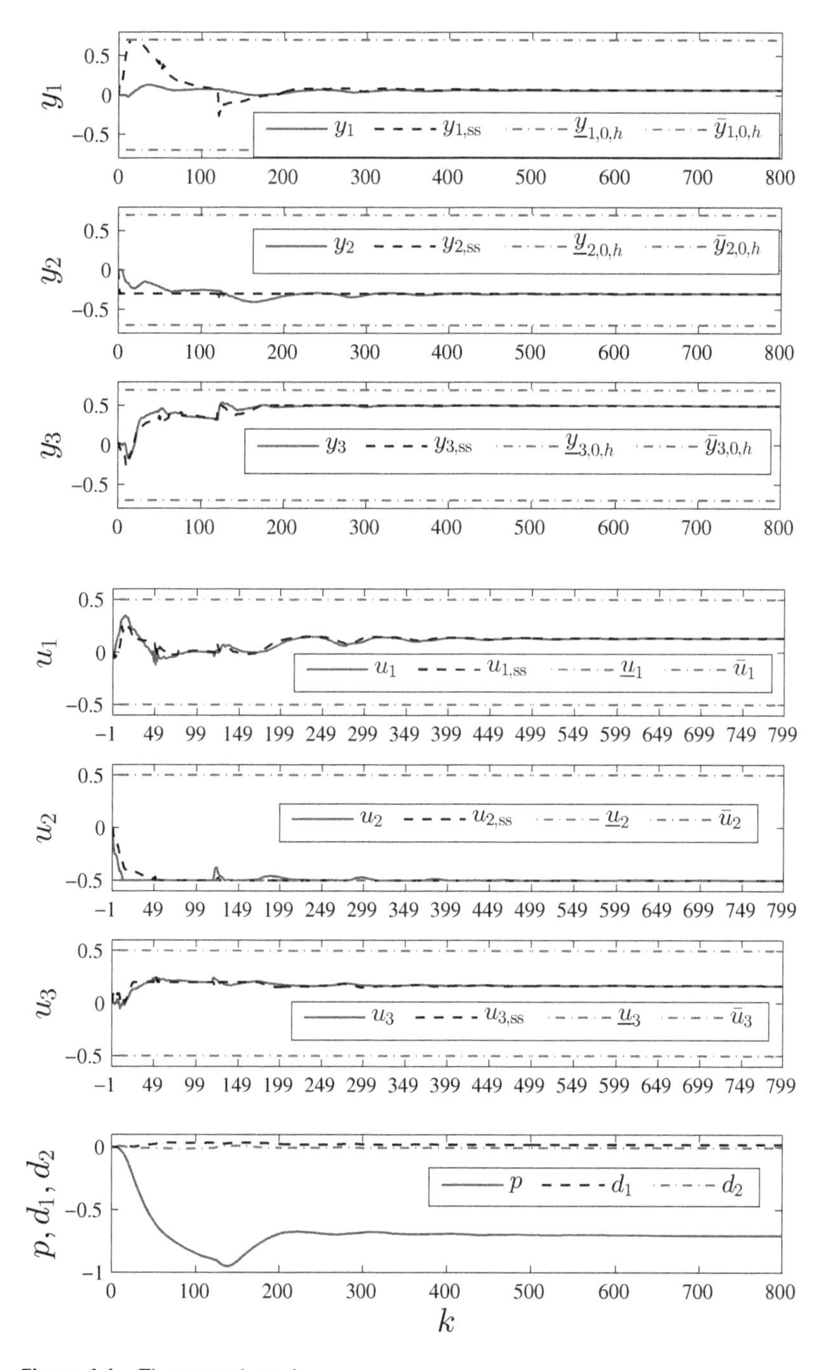

Figure 6.1 The control result.

7

Offset-Free, Nonlinearity and Variable Structure in Two-Layered MPC

Chapters 4, 5, and 6 propose the two-layered DMCs for sCV, iCV and the state-space model, respectively. They have the same idea, but are different in details.

This chapter discusses the offset-free property. The model utilizes (6.4) in Chapter 6, but the artificial disturbance $\{d_k, p_k\}$ is introduced. There are two important papers by [Rao and Rawlings, 1999, Muske and Badgwell, 2002] applying the state-space model and d_k, p_k. Section 7.1 expounds the idea of [Rao and Rawlings, 1999, Muske and Badgwell, 2002] in a unified way, in which the symbol ∇ and the bounded disturbance f are added; the conclusions of [Rao and Rawlings, 1999, Muske and Badgwell, 2002] are integrated. Section 7.2 summarizes the offset-free property, which is especially described more clearly in Rao and Rawlings [1999]. The so-called offset-free is $y(\infty) = y_{ss}(\infty)$, i.e., the actual value of y equals to its ss. Note that $y(\infty)$ may not be equal to y_t. Applying the offset-free analysis in Rao and Rawlings [1999] and Muske and Badgwell [2002] to the state space framework in Chapter 6, we can draw a conclusion, i.e., the two-layered DMC approaches proposed in Chapters 4–6 are offset-free.

7.1 State Space Steady-State Target Calculation with Target Tracking

The so-called offset-free means that the controller can drive CV to reach its ss. The offset-free property is very important, because the controller always attempts to eliminate the offset, an if it fails, will cause the large system fluctuations, thus affecting ET tracking. Compared with the single-layered MPC, the two-layered MPC is easier to ensure the dynamic stability and the offset-free.

The elimination of the offset is a major advantage of the two-layered MPC. In the single-layer MPC (i.e, MPC without SSTC), it is not easy to eliminate the offset, because the time-varying working conditions and the disturbances in the production process often make CV unable to reach its ss. In the two-layered MPC, ET is unnecessarily the targets to be tracked in DC; the actual ss to be tracked by DC are planned in SSTC, satisfying various constraints in the steady state. Since SSTC has been planned for the steady state, it is much easier to achieve the offset-free in DC.

Model Predictive Control, First Edition. Baocang Ding and Yuanqing Yang.

Consider LTI model in Chapter 6, i.e.,

$$\begin{cases} \nabla x_{k+1} = A\nabla x_k + B\nabla u_k + F\nabla f_k \\ \nabla y_k = C\nabla x_k \end{cases}, \quad k \geq 0, \tag{7.1}$$

where the output $\nabla y \in \mathbb{R}^{n_y}$, the control input $\nabla u \in \mathbb{R}^{n_u}$, the state $\nabla x \in \mathbb{R}^{n_x}$, and the measurable disturbance $\nabla f \in \mathbb{R}^{n_f}$. Assume that (A, B) is stabilizable, and (C, A) is detectable. Equation (7.1) describes the system approximately. In order to deal with, e.g., the unmeasurable disturbance and modeling error, we use the following state space model with the disturbance:

$$\begin{cases} \tilde{x}_{k+1} = \tilde{A}\tilde{x}_k + \tilde{B}\nabla u_k + \tilde{F}\nabla f_k \\ \nabla y_k = \tilde{C}\tilde{x}_k \end{cases}, \tag{7.2}$$

where the augmented state variable and system matrices are

$$\tilde{x}_k = \begin{bmatrix} \nabla x_k \\ d_k \\ p_k \end{bmatrix}, \quad \tilde{A} = \begin{bmatrix} A & G_d & 0 \\ 0 & I & 0 \\ 0 & 0 & I \end{bmatrix},$$
$$\tilde{B} = \begin{bmatrix} B \\ 0 \\ 0 \end{bmatrix}, \quad \tilde{F} = \begin{bmatrix} F \\ 0 \\ 0 \end{bmatrix}, \quad \tilde{C} = \begin{bmatrix} C & 0 & G_p \end{bmatrix}. \tag{7.3}$$

For the augmented model in (7.2) and (7.3), the estimate of the augmented state can be obtained using the following method:

$$\begin{bmatrix} \nabla \hat{x}_{k|k-1} \\ \hat{d}_{k|k-1} \\ \hat{p}_{k|k-1} \end{bmatrix} = \tilde{A} \begin{bmatrix} \nabla \hat{x}_{k-1|k-1} \\ \hat{d}_{k-1|k-1} \\ \hat{p}_{k-1|k-1} \end{bmatrix} + \tilde{B}\nabla u_{k-1} + \tilde{F}\nabla f_{k-1}$$
$$\begin{bmatrix} \nabla \hat{x}_{k|k} \\ \hat{d}_{k|k} \\ \hat{p}_{k|k} \end{bmatrix} = \begin{bmatrix} \nabla \hat{x}_{k|k-1} \\ \hat{d}_{k|k-1} \\ \hat{p}_{k|k-1} \end{bmatrix} + \begin{bmatrix} L_x \\ L_d \\ L_p \end{bmatrix} \left(\nabla y_k - \tilde{C} \begin{bmatrix} \nabla \hat{x}_{k|k-1} \\ \hat{d}_{k|k-1} \\ \hat{p}_{k|k-1} \end{bmatrix} \right). \tag{7.4}$$

The state filter gain L is divided into the state filter gain L_x of the process model, the state disturbance filter gain L_d, and the output disturbance filter gain L_p.

Based on (7.4), ss satisfy the following conditions:

$$\begin{cases} \nabla \hat{x}_{ss}(k) = A\nabla \hat{x}_{ss}(k) + B\nabla u_{ss}(k) + G_d\hat{d}(k|k) + F\nabla f(k) \\ \nabla y_{ss}(k) = C\nabla \hat{x}_{ss}(k) + G_p\hat{p}(k|k) \end{cases}, \tag{7.5}$$

where $\hat{d}(k|k)$ and $\hat{p}(k|k)$ come from (7.4). Denote y_t as CVET, and u_t as MVET. Differently from Chapter 6, all MVs and CVs have ETs.

7.1.1 Case all External Targets Having Equal Importance

Refer to Muske and Badgwell [2002], with some changes. The ss $\{u_{ss}, \triangledown\hat{x}_{ss}\}(k)$ are determined by solving the following QP:

$$\{\triangledown\hat{x}_{ss}(k), u_{ss}(k)\} \triangleq \arg\min \left[\|y_{ss}(k) - y_t\|^2_{Q_s} + \|u_{ss}(k) - u_t\|^2_{R_s}\right] \tag{7.6}$$

$$\text{s.t. } [(I - A) \quad -B] \begin{bmatrix} \triangledown\hat{x}_{ss}(k) \\ u_{ss}(k) \end{bmatrix} = G_d\hat{d}(k|k) + F\triangledown f(k) - Bu_{eq}, \tag{7.7}$$

$$\underline{u} \leq u_{ss}(k) \leq \bar{u}, \tag{7.8}$$

$$\underline{y}_0 \leq y_{ss}(k) \leq \bar{y}_0, \tag{7.9}$$

where R_s and Q_s are the symmetric matrices, and $y_{ss}(k)$ should be replaced by $y_{eq} + C\triangledown\hat{x}_{ss}(k) + G_p\hat{p}(k|k)$. This optimization problem amounts to a specific rank in the multi-priority-rank SSTC, i.e., this rank is "ET tracking" where all MVs and CVs have ETs at the same rank. Equation (7.7) is the first equation of (7.5). If (7.8) and (7.9) are inactive, then solving the optimization problem (7.6)–(7.9) is equivalent to solving (7.6) and (7.7).

If $\{\triangledown f_k, \hat{p}_{k|k}, \hat{d}_{k|k}\}$ is larger, the constraint (7.9) may cause the optimization problem (7.6)–(7.9) infeasible. One can soften (7.9), i.e., introduce the slack variables. See Chapter 6.

7.1.2 Case CV External Target Being More Important Than MV External Target

Refer to Rao and Rawlings [1999], with some changes. Assumed the realization of y_t is more important than that of u_t. When $n_u = n_y$, the model has an invertible steady-state gain matrix (at least, the system does not contain an integral dynamics), and the solution to the unconstrained target tracking problem can be obtained directly based on the steady-state gain matrix. However, mostly $n_u \neq n_y$, sometimes the integral dynamics is included, and there are usually inequality constraints on ss. Hence, the target tracking problem is described as mathematical programming. When $n_u > n_y$, often a variety of $u_{ss}(k)$ produce the same y_t, for which QP can be constructed to find the best $u_{ss}(k)$. When $n_u < n_y$, often there is no $u_{ss}(k)$ guaranteeing the tracking of y_t, and an optimization problem can be designed so that $y_{ss}(k)$ gets as close as possible to y_t in the sense of LS.

Instead of applying the two optimizations, it is better to adopt a unified optimization to deal with various situations. The soft constraints of the output target are

$$\begin{cases} y_t - y_{eq} - C\triangledown\hat{x}_{ss}(k) - G_p\hat{p}(k|k) \leq \varepsilon(k) \\ y_t - y_{eq} - C\triangledown\hat{x}_{ss}(k) - G_p\hat{p}(k|k) \geq -\varepsilon(k) \end{cases}. \tag{7.10}$$

In this way, the target tracking problem can be described as the following QP:

$$\min_{\triangledown\hat{x}_{ss}(k), u_{ss}(k), \varepsilon(k)} \frac{1}{2}\left[\|\varepsilon(k)\|^2_{Q_s} + \|u_{ss}(k) - u_t\|^2_{R_s}\right] + q_s^{\mathrm{T}}\varepsilon(k)$$

$$\text{s.t. } (7.7), (7.8), (7.9), (7.10), \tag{7.11}$$

where q_s is a vector composed of nonnegative elements, $y_{ss}(k)$ should be replaced by $y_{eq} + C\triangledown\hat{x}_{ss}(k) + G_p\hat{p}(k|k)$. This optimization cannot be equivalent to a rank in the multi-priority-rank SSTC. Since R_s and Q_s are symmetric positive-definite, due to the feature of QP, $\varepsilon(k)$ and $u_{\text{ss}}(k)$ will be uniquely determined.

The linear term $q_s^{\text{T}}\varepsilon(k)$ and the quadratic term $\varepsilon(k)^{\text{T}}Q_s\varepsilon(k)$ penalize the adjustment of the soft constraint. If q_s is large enough, the soft constraints will be tuned to a minimum. However, the lower bound of q_s to ensure this minimum is not easy to calculate accurately in advance. In practice, it is rarely required that the soft constraints be adjusted to their minimum, and q_s can be approximately chosen by experience. In light of the minimum adjustment of the soft constraints, the quadratic penalty term in the square brackets seems redundant. However, the quadratic term not only adds more tunable parameters, but also is necessary to guarantee the uniqueness of solution $u_{\text{ss}}^*(k)$.

In the two cases in Sections 7.1.1 and 7.1.2, since (A, B) is stabilizable, the matrix $\left[(I-A)\ -B\right]$ is nonsingular. According to (7.7), this is a sufficient condition to guarantee the existence of a feasible solution, as stated in the following conclusions.

Lemma 7.1 Since (A, B) is stabilizable, the target tracking problems in the two cases, without constraints (7.8) and (7.9), have feasible solutions.

For an arbitrary $\varepsilon^*(k)$, there exists the unique, nonnegative $\{\varepsilon_+, \varepsilon_-\}^*(k)$ such that $\varepsilon^*(k) = \varepsilon_+^*(k) + \varepsilon_-^*(k)$ and $y_{\text{ss}}^*(k) = y_t + \varepsilon_+^*(k) - \varepsilon_-^*(k)$. When QP is feasible, the unique $\{u_{\text{ss}}, \varepsilon\}^*(k)$ is obtained. According to (7.5), we have

$$\begin{bmatrix} I-A \\ C \end{bmatrix} \nabla\hat{x}_{\text{ss}}^*(k) = \begin{bmatrix} Bu_{\text{ss}}^*(k) - Bu_{\text{eq}} + G_d\hat{d}(k|k) + F\nabla f(k) \\ y_t + \varepsilon_+^*(k) - \varepsilon_-^*(k) - y_{\text{eq}} - G_p\hat{p}(k|k) \end{bmatrix}. \tag{7.12}$$

Obviously, when QP is feasible, the necessary and sufficient condition for determining the unique $\nabla\hat{x}_{\text{ss}}^*(k)$ by (7.12) is $\begin{bmatrix} I-A \\ C \end{bmatrix}$ being nonsingular. The detectability guarantees $\begin{bmatrix} I-A \\ C \end{bmatrix}$ nonsingular. In a word, since (C, A) is detectable, when QP is feasible, the solutions of the above two cases in Sections 7.1.1 and 7.1.2 are unique. If the inequality constraints (7.8) and (7.9) are removed, the feasible region must be non-empty, hence the detectability of (C, A) and stabilizability of (A, B) ensure the existence and uniqueness of the solution to the target tracking problem.

Consider the case that $\{y_t, u_t\}$ are accurately tracked, i.e, the state target $\nabla\hat{x}_{\text{ss}}(k)$ needs to meet

$$\begin{bmatrix} I-A \\ C \end{bmatrix} \nabla\hat{x}_{\text{ss}}(k) = \begin{bmatrix} Bu_t - Bu_{\text{eq}} + G_d\hat{d}(k|k) + F\nabla f(k) \\ y_t - y_{\text{eq}} - G_p\hat{p}(k|k) \end{bmatrix}. \tag{7.13}$$

The necessary condition for determining the unique $\nabla\hat{x}_{\text{ss}}(k)$ by (7.13) is $\begin{bmatrix} I-A \\ C \end{bmatrix}$ being full rank.

Remark 7.1 The nonuniqueness of $\nabla\hat{x}_{\text{ss}}(k)$ implies that it jumps irregularly; hence, if $\nabla\hat{x}_{\text{ss}}(k)$ is used for DC, the closed-loop system is unstable. In the light of guaranteeing the uniqueness of QP solution, the detectability (C, A) is not a necessary condition; when A does not contain integral mode, $\nabla\hat{x}_{\text{ss}}(k) = (I-A)^{-1}[B\nabla u_{\text{ss}}(k) + G_d\hat{d}_{k|k} + F\nabla f(k)]$ is obtained by $u_{\text{ss}}(k)$, which guarantees the uniqueness of the solution in the two cases. However, the detectability of (C, A) is a necessary condition for using (7.4).

In the aforementioned, we only discuss the target tracking problem. Compared with SSTC in Chapters 4–6, the target tracking is simpler.

7.2 QP-Based Dynamic Control and Offset-Free

In this section, the cost function is different from Chapter 6, but this is not the key to proving offset-free, i.e., using the cost function of Chapter 6, the offset-free is also guaranteed. Moreover, this section will not use the pre-stabilization control law in Chapter 6, but, as in Chapter 6, it can be used for unstable models.

Refer to Rao and Rawlings [1999], with some changes. At time k, with $\nabla\hat{x}_{k|k}$ and $\{\nabla\hat{x}_{ss}(k), u_{ss}(k)\}$ being known, adopt the following method to calculate the current control input:

$$u_{k|k} = v^*_{k|k} + u_{ss}(k), \tag{7.14}$$

where $v^*_{k|k}$ is the solution to the following QP:

$$\min_{\{v_{k+j|k},\ j=0,\dots,N-1\}} \sum_{j=0}^{N-1} \left[\|z_{k+j+1|k}\|^2_{C^{\mathrm{T}}QC} + \|v_{k+j|k}\|^2_R + \|\Delta v_{k+j|k}\|^2_S \right] \tag{7.15}$$

$$\text{s.t.} \begin{cases} z_{k+j+1|k} = Az_{k+j|k} + Bv_{k+j|k},\ j = 0,\dots,N-1 \\ z_{k|k} = \nabla\hat{x}_{k|k} - \nabla\hat{x}_{ss}(k) \\ v_{k-1|k} = u_{k-1} - u_{ss}(k) \\ \underline{y}_0 \le y_{eq} + C\left[z_{k+j|k} + \nabla\hat{x}_{ss}(k)\right] + G_p\hat{p}_{k|k} \le \bar{y}_0,\ j = j_1,\dots,N \\ \underline{u} \le v_{k+j|k} + u_{ss}(k) \le \bar{u},\ j = 0,\dots,N-1 \\ -\Delta\bar{u} \le \Delta v_{k+j|k} \le \Delta\bar{u},\ j = 0,\dots,N-1 \end{cases}, \tag{7.16}$$

where the weight matrices Q, R, S guarantee that Riccati iteration corresponding to LQR is feasible.

Theorem 7.1 Consider the output feedback MPC with the cost function (7.15), the constraint (7.16), the state estimator (7.4), and SSTC being the target tracking problem (7.6)–(7.9) or (7.11). Denote $\{y_\infty, \nabla\hat{x}_\infty, \hat{d}_\infty, \hat{p}_\infty\} = \lim_{k\to\infty}\{y_k, \nabla\hat{x}_{k|k}, \hat{d}_{k|k}, \hat{p}_{k|k}\}$. The controller achieves offset-free under the following conditions.

- (i) The closed-loop system is asymptotically stable, and $\{y_\infty, \nabla\hat{x}_\infty, \hat{p}_\infty, \hat{d}_\infty\}$ keep unchanged in the steady state;
- (ii) The process model in (7.1) is stabilizable and detectable;
- (iii) $n_d + n_p = n_y$;
- (iv) The augmented model shown in (7.2) is detectable;
- (v) The inequality constraints on inputs and outputs are inactive in the steady state.

The condition (i) requires the closed-loop system to actually reach a steady state, which includes that the optimization problem is always feasible. The condition (ii) makes possible the stable controller, the detectable augmented model, and the unique $\{\nabla\hat{x}_{ss}, u_{ss}\}(k)$. According to the conclusions in Chapter 6, the condition (iii) guarantees the existence of a detectable augmented model. Condition (iv) guarantees that stable observers can be constructed. The condition (v) guarantees that, at the steady state, we can utilize the control input determined by the unconstrained controller.

In Theorem 7.1, the satisfaction of the condition (i) needs in-depth study, which remains an open problem in various complex situations; the condition (v) requires that various inequality constraints (input and output amplitudes) be inactive in the steady state, which is difficult for a general SSTC. Hence, it needs further research to delete (v).

Example 7.1 We apply the heavy oil fractionator model in Chapter 3, whose continuous-time transfer function matrix around the equilibrium point is

$$G^u(s) = \begin{bmatrix} \frac{4.05e^{-27s}}{50s+1} & \frac{1.77e^{-28s}}{60s+1} & \frac{5.88e^{-27s}}{50s+1} \\ \frac{5.39e^{-18s}}{50s+1} & \frac{5.72e^{-14s}}{60s+1} & \frac{6.90e^{-15s}}{40s+1} \\ \frac{4.38e^{-20s}}{33s+1} & \frac{4.42e^{-22s}}{44s+1} & \frac{7.20}{19s+1} \end{bmatrix}, \quad G^f(s) = \begin{bmatrix} \frac{1.20e^{-27s}}{45s+1} & \frac{1.44e^{-27s}}{40s+1} \\ \frac{1.52e^{-15s}}{25s+1} & \frac{1.83e^{-15s}}{20s+1} \\ \frac{1.14}{27s+1} & \frac{1.26}{32s+1} \end{bmatrix}.$$

The sampling time is 4. The state space model is obtained by the subspace identification method, with $n_x = 20$ (see the example in Chapter 6). First, choose $y_{\text{eq}} = 0$, $u_{\text{eq}} = 0$ and $f_{\text{eq}} = 0$. $u(-1) = u_{\text{eq}}$, $y(0) = y_{\text{eq}}$, $\triangledown\hat{x}(0|0) = \triangledown x(0) = 0$. The constraints on MV, CV are $\underline{u}_i = u_{\text{eq}} - 0.5$, $\bar{u}_i = u_{\text{eq}} + 0.5$, $\Delta\bar{u}_i = 0.1$; $\underline{y}_{j,0} = y_{\text{eq}} - 0.5$, $\bar{y}_{j,0} = y_{\text{eq}} + 0.5$. The external targets are $y_t = y_{\text{eq}} + [0.5, -0.5, 0.5]^{\text{T}}$ and $u_t = u_{\text{eq}} + [0.5, -0.5, 0.5]^{\text{T}}$.

Choose $G_d = [I_2, 0]^{\text{T}}$, $G_p = [1, 0, 0]^{\text{T}}$. In SSTC, ss at each time is obtained by solving the optimization (7.11). The parameters are selected as $Q_s = I_3$, $R_s = I_3$, $q_s = [0.5, 0.5, 0.5]^{\text{T}}$. In DC, choose $N = 10$, $j_1 = 1$, $Q = Q_s$, $R = S = R_s$, and the controlled system is parameterized as $A_r = 0.8A$, $B_r = 0.8B$, $C_r = 0.8C$, $F_r = 0.8F$. The measurable disturbance f_k varies at time $k = 158 \sim 168$ and $k = 228 \sim 238$, and the amplitudes are $f_{\text{eq}} + [0.05, 0.05]^{\text{T}}$ and $f_{\text{eq}} - [0.05, 0.05]^{\text{T}}$, respectively. At the other time intervals $f_k = f_{\text{eq}}$. We solve the optimization problem (7.15)–(7.16), and using (7.14) to obtain the dynamic value of MV. The control result is shown in Figure 7.1, which implies MV and CV can track ss offset-free.

If there is no model-system mismatch (i.e., $A_r = A, B_r = B, C_r = C, F_r = F$), and the equilibrium point $\{x, y, u, f\}_{\text{eq}}$ satisfies

$$(I - A)x_{\text{eq}} = Bu_{\text{eq}} + Ff_{\text{eq}}, \; y_{\text{eq}} = Cx_{\text{eq}},$$

then any move of the equilibrium point affects the value of each variable but does not change the relative difference (i.e., the shape of the curves). Choose $\triangledown\hat{x}_i(0|0) = -0.05$, $\triangledown x_i(0) = 0$. For two groups of equilibrium points $\{u_{i,\text{eq}} = 23, f_{i,\text{eq}} = 17\}$ and $\{u_{i,\text{eq}} = -123, f_{i,\text{eq}} = 32\}$, the control results are shown in Figures 7.2 and 7.3, respectively. Note that if there is no model-plant mismatch, then ss of d and p are zeros.

When we take another equilibrium point (whether or not $(I - A)x_{\text{eq}} = Bu_{\text{eq}} + Ff_{\text{eq}}$ and $y_{\text{eq}} = Cx_{\text{eq}}$), the result is still available. Choose $u_{i,\text{eq}} = 0.1, f_{i,\text{eq}} = 0.09, x_{i,\text{eq}} = \xi_i(I - A)^{-1}(Bu_{\text{eq}} + Ff_{\text{eq}}) - 0.1$, $y_{i,\text{eq}} = \xi_i Cx_{\text{eq}} + 0.05$ where ξ_i is the ith row of the identity matrix. Choose ET as $y_t = y_{\text{eq}} + [0.5, -0.2, 0.4]^{\text{T}}$, $u_t = u_{\text{eq}} + [0.5, -0.5, 0.5]^{\text{T}}$, and choose $\triangledown\hat{x}_i(0|0) = -0.05$, $\triangledown x_i(0) = 0.05$. The controlled system is parameterized as $A_r = 0.95A$, $B_r = 0.95B$, $C_r = 0.95C, F_r = 0.95F$, and the other parameters are the same as before. The control result is shown in Figure 7.4.

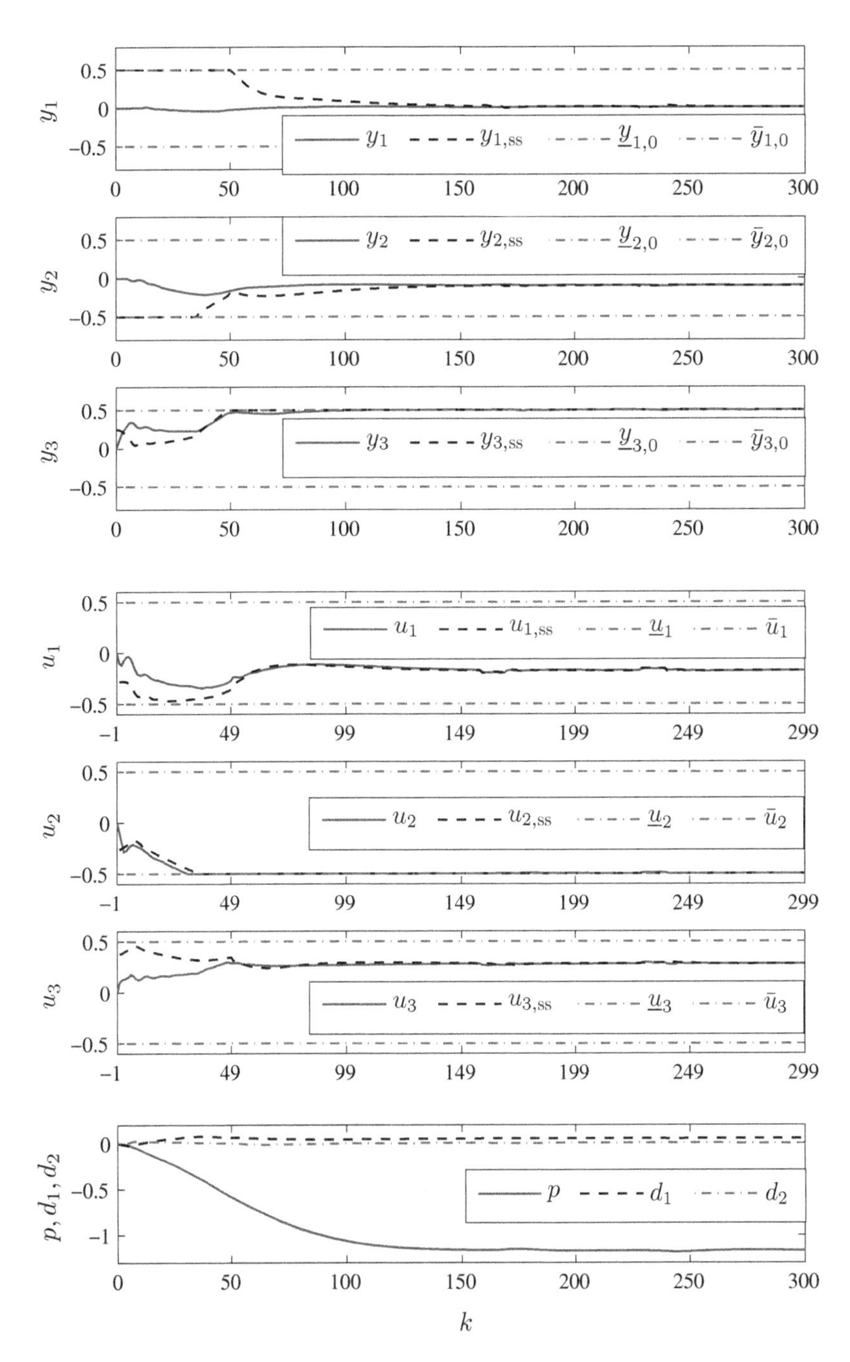

Figure 7.1 The control result.

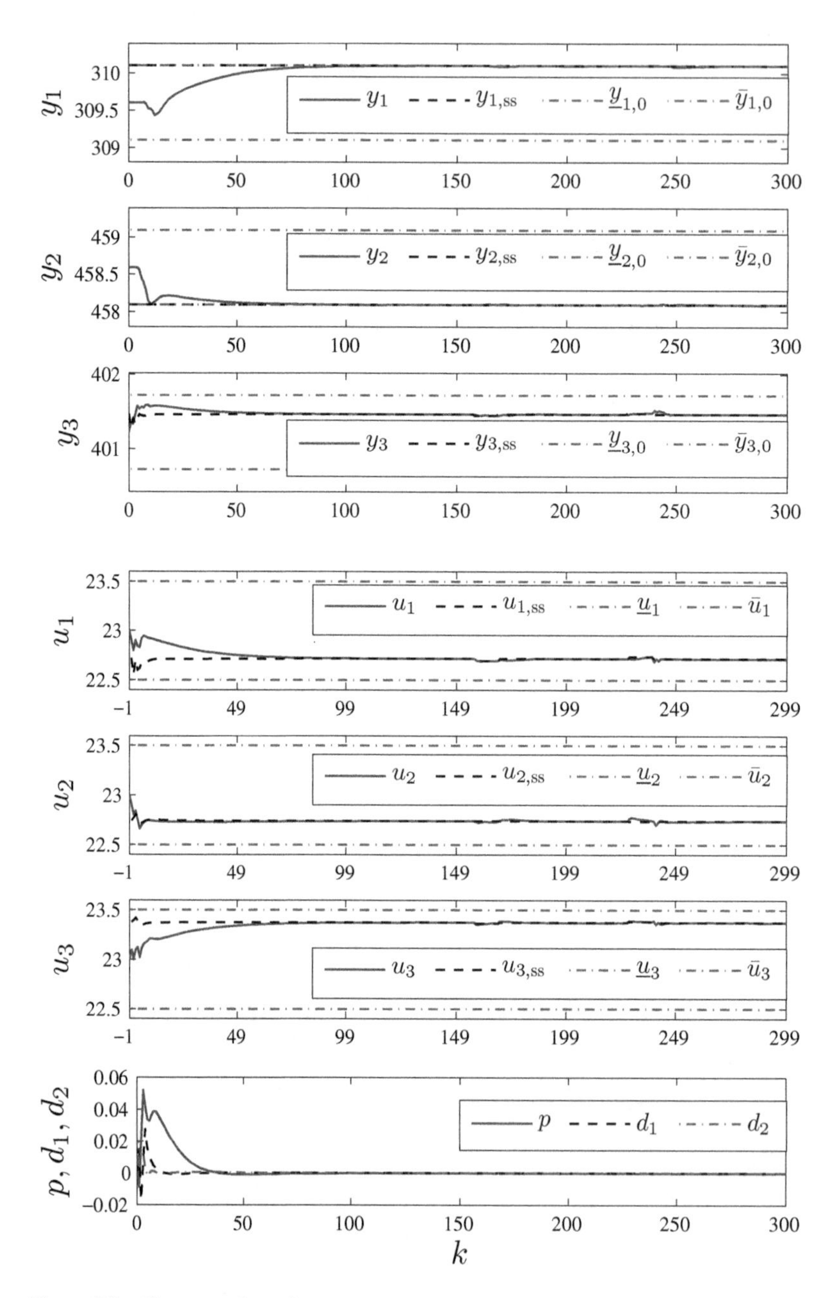

Figure 7.2 The control result.

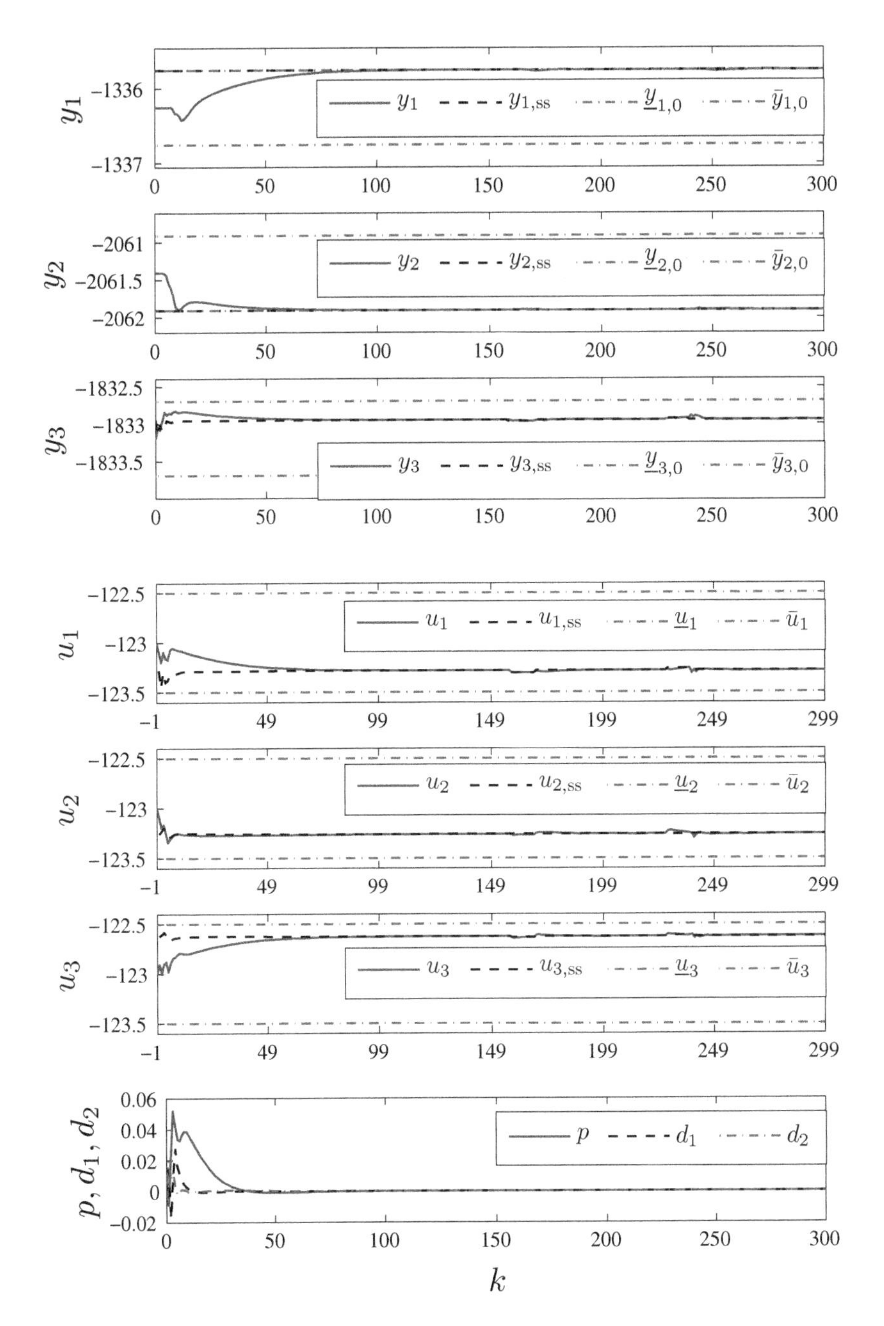

Figure 7.3 The control result.

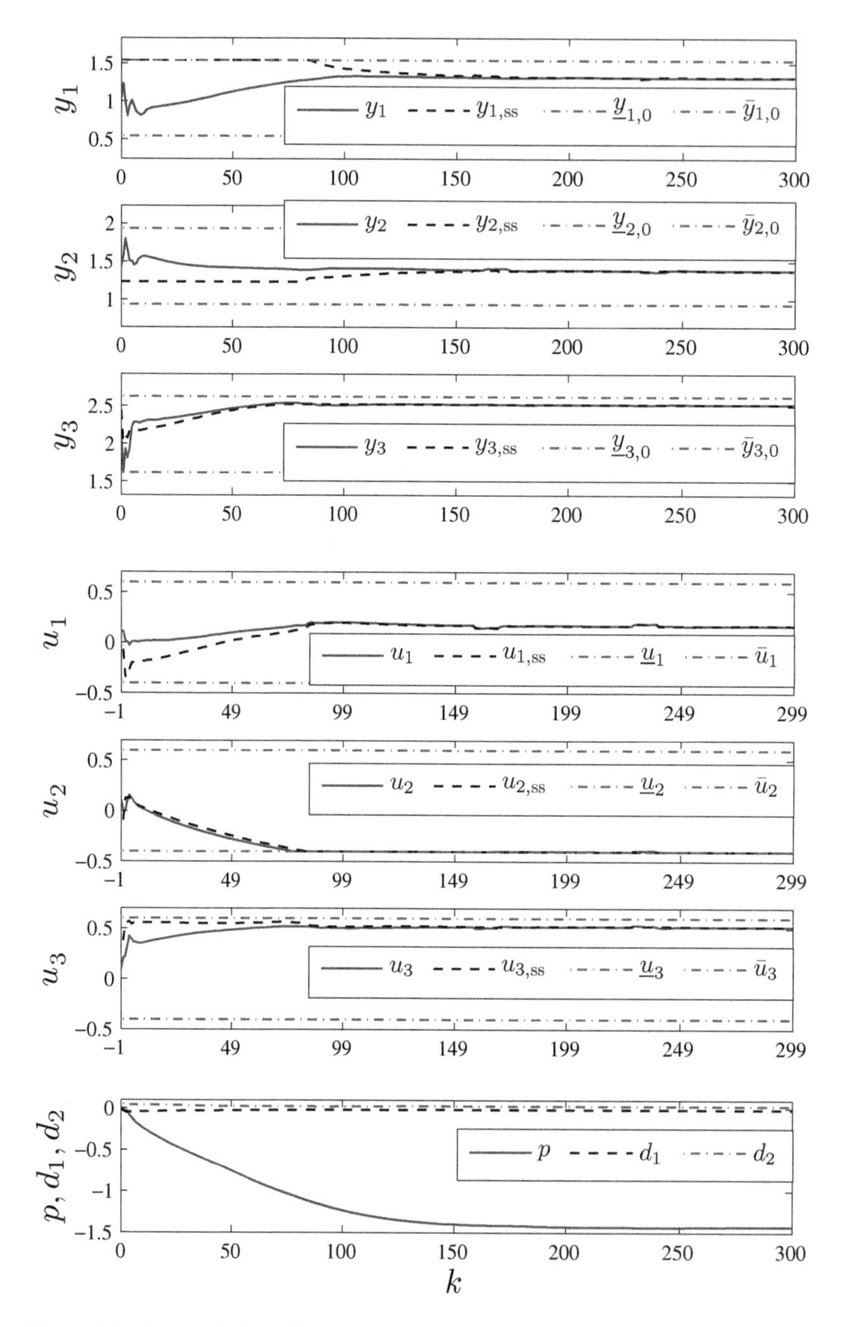

Figure 7.4 The control result.

In the two-layered MPC introduced in Sections 7.1 and 7.2, $-Bu_{\mathrm{eq}} + F\nabla f(k)$ always appears with $G_d\hat{d}(k|k)$, and y_{eq} always appears with $G_p\hat{p}(k|k)$. $\{y, u, f\}_{\mathrm{eq}}$ does not need to be known exactly; it can even be omitted. Certainly, using more accurate $\{y, u, f\}_{\mathrm{eq}}$ generally receives better control effects; using $G_d\hat{d}(k|k)$ and $G_p\hat{p}(k|k)$ to include the equilibrium point will affect the control performance.

The offset-free in Theorem 7.1 can be appropriately extended to the controllers in Chapters 4–6.

Note that all previous analyses of the offset-free of the two-layered MPC do not rely on the accuracy of the model. Indeed, the accuracy of the model is no longer required under the conditions for the above offset-free, but the inaccuracy of the model may destroy these conditions. Therefore, for the practical two-layered MPC, the accuracy of the model is the most important for the dynamic stability and offset-free.

7.3 Static Nonlinear Transformation

The linear MPC (i.e., MPC based on the linear model) technology accounts for more than 90% of MPC applications (see [Qin and Badgwell, 2003]), but the complexity of the industrial processes often exceeds the scope of linear MPC. The approaches to handle nonlinearity include, e.g., the nonlinear transformation and the nonlinear MPC (i.e., MPC based on a nonlinear model). The nonlinear MPC contains extremely rich contents which are not in the scope of this book. This chapter only introduces the nonlinear transformation.

Although the industrial processes are nonlinear in nature, so far, most MPCs in practice use the linear dynamic models, most of which are non-parametric models (FIR, FSR). The potential reasons are as follows.

- The linear empirical model of the process can be obtained directly by the test data of the process;
- Most of the MPCs are applied in the petrochemical and chemical industries, the goal of which is to maintain the process near the desired equilibrium point, so the linear model is accurate enough in the region around the equilibrium point;
- The linear MPC algorithm gets mature.

For the aforementioned reasons, the MPC algorithm based on the linear model can receive good results in many applications. However, sometimes, the operating ranges of a process variable are not limited to the neighborhood of the equilibrium point, i.e., it requires a wider range of change (e.g., a control valve or a product composition). In order to control such nonlinear processes, the linear MPC requires further treatment.

7.3.1 Principle of Nonlinear Transformation

The control system can be divided into four parts, i.e., the linear MPC, the input, the controlled process and the output. The input and output variables are the interface between MPC and the process. In practice, the nonlinearity of the process may appear in the input (MV and DV), the output (CV), or both.

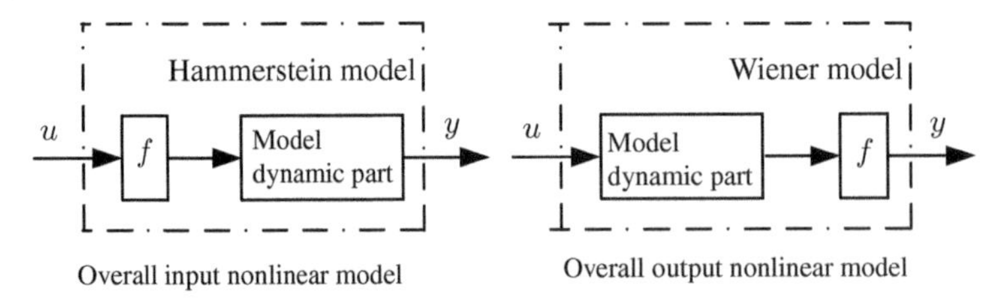

Figure 7.5 Hammerstein–Wiener nonlinear model.

The controlled process containing nonlinearity can be described by Hammerstein-Wiener model is shown in Figure 7.5, where f is the static nonlinearity. There are at least two methods to specify the nonlinearity of the process, i.e.,

- the mechanism method, where the dynamics of a unit in the industrial process have been thoroughly studied, so the static nonlinearity of this unit can be directly defined;
- the identification method, in which by fitting (regression), we can judge whether or not the process data has good consistency with the specified nonlinear function, and obtain the parameters of this nonlinear function.

The identification of Hammerstein–Wiener model has received wide research.

We can use the nonlinear transformation method to eliminate the static nonlinearity of the model, and then the linear MPC to control the process. Figure 7.6 shows the schematic diagram of the nonlinear transformation method. The linear MPC becomes more effective by the nonlinear transformation of input/output variables. The linear MPC sends MV, after the inverse nonlinear transformation, to the actuator, and displays CV, after the inverse nonlinear transformation, to users.

The implementation of nonlinear transformation is as follows.

- Specify the input and output static nonlinearities.
- Nonlinear transform the input and output data, which is applied to identify the linear model.
- When the linear MPC runs online, the input and output data and some controller parameters (such as the constraints on input and output) are nonlinearly transformed, prior to the controller optimization.
- The linear MPC is applied to a linear model. Hence, the calculated ss and MV moves are for the linear model rather than the practical nonlinear process. Some results of

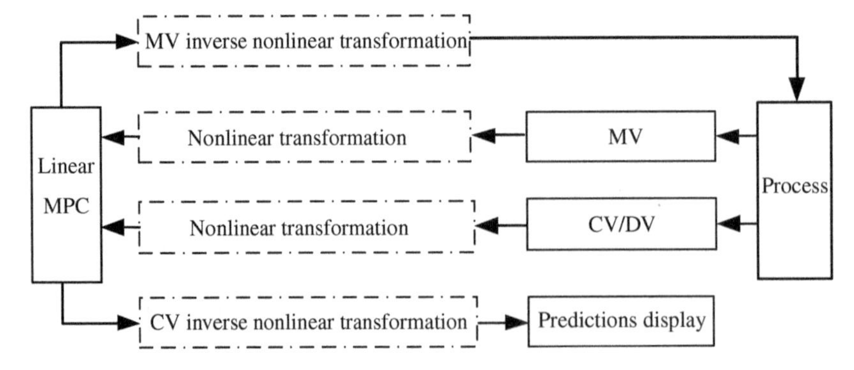

Figure 7.6 The schematic diagram of the nonlinear transformation method.

linear MPC require an inverse nonlinear transformation in order to be implemented and displayed.

Since there are many parameters in the implementation of linear MPC, it should carefully consider, together with the specific MPC algorithm, which parameters should be nonlinear transformed and which inverse nonlinear transformed.

7.3.2 Usual Nonlinear Transformations

The usual nonlinear transformations include, e.g., the control valve, the logarithmic, the exponential, and the piecewise linear. The two most useful nonlinear transformations are introduced.

7.3.2.1 Nonlinear Transformation of Valve Output

The pneumatic valve controls the liquid flow by moving the valve stem (0–100%), which accounts for the majority of the process control. In MPC, the actuator is usually a pneumatic control valve, and the so-called valve opening refers to the position of the valve stem. The relationship between flow and valve opening is not simply proportional, but nonlinearly related to the valve structures. If MV of the linear MPC refers to the flow rate in the valve, and MV of the overall MPC (i.e., linear MPC + nonlinear and inverse transformations) refers to the valve opening, then nonlinear transformation is required so that the flow rate corresponds to the valve opening. Moreover, sometimes, the valve opening is directly used as CV. The reason for using the flow rate as MV is that the obtained linear model may be more accurate. If MV of linear MPC is sp of PID with valve actuator, then it is unnecessary to perform the valve nonlinearly transformation. Due to PID, the loop model, with PID and valve actuator, may be closer to linearity. This "feedback linearization" effect of PID is common in the applications of MPC, but is not everywhere available.

The relationship between the movement of the valve stem and the flow rate is called the flow characteristic of the valve, which is related to not only the valve opening, but also the pressure drop in the valve. In practice, it is often to distinguish between the inherent flow characteristics and the installed flow characteristics, according to the pressure drop in the valve. The flow characteristic, obtained by a constant pressure drop in the valve, is called the inherent flow characteristic. There are several typical flow characteristics of valves, i.e., linear, equal percentage (logarithmic), parabolic, and quick opening. However, in practice, the pressure drop in the valve cannot keep constant, so the inherent flow characteristic cannot fully show the relationship between the flow rate and valve opening; due to the numerous factors, it is difficult to obtain the exact relationship; this flow characteristic is called the installed flow characteristic. In MPC applications, the linear, the equal percentage, and the parabolic valves are usually applied. The simple method is to use an empirical formula to describe the installed flow characteristic.

In the linear valve, we adopt the following nonlinear transformation:

$$r_L = \frac{L - L_{\min}}{L_{\max} - L_{\min}}, \quad r_F = \frac{(r_L - r_{L,\text{offset}})^2}{\sqrt{a + (1-a)(r_L - r_{L,\text{offset}})^2}}, \quad F = r_F F_{\max}, \tag{7.17}$$

where $L, L_{\max}, L_{\min}$ are the actual opening, the maximum opening and the minimum opening of the valve, respectively. r_L is the opening ratio; $r_{L,\text{offset}}$ is the move value; a is the pressure drop variation factor, i.e., the ratio of the pressure drop at the maximum flow to that at the minimum; F, F_{max} are the actual flow rate and the maximum flow rate, respectively; r_F is the flow rate ratio. The inverse transformation for the linear valve is

$$r_F = \frac{F}{F_{max}}, \quad r_L = \sqrt{\frac{(1-a)r_F^2 + r_F\sqrt{(1-a)^2 r_F^2 + 4a}}{2}} + r_{L,\text{offset}},$$
$$L = r_L(L_{max} - L_{min}) + L_{min}. \tag{7.18}$$

In the parabolic (equal percentage) valve, we use the following nonlinear transformation:

$$r_L = \frac{L - L_{min}}{L_{max} - L_{min}}, \quad r_F = \frac{(r_L - r_{L,\text{offset}})^4}{\sqrt{a + (1-a)(r_L - r_{L,\text{offset}})^4}}, \quad F = r_F F_{max}, \tag{7.19}$$

where the physical meaning of each variable is the same as that of the linear valve. The inverse transformation for the parabolic valve is

$$r_F = \frac{F}{F_{max}}, \quad r_L = \sqrt[4]{\frac{(1-a)r_F^2 + r_F\sqrt{(1-a)^2 r_F^2 + 4a}}{2}} + r_{L,\text{offset}},$$
$$L = r_L(L_{max} - L_{min}) + L_{min}. \tag{7.20}$$

7.3.2.2 Piecewise Linear Transformation

The piecewise linearity is a usual nonlinearity whose essence is the linear interpolation. A piecewise linear curve is the curve by connecting a series of points $\{(x_1, y_1), (x_2, y_2), \ldots, (x_n, y_n)\}$, satisfying $x_i < x_{i+1}$ and $y_i < y_{i+1}$. The nonlinear transformation is

$$v' = \begin{cases} y_1 + \frac{y_2 - y_1}{x_2 - x_1}(v - x_1), & v \le x_2 \\ y_i + \frac{y_{i+1} - y_i}{x_{i+1} - x_i}(v - x_i), & x_i \le v \le x_{i+1}, 2 \le i \le n-2. \\ y_{n-1} + \frac{y_n - y_{n-1}}{x_n - x_{n-1}}(v - x_{n-1}), & v \ge x_{n-1} \end{cases} \tag{7.21}$$

The corresponding inverse nonlinear transformation is

$$v = \begin{cases} x_1 + \frac{x_2 - x_1}{y_2 - y_1}(v' - y_1), & v' \le y_2, \\ x_i + \frac{x_{i+1} - x_i}{y_{i+1} - y_i}(v' - y_i), & y_i \le v' \le y_{i+1}, 2 \le i \le n-2, \\ x_{n-1} + \frac{x_n - x_{n-1}}{y_n - y_{n-1}}(v' - y_{n-1}), & v' \ge y_{n-1}. \end{cases} \tag{7.22}$$

In the aforementioned, it briefly introduces, in the linear MPC, how to handle the process with some nonlinear characteristics. However, we should pay attention to two points in practice.

- The static nonlinear characteristic of some processes may be complex, which is beyond the above nonlinear transformation methods. In the applications, it is necessary to find more effective nonlinear transformation methods;
- Some nonlinearity of the processes has not only static nonlinearity but also strong dynamic nonlinearity. In this case, the MPC method based on dynamic nonlinear model is recommended.

7.4 Two-Layered MPC with Varying Degree of Freedom

The two-layered MPC has varying degrees of freedom. This controller will not vary either from two-layer to multilayer, or from two-layer to single-layer, but its number of CVs, MVs and DVs may vary. The so-called degree of freedom is defined as the "effective numbers of MVs – effective numbers of CVs", which implies the control ability of MPC. The higher the degree of freedom, the stronger the control ability of MPC. This is obvious since it is more difficult to control more CVs with less MVs and easier to control less CVs with more MVs. DV may make some MVs saturated, which reduces the degree of freedom. During the running of MPC, the degree of freedom vary due to, e.g., the manual operation, the numerical judgment, the change in process conditions. In general, the varying degree of freedom may cause varying in the model; in this sense, the two-layer MPC is also a variable structure control. The variable structure control is a general term for a large class of control algorithms, in which the "variable structure" implies the "variable model" (including the open-loop model and closed-loop model).

When the degree of freedom varies, the principle algorithm of the two-layer MPC does not change. Some usual variations of degree of freedom, with corresponding treatments, are given as follows.

(1) If an MV lies on a boundary of hard constraint, i.e., both the current value and ss are on this boundary, then this MV can be changed to DV, and the corresponding MV constraints are no longer considered.
(2) If an MV is removed by the operator, then this MV becomes a DV, and the corresponding MV constraints are no longer considered.
(3) If a CV is removed by the operator, then the corresponding CV constraints are no longer considered.
(4) If SSTC cannot make the feasible region nonempty by relaxing constraints at the highest rank, then find out the CV that causes the infeasibility and remove it, and the corresponding CV constraints are no longer considered. The method is, offline, to rank CV hard constraints according to the priority (the upper and lower limit of the hard constraints of each CV are at the same rank); online, if the highest rank of SSTC is infeasible, then judge the compatibility of hard constraint set according to the priority-number (using ascending-number strategy), and CV whose hard constraints cause incompatibility will be removed. If the steady-state constraint on CV rate causes incompatibility, then remove the constraint instead of CV.
(5) After the entry or exit of MV and CV, determine whether or not it is required to further exclude CV by judging the collinearity between the weighted CVs.

The large disturbances may cause (4). The fact that, CV cannot be controlled because of the large disturbance, is tested in both SSTC and DC; in this situation, in SSTC, the feasible region is empty, while in DC, the dynamic prediction violates the constraint. If some CVs cannot be controlled for a long time, it should be regarded as an exception, and MPC should be stopped. In fact, when $y_{ss}(k)$ from SSTC continuously violates the operational constraints, it is suggested as an exception in Rao and Rawlings [1999] and Muske and Badgwell [2002], because it implies that MPC cannot suppress the effects of the large disturbances.

For (5), by calculating/changing the condition number of Hessian matrix of the unconstrained LS problem, the optimization problem of DC obtains an appropriate solution. Take Chapter 4 as an example, which is equivalent to considering the condition number of the matrix $\mathcal{S}^{\mathrm{T}}\tilde{Q}(k)\mathcal{S} + \tilde{\Lambda}$ ($\tilde{Q}(k)$ and $\tilde{\Lambda}$ being diagonal matrices composed of the weight matrices); if the condition number is over large, then some MVs may vary sharply, which is very harmful to the system. Assuming that the condition number gets too large, there are two methods. One is to temporarily increase/change $\tilde{\Lambda} = \mathrm{diag}\{\Lambda, \Lambda, \ldots, \Lambda\}$ to reduce the condition number. The other is to restructure the controlled variables. We consider the second in the following.

At time k,

$$y_{ss}(k) - G^f f(k) = G^u u_{ss}(k), \tag{7.23}$$

where $G^u \in \mathbb{R}^{n_y \times n_u}$. Since different CVs have different weights, we handle the following equation:

$$Q(k)^{1/2} y_{ss}(k) - Q(k)^{1/2} G^f f(k) = Q(k)^{1/2} G^u u_{ss}(k). \tag{7.24}$$

By SVD decomposition of $Q(k)^{1/2} G^u$, it yields

$$\begin{aligned} Q(k)^{1/2} y_{ss}(k) - Q(k)^{1/2} G^f f(k) &= \begin{bmatrix} \Sigma_1 & \Sigma_2 \end{bmatrix} \begin{bmatrix} U_1 & 0 \\ 0 & U_2 \end{bmatrix} \begin{bmatrix} V_1^{\mathrm{T}} \\ V_2^{\mathrm{T}} \end{bmatrix} u_{ss}(k) \\ &\approx \Sigma_1 U_1 V_1^{\mathrm{T}} u_{ss}(k), \end{aligned} \tag{7.25}$$

where $\mathrm{diag}\{U_1, U_2\} \in \mathbb{R}^{n_y \times n_u}$. If $n_y \neq n_u$, then choose U_2 as the square matrix. By applying (7.25) and further approximation, obtains

$$\Sigma_1^{\mathrm{T}} Q(k)^{1/2} y_{ss}(k) - \Sigma_1^{\mathrm{T}} Q(k)^{1/2} G^f f(k) = U_1 V_1^{\mathrm{T}} u_{ss}(k). \tag{7.26}$$

Writing (7.26) equivalently as

$$y'_{ss}(k) - G'^f(k) = G'^u u_{ss}(k). \tag{7.27}$$

Then, CV becomes $y' = \Sigma_1^{\mathrm{T}} Q(k)^{1/2} y$, so all ss and the constraints corresponding to y are converted into those to y'; $\{G^u, G^f\}$ is substituted with $\{G'^u, G'^f\}$ (for DMC, all $\{S_i^u, S_i^f\}$ are substituted with $\{\Sigma_1^{\mathrm{T}} Q(k)^{1/2} S_i^u, \Sigma_1^{\mathrm{T}} Q(k)^{1/2} S_i^f\}$). If a part of $y_{ss}(k)$ has ET and the remaining does not, then $y'_{ss}(k) = \Sigma_1^{\mathrm{T}} Q(k)^{1/2} y_{ss}(k)$ no longer has ET. In addition, DC weight for y' is temporarily taken as the identity matrix. This method keeps the condition number of the real-time weighted steady-state gain matrix $Q(k)^{1/2} G^u$ within an appropriate range. If the condition number is overly large, the small differences in sps between some CVs may cause some MVs to vary sharply, which is harmful to the system.

Strictly, the condition number by treating the matrix $Q(k)^{1/2} G^u$ is not equivalent to that by treating $\mathcal{S}^{\mathrm{T}}\tilde{Q}(k)\mathcal{S} + \tilde{\Lambda}$, so it is unnecessary to observe the condition number of $\mathcal{S}^{\mathrm{T}}\tilde{Q}(k)\mathcal{S} + \tilde{\Lambda}$ in advance.

7.4.1 Numerical Example Without Varying Structure

Apply the previous heavy oil fractionator model. The sampling time is 4, the model horizon 100. Take $\underline{u} = [-0.5; -0.5; -0.5]$, $\bar{u} = [0.5; 0.5; 0.5]$, $\Delta\bar{u}_i = \delta\bar{u}_{i,ss} = 0.1$; $\underline{y}_{0,h} = [-0.7; -0.7;$

Table 7.1 Parameters of multi-priority-rank SSTC.

Rank no.	Type of constraint	Variable	Soft constraint	$\overline{\varepsilon}$
1	Inequality	$y_{2,\text{ss}}$	CV lower bound	0.20
1	Inequality	$y_{3,\text{ss}}$	CV upper bound	0.20
2	Eequality	$y_{2,\text{ss}}$	−0.3	0.25
3	Inequality	$y_{2,\text{ss}}$	CV upper bound	0.20
3	Inequality	$y_{3,\text{ss}}$	CV lower bound	0.20
4	Equality	$u_{3,\text{ss}}$	−0.2	0.25
4	Equality	$y_{1,\text{ss}}$	0.3	0.25
5	Inequality	$y_{1,\text{ss}}$	CV upper, lower bound	0.20

$-0.7]$, $\overline{y}_{0,h} = [0.7; 0.7; 0.7]$, $\underline{y}_0 = [-0.5; -0.5; -0.5]$, $\overline{y}_0 = [0.5; 0.5; 0.5]$, $\delta\overline{y}_{\text{ss}} = [0.2; 0.2; 0.3]$; where y_1, y_2, u_3 have ETs, and their expected allowable range is 0.5.

The parameters of SSTC are shown in Table 7.1. In the economic stage, u_2 is the minimum-move MV, take $h = [-2, -1, 2]$, $J_{\min} = -0.4$. At $k \in [65, 80]$, there exists the disturbance with the value $[0.2; 0.1]$; at $k \geq 120$, there exists the disturbance with the value $f_{\text{eq}} + [1; -1]$.

The parameters for DC are $P = 15$, $M = 8$, $\Lambda = \text{diag}\{3, 5, 3\}$, $\overline{z} = [0.4; 0.4; 0.4]$; $\underline{z} = [-0.4; -0.4; -0.4]$, $\underline{q}_1 = 2.0$, $\check{q}_1 = 0.5$, $\overline{q}_1 = 2.0$, $\underline{q}_2 = 2.0$, $\check{q}_2 = 1.0$, $\overline{q}_2 = 2.0$, $\underline{q}_3 = 2.5$, $\check{q}_3 = 2.5$, $\overline{q}_3 = 4.0$, $\rho = 0.2$. $\Delta u(k) = \Delta u(k|k)$. The true system output is produced using the above transfer function matrix, being multiplied by 0.9 to represent the model mismatch.

The control results are shown in Figures 7.7 and 7.8, which demonstrates that ss can be tracked by DC with offset-free. Finally $y_1 = y_{1,\text{ss}} = 0.299 \neq y_{1,t}$, $y_2 = y_{2,\text{ss}} = -0.3 \neq y_{2,t}$, $u_3 = u_{3,\text{ss}} = -0.163 \neq u_{3,t}$.

7.4.2 Numerical Example with Varying Number of Manipulated Variables

One of the important features of two-layered DMC is that the current value and ss are often both on the constraint boundary. However, when an MV runs at the hard constraint boundary, due to e.g., the large disturbance or equipment failure, it may violate the constraint so that MPC has to stop running. When some MVss runs at the hard constraint boundary, it can be changed as a constant valued DV, and the corresponding constraints are no longer considered. The transfer function matrix of the original model is

$$G^u(s) = \begin{bmatrix} G^u_{1,1}(s) & G^u_{1,2}(s) & \cdots & G^u_{1,n_u}(s) \\ \vdots & \vdots & \ddots & \vdots \\ G^u_{n_y,1}(s) & G^u_{n_y,2}(s) & \cdots & G^u_{n_y,n_u}(s) \end{bmatrix}_{n_y \times n_u},$$

$$G^f(s) = \begin{bmatrix} G^f_{1,1}(s) & G^f_{1,2}(s) & \cdots & G^f_{1,n_f}(s) \\ \vdots & \vdots & \ddots & \vdots \\ G^f_{n_y,1}(s) & G^f_{n_y,2}(s) & \cdots & G^f_{n_y,n_f}(s) \end{bmatrix}_{n_y \times n_f},$$

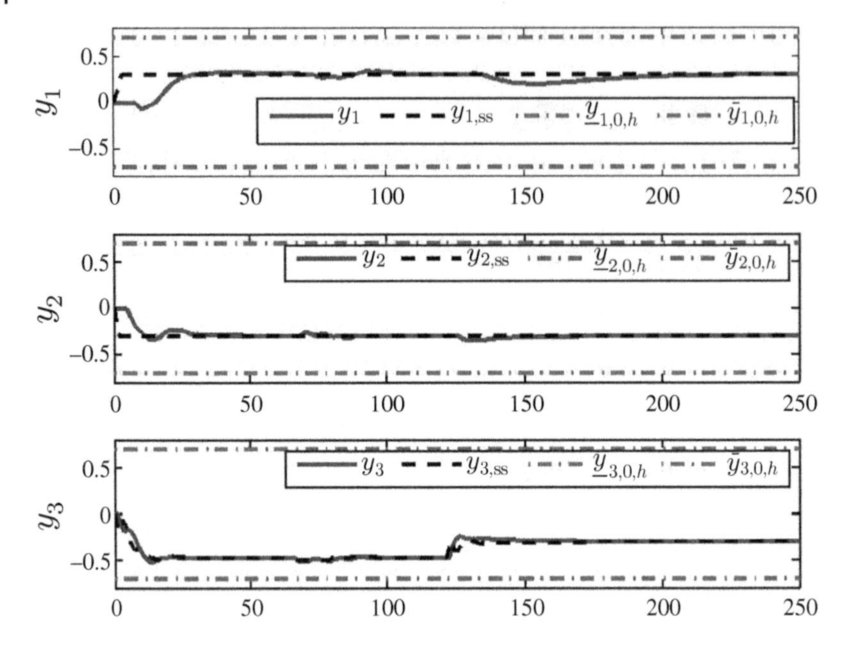

Figure 7.7 The control result of CV.

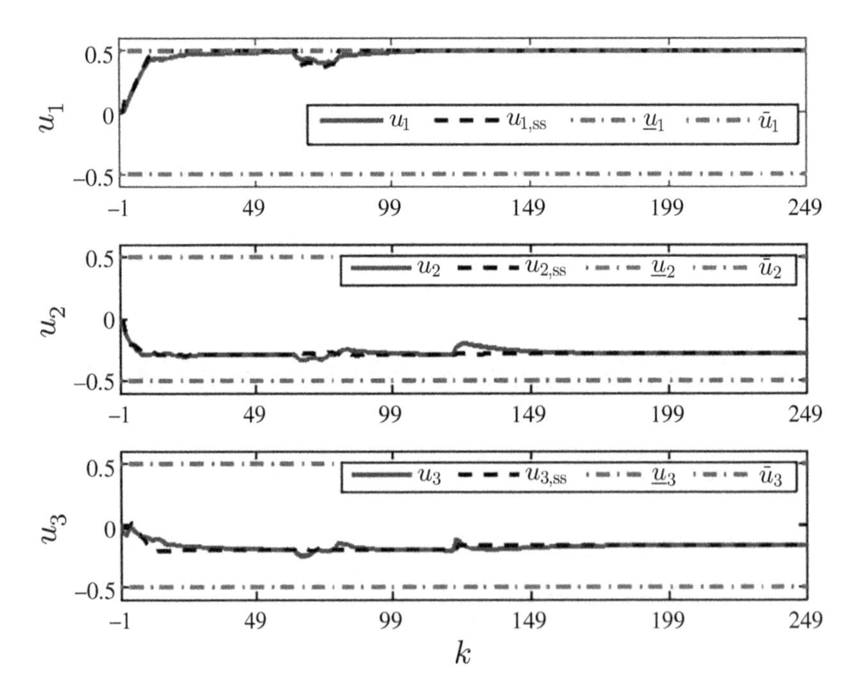

Figure 7.8 The control result of MV.

where n_u, n_y, and n_f stand for the dimensions of MV, CV, and DV, respectively. $G^u_{i,j}(s)$ denotes the transfer function of the jth input to the ith output, where $1 \leq i \leq n_y$ and $1 \leq j \leq n_u$; $G^f_{m,n}(s)$ denotes the transfer function of the nth disturbance to the mth output, where $0 \leq m \leq n_y$ and $0 \leq n \leq n_f$.

Assuming that the lth MVss runs on the constraint boundary, then the lth MV becomes the constant DV, and the model becomes

$$G^{u'}(s) = \begin{bmatrix} G^u_{1,1}(s) & G^u_{1,2}(s) & \cdots & G^u_{1,l-1}(s) & G^u_{1,l+1}(s) & \cdots & G^u_{1,n_u}(s) \\ \vdots & \vdots & \ddots & \vdots & \vdots & \ddots & \vdots \\ G^u_{n_y,1}(s) & G^u_{n_y,2}(s) & \cdots & G^u_{n_y,l-1}(s) & G^u_{n_y,l+1}(s) & \cdots & G^u_{n_y,n_u}(s) \end{bmatrix}_{n_y \times (n_u-1)},$$

$$G^{f'}(s) = \begin{bmatrix} G^f_{1,1}(s) & G^f_{1,2}(s) & \cdots & G^f_{1,n_f}(s) & G^u_{1,l}(s) \\ \vdots & \vdots & \ddots & \vdots & \vdots \\ G^f_{n_y,1}(s) & G^f_{n_y,2}(s) & \cdots & G^f_{n_y,n_f}(s) & G^u_{n_y,l}(s) \end{bmatrix}_{n_y \times (n_f+1)}.$$

The structure of the controller now becomes $\{n_u - 1 \text{ MVs}, n_y \text{ CVs}, n_f + 1 \text{ DVs}\}$, so the degree of freedom is reduced. After an MV on the constraint boundary becomes a DV, the DV remains at its boundary value. In the variable structure control of MV, for the open-loop prediction module, SSTC module and DC module, only the number of MV and DV varies. When the number of MV is changed, the overall DMC algorithm does not change.

Consider the previous heavy oil fractionator model, and all parameters are the same as in Section 7.4.1. As shown in Figure 7.8, at time $k = 150$, ss and the actual value of u_1 are both at the constraint upper bound, 0.5. Hence, at time $k = 150$, it reduces the number of MVs, and u_1 becomes the disturbance; the corresponding model changes, i.e.,

$$G^{u'}(s) = \begin{bmatrix} \frac{1.77e^{-28s}}{60s+1} & \frac{5.88e^{-27s}}{50s+1} \\ \frac{5.72e^{-14s}}{60s+1} & \frac{6.90e^{-15s}}{40s+1} \\ \frac{4.42e^{-22s}}{44s+1} & \frac{7.20}{19s+1} \end{bmatrix},$$

$$G^{f'}(s) = \begin{bmatrix} \frac{1.20e^{-27s}}{45s+1} & \frac{1.44e^{-27s}}{40s+1} & \frac{4.05e^{-27s}}{50s+1} \\ \frac{1.52e^{-15s}}{25s+1} & \frac{1.83e^{-15s}}{20s+1} & \frac{5.39e^{-18s}}{50s+1} \\ \frac{1.14}{27s+1} & \frac{1.26}{32s+1} & \frac{4.38e^{-20s}}{33s+1} \end{bmatrix}.$$

At this interval, all the corresponding constraints and target for u_1 are no longer considered. However, a constant disturbance with the value 0.5 (constraint upper bound) is added to the disturbance channel, and the model varies from a square system with three inputs and three outputs to a thin system with two inputs and three outputs. The control results with the reduced number of MVs are shown in Figures 7.9 and 7.10. Since u_1 is changed as a constant disturbance at $k = 150$, the control curve begins to fluctuate, but tracks quickly and tends to be stable. Comparing Figures 7.9 and 7.10 with Figures 7.7 and 7.8, the trend of the curves do not change, and finally $y_1 = y_{1,ss} = 0.298 \neq y_{1,t}$, $y_2 = y_{2,ss} = -0.3 \neq y_{2,t}$,

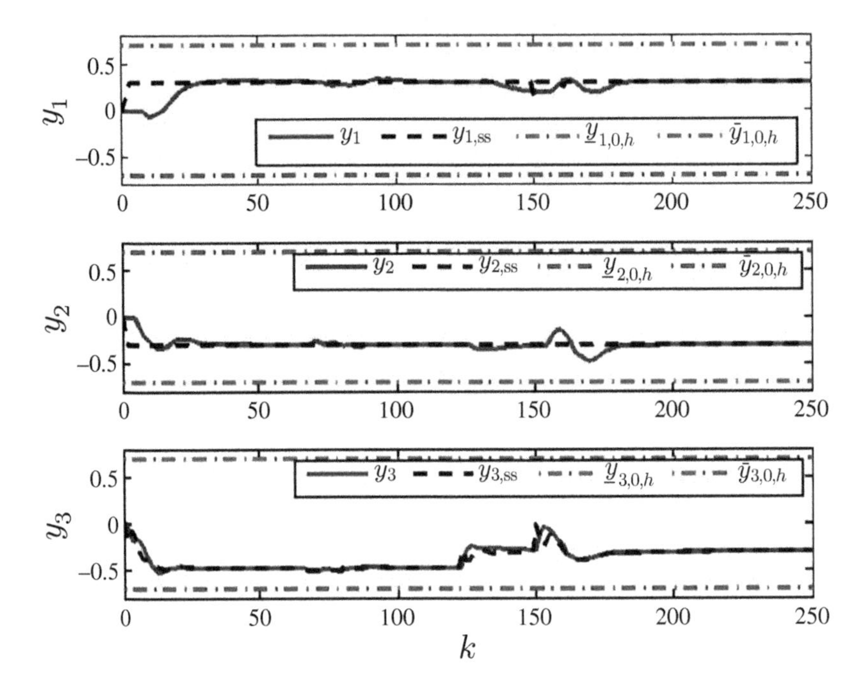

Figure 7.9 The control result of CV.

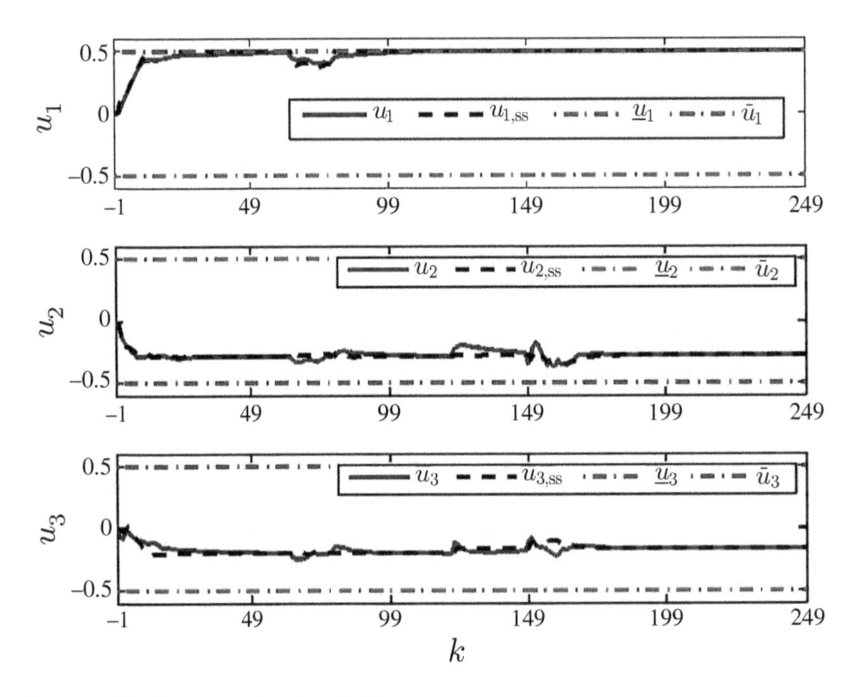

Figure 7.10 The control result of MV.

$u_3 = u_{3,\mathrm{ss}} = -0.183 \neq u_{3,t}$; the control results of this section are roughly similar to Section 7.4.1, which verifies the effectiveness of the algorithm.

7.5 Numerical Example with Output Collinearity

Let us modify the previous heavy oil fractionator model, i.e.,

$$G^u(s) = \begin{bmatrix} \frac{4.05e^{-27s}}{50s+1} & \frac{1.77e^{-28s}}{60s+1} & \frac{5.88e^{-27s}}{50s+1} & \frac{4.00e^{-20s}}{50s+1} \\ \frac{5.39e^{-18s}}{50s+1} & \frac{5.72e^{-14s}}{60s+1} & \frac{6.90e^{-15s}}{40s+1} & \frac{5.30e^{-15s}}{50s+1} \\ \frac{4.38e^{-20s}}{33s+1} & \frac{4.42e^{-22s}}{44s+1} & \frac{7.20}{19s+1} & \frac{4.30e^{-20s}}{33s+1} \\ \frac{3.66e^{-23s}}{38s+1} & \frac{4.55}{30s+1} & \frac{5.21e^{-12s}}{15s+1} & \frac{3.60e^{-19s}}{30s+1} \end{bmatrix},$$

$$G^f(s) = \begin{bmatrix} \frac{1.20e^{-27s}}{45s+1} & \frac{1.44e^{-27s}}{40s+1} \\ \frac{1.52e^{-15s}}{25s+1} & \frac{1.83e^{-15s}}{20s+1} \\ \frac{1.14}{27s+1} & \frac{1.26}{32s+1} \\ \frac{1.04e^{-15s}}{23s+1} & \frac{1.22}{30s+1} \end{bmatrix}.$$

The steady-state gain matrix of the MV channel is

$$G^u = \begin{bmatrix} 4.05 & 1.77 & 5.88 & 4.00 \\ 5.39 & 5.72 & 6.90 & 5.30 \\ 4.38 & 4.42 & 7.20 & 4.30 \\ 3.66 & 4.55 & 5.21 & 3.60 \end{bmatrix}.$$

The sampling time is 4, and the model horizon $N = 100$. Take $\underline{u} = [-0.5; -0.5; -0.5; -0.5]$, $\bar{u} = [0.5; 0.5; 0.5; 0.5]$, $\Delta\bar{u}_i = \delta\bar{u}_{i,\mathrm{ss}} = 0.1$; $\underline{y}_{0,h} = [-0.7; -0.7; -0.7; -0.7]$, $\bar{y}_{0,h} = [0.7; 0.7; 0.7; 0.7]$, $\underline{y}_0 = [-0.5; -0.5; -0.5; -0.5]$, $\bar{y}_0 = [0.5; 0.5; 0.5; 0.5]$, $\delta\bar{y}_{\mathrm{ss}} = [0.1; 0.3; 0.1; 0.3]$. ETs are $y_t = [0.3, 0.3, 0.2, 0]$, $u_{3,t} = -0.2$, and their expected allowable range is 0.5.

The parameters of SSTC are shown in Table 7.2. There is no minimum-move MV in the economic optimization. Take $h = [-2, 1, 2, 1]$, $J_{\min} = -0.4$. At $k \in [65, 80]$, there exists the disturbance with value $[0.2; 0.1]$; at $k \geq 120$, there exists the disturbance with value $f_{\mathrm{eq}} + [-0.1; -0.2]$.

The parameters of the dynamic controller are $P = 15$, $M = 8$, $\Lambda = \mathrm{diag}\{3, 5, 3, 5\}$, $\bar{z} = [0.4; 0.4; 0.4; 0.4]$; $\underline{z} = [-0.4; -0.4; -0.4; -0.4]$, $Q = I_4$, $\rho = 0.2$. $\Delta u(k) = \Delta u(k|k)$. The control results are shown in Figures 7.11 and 7.12.

By calculating $[\Sigma \ U \ V] = \mathrm{SVD}(Q^{1/2}G^u)$, we have $U = \begin{bmatrix} U_1 & 0 \\ 0 & U_2 \end{bmatrix} = \mathrm{diag}\{19.61, 2.30, 0.82, 0.006\}$, where $U_1 = \mathrm{diag}\{19.61, 2.30, 0.82\}$, $U_2 = [0.006]$. Since each eigenvalue of U_1

Table 7.2 Parameters of multi-priority-rank SSTC.

Rank no.	Type of constraint	Variable	Soft constraint	$\bar{\varepsilon}$
1	Inequality	$y_{2,\mathrm{ss}}$	CV lower bound	0.20
1	Inequality	$y_{3,\mathrm{ss}}$	CV upper bound	0.20
2	Equality	$y_{2,\mathrm{ss}}$	0.3	0.25
2	Equality	$y_{4,\mathrm{ss}}$	0	0.25
3	Inequality	$y_{2,\mathrm{ss}}$	CV upper bound	0.20
3	Inequality	$y_{3,\mathrm{ss}}$	CV lower bound	0.20
4	Equality	$u_{3,\mathrm{ss}}$	−0.2	0.25
4	Equality	$y_{1,\mathrm{ss}}$	0.3	0.25
4	Equality	$y_{3,\mathrm{ss}}$	0.2	0.25
5	Inequality	$y_{1,\mathrm{ss}}$	CV upper, lower bound	0.20

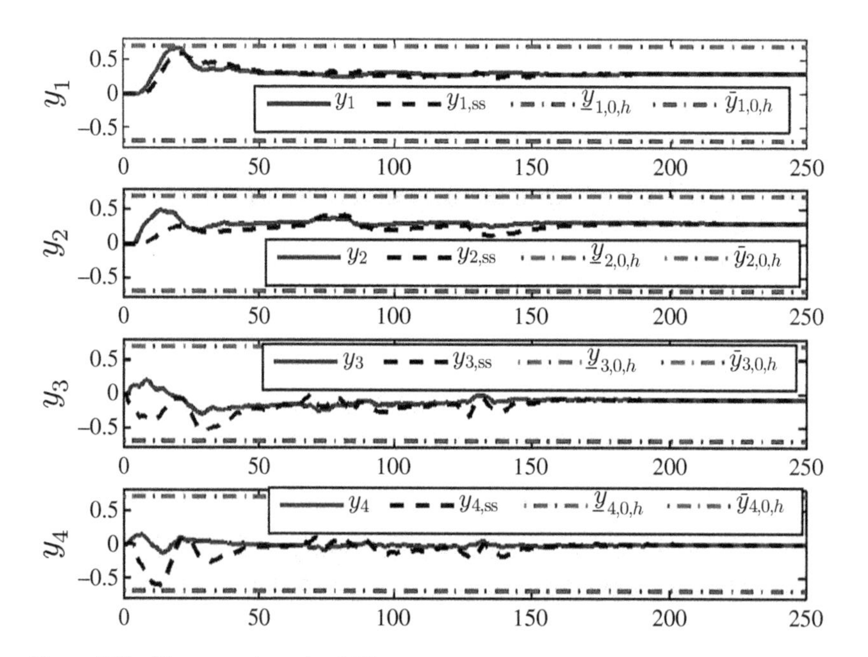

Figure 7.11 The control result of CV.

is much larger than that of U_2, the variable structure control is used to eliminate the collinearity in CVs. After SVD transformation, the corresponding constraints are changed, and the specific control parameters are as follows.

The sampling time is 4, and the model horizon $N = 100$. Take $\underline{u} = [-0.5;-0.5;-0.5;-0.5]$, $\bar{u} = [0.5;0.5;0.5;0.5]$, $\Delta\bar{u}_i = \delta\bar{u}_{i,\mathrm{ss}} = 0.1$; $\underline{y}_{0,h} = [-1.385;-1.385;-1.385]$, $\bar{y}_{0,h} = [1.385;1.385;1.385]$, $\underline{y}_0 = [-0.99;-0.99;-0.99]$, $\bar{y}_0 = [0.99;0.99;0.995]$, $\delta\bar{y}_{\mathrm{ss}} = [0.4043;0.1369;0.0906]$. ETs are $y_t = [-0.4095, 0.1612, 0.0999]$, $u_{3,t} = -0.2$ and their expected allowable range is 0.5.

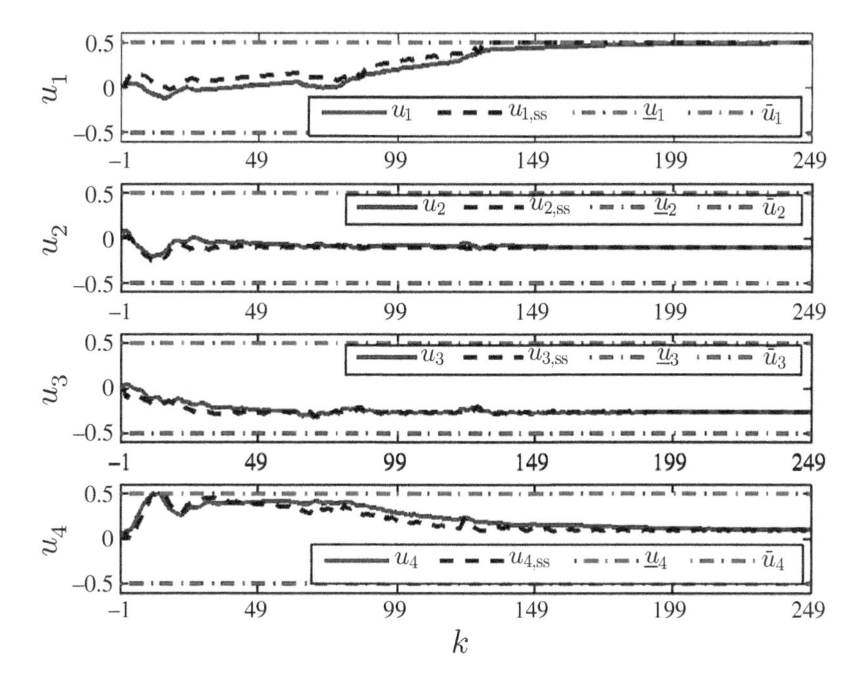

Figure 7.12 The control result of MV.

Table 7.3 Parameters of multi-priority-rank SSTC.

Rank no.	Type of constraint	Variable	Soft constraint	$\bar{\varepsilon}$
1	Inequality	$y_{2,\mathrm{ss}}$	CV lower bound	0.20
1	Inequality	$y_{3,\mathrm{ss}}$	CV upper bound	0.20
2	Equality	$y_{2,\mathrm{ss}}$	0.1624	0.25
3	Inequality	$y_{2,\mathrm{ss}}$	CV upper bound	0.20
3	Inequality	$y_{3,\mathrm{ss}}$	CV lower bound	0.20
4	Equality	$u_{3,\mathrm{ss}}$	−0.2	0.25
4	Equality	$y_{1,\mathrm{ss}}$	−0.4095	0.25
4	Equality	$y_{1,\mathrm{ss}}$	0.0999	0.25
5	Inequality	$y_{1,\mathrm{ss}}$	CV upper, lower bound	0.20

The parameters of SSTC are shown in Table 7.3. There is no minimum-move MV for economic optimization. Take $h = [-2, 1, 2, 1]$, $J_{\min} = -0.4$. At $k \in [65, 80]$, there exists the disturbance with value $[0.2; 0.1]$; at $k \geq 120$, there exists the disturbance with value $f_{\mathrm{eq}} + [-0.1; -0.2]$.

The parameters of DC are $P = 15$, $M = 8$, $\Lambda = \mathrm{diag}\{3, 5, 3, 5\}$, $\bar{z} = [0.4; 0.4; 0.4]$; $\underline{z} = [-0.4; -0.4; -0.4]$, $Q = I_3$, $\rho = 0.2$. $\Delta u(k) = \Delta u(k|k)$.

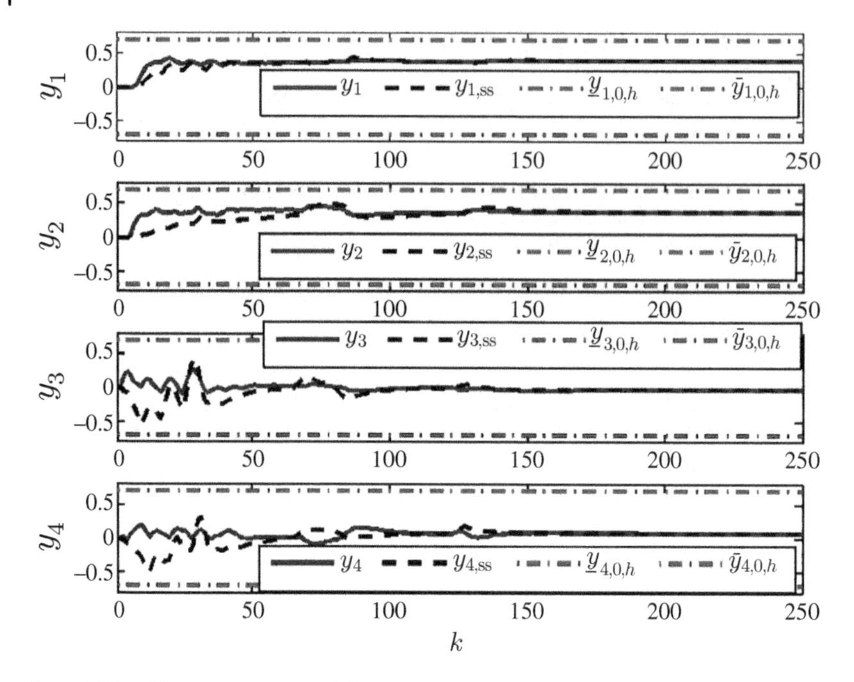

Figure 7.13 The control result of CV.

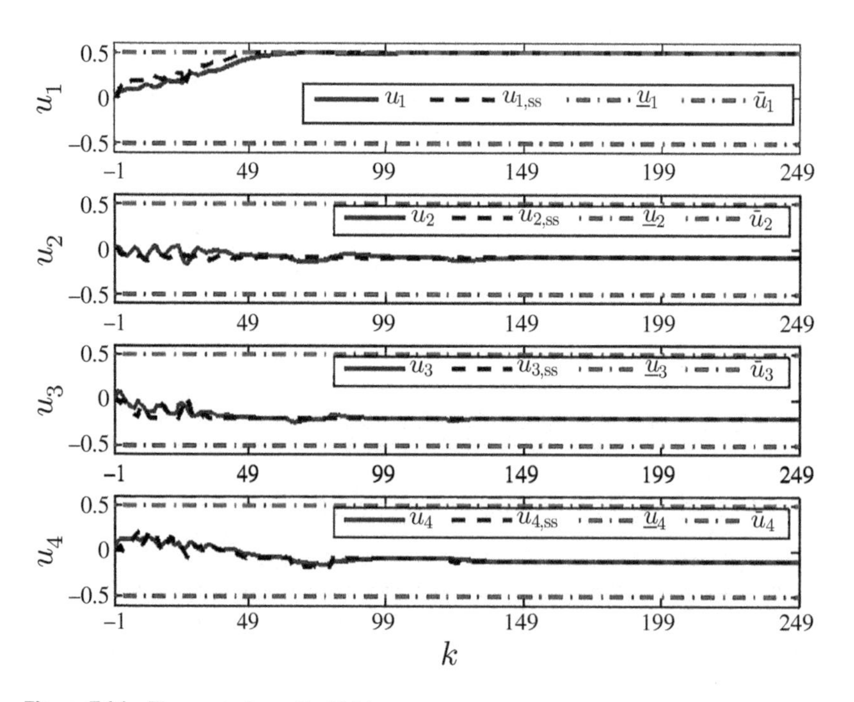

Figure 7.14 The control result of MV.

The control results are shown in Figures 7.13 and 7.14. The true CV and MV can finally track ss given by SSTC. Comparing Figures 7.13 and 7.14 with Figures 7.11 and 7.12, their control results are roughly similar; for the variable structure control, MV varies more smoothly and tracking is faster.

8

Two-Step Model Predictive Control for Hammerstein Model

Here, the two-step control applies to a class of special systems, i.e., systems with input nonlinearities. Input nonlinearities include, e.g., input saturation, dead zone. Moreover, a system represented by the Hammerstein model is often seen as an input nonlinear system. The Hammerstein model consists of a static nonlinear part followed by a dynamic linear part; see Pearson and Pottmann [2000]. There are many research on the identification of Hammerstein model, e.g., Yu et al. [2013]. Model predictive control (MPCs) for input nonlinear systems (mainly referred to as input saturation and Hammerstein nonlinearity) can be classified into two categories.

One category takes the nonlinear model as a whole (overall category, e.g., [Bloemen et al., 2001]), incorporates the nonlinear part into the objective function and directly solves the control moves. Note that the input saturation is usually taken as the constraint in the optimization. For this category, the control law calculation is rather complex and it is more difficult for real applications.

The other category utilizes nonlinear separation technique (separation category, e.g., [Fruzzetti et al., 1997]), i.e., first calculates the intermediate variable utilizing the linear submodel and MPC, then computes the actual control move via nonlinear inversion. The input saturation can be regarded as a kind of input nonlinearity or as a constraint in the optimization. The nonlinear separation with respect to the Hammerstein model invokes the special structure of the Hammerstein model, and classifies the controller designing problem as the linear control, which is much simpler than the overall category. In many control problems, the target is to make the system output track sp as soon as possible (or drive the system's state to the origin as soon as possible); the weight on the control move is to restrict the magnitude or the variation of the control move. Hence, although the control moves are not directly incorporated into the optimization, the separation strategy is often more practical then the overall strategy.

We are especially talking about the two-step model predictive control (TSMPC). For the Hammerstein model with input saturation, first utilize the linear submodel and unconstrained MPC to compute the desired intermediate variable, then solve the nonlinear algebraic equation (group) (represented by Hammerstein nonlinearity) to obtain the control action, and utilize desaturation to satisfy the input saturation constraint. Since the computational time can be greatly reduced, TSMPC is very suitable for the fast control requirement, especially for the actual system where the model is on-line identified. The reserved nonlinear item in the closed-loop of TSMPC is static.

Model Predictive Control, First Edition. Baocang Ding and Yuanqing Yang.

In TSMPC, if the intermediate variable is exactly implemented by the actual control moves passing the nonlinear part of the system, then stability of the whole system is guaranteed by stability of the linear subsystem. However, in a real application, this ideal case is hard to ensure. The control move may saturate, and the solution of the nonlinear algebraic equation (group) unavoidably introduces solution error.

Sections 8.1–8.3 are referred to Ding et al. [2003b] and Ding and Xi [2006]; Section 8.4 is referred to Ding et al. [2004]; Section 8.5 is referred to Ding and Xi [2004b]; Section 8.6 is referred to Ding et al. [2003a]; Section 8.7 is referred to Ding and Xi [2004a].

8.1 Two-Step State Feedback MPC

Consider the following discrete-time system:

$$x(k+1) = Ax(k) + Bv(k),\ y(k) = Cx(k),\ v(k) = \phi(u(k)), \tag{8.1}$$

where $x \in \mathbb{R}^n$, $v \in \mathbb{R}^m$, $y \in \mathbb{R}^p$, $u \in \mathbb{R}^m$ are the state, the intermediate variable, the output and the input, respectively; ϕ represents the relationship between the input and the intermediate variable with $\phi(0) = 0$. Moreover, the following assumptions are given.

Assumption 8.1 The state x is measurable.

Assumption 8.2 The pair (A, B) is stabilizable.

Assumption 8.3 $\phi = f \circ \text{sat}$, where f is the invertible static nonlinearity and sat represents the following input saturation (physical) constraint:

$$-\underline{u} \leq u(k) \leq \bar{u}, \tag{8.2}$$

where $\underline{u} := [\underline{u}_1, \underline{u}_2, \ldots, \underline{u}_m]^T$, $\bar{u} := [\bar{u}_1, \bar{u}_2, \ldots, \bar{u}_m]^T$, $\underline{u}_i > 0, \bar{u}_i > 0, i \in \{1, 2, \ldots, m\}$.

In TSMPC, $v(k)$ usually has some form of uncertainty. The closed-loop system is

$$x(k+1) = Ax(k) + Bv(k) = Ax(k) + Bv^L(k) + B[v(k) - v^L(k)], \tag{8.3}$$

where the desired intermediate variable $v^L(k) = v^L(k|k)$ is the solution to the optimization

$$\begin{aligned} &\min_{\tilde{v}^L(k|k)} J(N, x(k)), \\ &\text{s.t. } x^L(k+i+1|k) = Ax^L(k+i|k) + Bv^L(k+i|k),\ i \geq 0,\ x^L(k|k) = x(k). \end{aligned} \tag{8.4}$$

In (8.4),

$$J(N, x(k)) = \sum_{i=0}^{N-1} \left[\|x^L(k+i|k)\|_Q^2 + \|v^L(k+i|k)\|_R^2 \right] + \|x^L(k+N|k)\|_{P_N}^2, \tag{8.5}$$

$$\tilde{v}^L(k|k) = \left[v^L(k|k)^T, v^L(k+1|k)^T, \ldots, v^L(k+N-1|k)^T\right]^T, \tag{8.6}$$

where $Q \geq 0$, $R > 0$ are the symmetric matrices; $P_N > 0$ is called the terminal state weight matrix. Equation (8.4) is a finite-horizon standard LQ problem, to which we can apply the following Riccati iteration:

$$P_j = Q + A^T P_{j+1} A - A^T P_{j+1} B \left(R + B^T P_{j+1} B\right)^{-1} B^T P_{j+1} A, \; 0 \leq j < N. \tag{8.7}$$

It obtains

$$v^L(k|k) = Kx(k) = -(R + B^T P_1 B)^{-1} B^T P_1 A x(k). \tag{8.8}$$

Then $\hat{u}(k)$ is obtained by solving the algebraic equation $v^L(k) - f(\hat{u}(k)) = 0$, which is denoted as $\hat{u}(k) = \hat{f}^{-1}\left(v^L(k)\right)$. The control input $u(k)$ can be obtained by desaturating $\hat{u}(k)$ with $u(k) = \text{sat}\{\hat{u}(k)\}$ such that (8.2) is satisfied, which is denoted as $u(k) = g\left(v^L(k)\right)$.

Thus, $v(k) = \phi(\text{sat}\{\hat{u}(k)\}) = (\phi \circ \text{sat} \circ \hat{f}^{-1})(v^L(k)) = (f \circ \text{sat} \circ g)(v^L(k))$. Denote $h = f \circ \text{sat} \circ g$. The control law in terms of $v(k)$ is

$$v(k) = h(v^L(k)), \tag{8.9}$$

and the closed-loop representation of the system is

$$x(k+1) = Ax(k) + Bv(k) = (A + BK)x(k) + B[h\left(v^L(k)\right) - v^L(k)]. \tag{8.10}$$

If the reserved nonlinear item $h = \tilde{1} = [1, 1, \ldots, 1]^T$, in (8.10) $[h\left(v^L(k)\right) - v^L(k)]$ will disappear and the system will become linear. However, this generally cannot be guaranteed, since h may include

- the solution error of nonlinear equation;
- the desaturation that makes $v^L(k) \neq v(k)$.

In real applications, it is usual that $h \neq \tilde{1}, v^L(k) \neq v(k)$. So, we make the following assumptions on h.

Assumption 8.4 The nonlinearity h satisfies

$$\|h(s)\| \geq b_1 \|s\|, \; \|h(s) - s\| \leq |b - 1| \, \|s\|, \; \forall \|s\| \leq \Delta, \tag{8.11}$$

where b and b_1 are scalars.

Assumption 8.5 For the decentralized f (i.e., one element of u has a relation with one and only one of the elements in v, and the relationship is in order), h satisfies

$$b_{i,1} s_i^2 \leq h_i(s_i) s_i \leq b_{i,2} s_i^2, \; i \in \{1, \ldots, m\}, \; \forall |s_i| \leq \Delta, \tag{8.12}$$

where $b_{i,2}$ and $b_{i,1}$ are positive scalars.

Since $h_i(s_i)$ and s_i has the same sign, it yields $|h_i(s_i) - s_i| = \left||h_i(s_i)| - |s_i|\right| \leq \max\{|b_{i,1} - 1|, |b_{i,2} - 1|\} \, |s_i|$. Denote

$$\begin{aligned} b_1 &= \min\left\{b_{1,1}, b_{2,1}, \ldots, b_{m,1}\right\}, \\ |b - 1| &= \max\{|b_{1,1} - 1|, \ldots, |b_{m,1} - 1|, |b_{1,2} - 1|, \ldots, |b_{m,2} - 1|\}. \end{aligned} \tag{8.13}$$

Then (8.11) can be deduced from (8.12). The lower the degree of desaturation, the smaller will be $b_{i,1}$ and b_1. Therefore, with b_1 given, $\|h(s)\| \geq b_1 \|s\|$ in (8.11) will mainly represent a restriction on the degree of desaturation.

8.2 Stability of Two-Step State Feedback MPC

Definition 8.1 A region Ω^{N} is the null controllable region (see [Hu et al., 1998]) of the system (8.1), if

(i) for all $x(0) \in \Omega^{\mathrm{N}}$, there exists an admissible control sequence ($\{u(0), u(1), \dots\}$, $-\underline{u} \le u(i) \le \bar{u}$ for all $i \ge 0$) such that $\lim_{k\to\infty} x(k) = 0$;
(ii) for all $x(0) \notin \Omega^{\mathrm{N}}$, there does not exist an admissible control sequence such that $\lim_{k\to\infty} x(k) = 0$.

According to Definition 8.1, for any setting of $\{\lambda, P_N, N, Q\}$ and any equation solution error, the region of attraction (RoA) of system (8.10) (denoted as Ω) satisfies $\Omega \subseteq \Omega^{\mathrm{N}}$.

In the following, for simplicity, we take $R = \lambda I$.

Theorem 8.1 ***(Exponential stability of TSMPC)*** Consider system (8.1) with TSMPC (8.8) and (8.9). Suppose

(i) the choices $\{\lambda, P_N, N, Q\}$ make $Q - P_0 + P_1 > 0$;
(ii) for all $x(0) \in \Omega \subset \mathbb{R}^n$ and all $k \ge 0$,

$$-\lambda h(v^L(k))^T h(v^L(k)) + [h(v^L(k)) - v^L(k)]^T(\lambda I + B^T P_1 B)[h(v^L(k)) - v^L(k)] \le 0. \quad (8.14)$$

Then the equilibrium $x = 0$ of the closed-loop system (8.10) is locally exponentially stable with RoA Ω.

Proof: Define a quadratic function $V(k) = x(k)^T P_1 x(k)$. For $x(0) \in \Omega$, applying (8.7), (8.8), and (8.10), we have (omitting time (k))

$$\begin{aligned}
&V(k+1) - V(k)\\
&\quad = x^T(A+BK)^T P_1 (A+BK)x - x^T P_1 x\\
&\qquad - 2\lambda x^T K^T \left[h\left(v^L\right) - v^L\right] + \left[h\left(v^L\right) - v^L\right]^T B^T P_1 B \left[h\left(v^L\right) - v^L\right]\\
&\quad = x^T(-Q + P_0 - P_1 - \lambda K^T K)x\\
&\qquad - 2\lambda x^T K^T \left[h\left(v^L\right) - v^L\right] + \left[h\left(v^L\right) - v^L\right]^T B^T P_1 B \left[h\left(v^L\right) - v^L\right]\\
&\quad = x^T\left(-Q + P_0 - P_1\right)x - \lambda \left(v^L\right)^T v^L\\
&\qquad - 2\lambda \left(v^L\right)^T \left[h\left(v^L\right) - v^L\right] + \left[h\left(v^L\right) - v^L\right]^T B^T P_1 B \left[h\left(v^L(k)\right) - v^L\right]\\
&\quad = x^T\left(-Q + P_0 - P_1\right)x - \lambda h\left(v^L\right)^T h\left(v^L\right)\\
&\qquad + \left[h\left(v^L\right) - v^L\right]^T \left(\lambda I + B^T P_1 B\right)\left[h\left(v^L\right) - v^L\right].
\end{aligned}$$

Note that in the above, we have utilized the following facts:

$$(A+BK)^T P_1 B = A^T P_1 B\left[I - \left(\lambda I + B^T P_1 B\right)^{-1} B^T P_1 B\right] = -\lambda K^T.$$

Under conditions (i)–(ii), it is clear that $V(k+1) - V(k) \leq -\sigma_{\min}(Q - P_0 + P_1)x(k)^T x(k) < 0$, $\forall x(k) \neq 0$, where $\sigma_{\min}(\cdot)$ denotes the minimum eigenvalue. Therefore, $V(k)$ is the Lyapunov function for the exponential stability.

The conditions in Theorem 8.1 just reflect the essential idea of two-step design. The condition (i) is a requirement on the linear control law (8.8), while the condition (ii) is an extra requirement on h. From the proof of Theorem 8.1, it is easy to know that (i) is a sufficient stability condition for unconstrained linear system because, with $h = \tilde{1}$, (8.14) becomes $-\lambda v^L(k)^T v^L(k) \leq 0$, i.e., the condition (ii) is always satisfied.

Since (8.14) is not easy to check, the following two corollaries are given.

Corollary 8.1 ***(Exponential stability of TSMPC)*** Consider system (8.1) with TSMPC (8.8) and (8.9). Suppose

(i) $Q - P_0 + P_1 > 0$;
(ii) whenever $x(0) \in \Omega \subset \mathbb{R}^n$, $\|v^L(k)\| \leq \Delta$ for all $\forall k \geq 0$;
(iii)

$$-\lambda\left[b_1^2 - (b-1)^2\right] + (b-1)^2 \sigma_{\max}\left(B^T P_1 B\right) \leq 0. \tag{8.15}$$

Then the equilibrium $x = 0$ of the closed-loop system (8.10) is locally exponentially stable with RoA Ω.

Corollary 8.2 ***(Exponential stability of TSMPC)*** Consider system (8.1) with TSMPC (8.8) and (8.9). Suppose

(i) $Q - P_0 + P_1 > 0$;
(ii) whenever $x(0) \in \Omega \subset \mathbb{R}^n$, $\left|v_i^L(k)\right| \leq \Delta$ for all $\forall k \geq 0$;
(iii) f is decentralized, and

$$-\lambda\left(2b_1 - 1\right) + (b-1)^2 \sigma_{\max}\left(B^T P_1 B\right) \leq 0 \tag{8.16}$$

Then the equilibrium $x = 0$ of the closed-loop system (8.10) is locally exponentially stable with RoA Ω.

Proposition 8.1 ***(Exponential stability of TSMPC)*** In Corollary 8.1, suppose the inequalities in conditions (i) and (iii) are substituted by

$$Q - P_0 + P_1 + \eta A^T P_1 B(\lambda I + B^T P_1 B)^{-2} B^T P_1 A > 0,$$

where $\eta = \lambda\left[b_1^2 - (b-1)^2\right] - (b-1)^2 \sigma_{\max}\left(B^T P_1 B\right)$ (Corollary 8.1) or $\eta = \lambda\left(2b_1 - 1\right) - (b-1)^2 \sigma_{\max}\left(B^T P_1 B\right)$ (Corollary 8.2). Then, the conclusion still holds.

Remark 8.1 The stability conclusion in Proposition 8.1 is less conservative than Corollaries 8.1 and 8.2, and is unnecessarily more conservative than Theorem 8.1. However, for the controller parameter tuning, applying Proposition 8.1 is not as straightforward as applying Corollaries 8.1 and 8.2.

Remark 8.2 If $f = \tilde{1}$, i.e., there is only input saturation, then $b_{i,2} = 1$, $(b-1)^2 = (b_1 - 1)^2$ and both (8.15) and (8.16) will become $-\lambda\left(2b_1 - 1\right) + \left(b_1 - 1\right)^2 \sigma_{\max}\left(B^T P_1 B\right) \le 0$.

Denote (8.15) and (8.16) as

$$-\lambda + \beta\sigma_{\max}\left(B^T P_1 B\right) \le 0, \tag{8.17}$$

where $\beta = (b-1)^2/[b_1^2 - (b-1)^2]$ for (8.15); $\beta = (b-1)^2/\left(2b_1 - 1\right)$ for (8.16).

As for RoA in Theorem 8.1, Corollaries 8.1 and 8.2, we give the following easily manipulable ellipsoidal one.

Corollary 8.3 ***(RoA of TSMPC)*** Consider system (8.1) with TSMPC (8.8) and (8.9). If

(i) $Q - P_0 + P_1 > 0$;
(ii) the choices $\left\{\Delta, b_1, b\right\}$ satisfy (8.15),

then RoA Ω for the equilibrium $x = 0$ of the closed-loop system (8.10) is not smaller than

$$S_c = \left\{x | x^T P_1 x \le c\right\},\ c = \frac{\Delta^2}{\left\|\left(\lambda I + B^T P_1 B\right)^{-1} B^T P_1 A P_1^{-1/2}\right\|^2}. \tag{8.18}$$

Proof: Transform the linear system (A, B, C) into $\left(\bar{A}, \overline{B}, \overline{C}\right)$ with nonsingular transformation $\overline{x} = P_1^{1/2} x$. Then, for all $x(0) \in S_c$, $\|\overline{x}(0)\| \le \sqrt{c}$ and

$$\begin{aligned}\|v^L(0)\| &= \left\|\left(\lambda I + B^T P_1 B\right)^{-1} B^T P_1 A x(0)\right\| = \left\|\left(\lambda I + B^T P_1 B\right)^{-1} B^T P_1 A P_1^{-1/2} \overline{x}(0)\right\| \\ &\le \left\|\left(\lambda I + B^T P_1 B\right)^{-1} B^T P_1 A P_1^{-1/2}\right\| \|\overline{x}(0)\| \le \Delta.\end{aligned}$$

Under (i) and (ii), all the conditions in Corollary 8.1 are satisfied at time $k = 0$ if $x(0) \in S_c$. Furthermore, according to the proof of Theorem 8.1, $x(1) \in S_c$ if $x(0) \in S_c$. Hence, for all $x(0) \in S_c$, $\|v^L(1)\| \le \Delta$. This shows that all the conditions in Corollary 8.1 are satisfied at time $k = 1$, and by analogy, they are also satisfied for any $k > 1$.

Apparently RoA for given the controller parameters may be too small if we have no desired RoA prior to the controller design. To design a controller with a desired RoA, the concept of "semi-global stabilization" could be technically involved.

If A has no eigenvalues outside of the unit circle, semi-global stabilization (see [Lin and Saberi, 1993, Lin et al., 1996]) means the design of a feedback control law that results in RoA that includes any prior given compact set in n dimensional space. If A has eigenvalues outside of the unit circle, semi-global stabilization (see [Hu et al., 2001]) means the design of a feedback control law that results in RoA that includes any a prior given compact subset of the null controllable region.

The Section 8.3 will derive the semi-global stabilization techniques for TSMPC. If A has no eigenvalues outside of the unit circle, TSMPC can be designed (tuning $\left\{\lambda, P_N, N, Q\right\}$) to have an arbitrarily large RoA. Otherwise, a set of $\left\{\lambda, P_N, N, Q\right\}$ can be chosen to obtain a set of RoAs with their union contained in the null controllable region.

8.3 Region of Attraction for Two-Step MPC: Semi-Global Stability

8.3.1 System Matrix Having No Eigenvalue Outside of Unit Circle

Theorem 8.2 ***(Semi-global stability of TSMPC)*** Consider system (8.1) with TSMPC (8.8) and (8.9). Suppose

(i) A has no eigenvalues outside of the unit circle;
(ii) $b_1 > |b-1| > 0$, i.e., $\beta > 0$.

Then, for any bounded set $\Omega \subset \mathbb{R}^n$, there exist $\{\lambda, P_N, N, Q\}$ such that the equilibrium $x = 0$ for the closed-loop system (8.10) is locally exponentially stable with RoA Ω.

Proof: We show how $\{\lambda, P_N, N, Q\}$ can be chosen to satisfy condition (i)-(iii) in Corollary 8.1. First of all, choose $Q > 0$ and an arbitrary N. In the following, we elaborate on how to choose λ and P_N.

- Step 1: Choose P_N to satisfy

$$P_N = Q_1 + Q + A^T P_N A - A^T P_N B \ \left(\lambda I + B^T P_N B\right)^{-1} B^T \ P_N A, \tag{8.19}$$

where $Q_1 \geq 0$ is an arbitrary symmetric matrix. Equation (8.19) means that Riccati iteration (8.7) satisfies $P_{N-1} \leq P_N$ and has monotonic decreasing property, so $P_0 \leq P_1$, $Q - P_0 + P_1 > 0$. Therefore, condition (i) in Corollary 8.1 can be satisfied for any λ. Changing λ, P_N is also changed corresponding to (8.19).
- Step 2: The fake algebraic Riccati equation (see [Poubelle et al., 1988]) of (8.7) is

$$P_{j+1} = \left(Q + P_{j+1} - P_j\right) + A^T P_{j+1} A - A^T P_{j+1} B\left(\lambda I + B^T P_{j+1} B\right)^{-1} B^T \ P_{j+1} A,$$
$$0 \leq j < N, \ P_N - P_{N-1} = Q_1. \tag{8.20}$$

Multiply both sides of (8.20) by λ^{-1}, then

$$\overline{P}_{j+1} = (\lambda^{-1} Q \ + \ \overline{P}_{j+1} - \overline{P}_j) + A^T \ \overline{P}_{j+1} A - A^T \overline{P}_{j+1} B\left(I + B^T \ \overline{P}_{j+1} B\right)^{-1} B^T \overline{P}_{j+1} A,$$
$$0 \leq j < N, \ \overline{P}_N - \overline{P}_{N-1} = \lambda^{-1} Q_1, \tag{8.21}$$

where $\overline{P}_{j+1} = \lambda^{-1} P_{j+1}, 0 \leq j < N$. As $\lambda \to \infty, \overline{P}_{j+1} \to 0, 0 \leq j < N$. Since $\beta > 0$, there exists a suitable λ_0^* such that whenever $\lambda \geq \lambda_0^*$, $\beta \sigma_{\max}\left(B^T \overline{P}_1 B\right) \leq 1$, i.e., condition (iii) in Corollary 8.1 can be satisfied.
- Step 3: Further, choose an arbitrary constant $\alpha > 1$, then there exists $\lambda_1^* \geq \lambda_0^*$ such that whenever $\lambda \geq \lambda_1^*$,

$$\overline{P}_1^{1/2} B\left(I + B^T \overline{P}_1 B\right)^{-1} B^T \overline{P}_1^{1/2} \leq (1 - 1/\alpha) I. \tag{8.22}$$

For $j = 0$, LHS and RHS multiplying both sides of (8.21) by $\overline{P}_1^{-1/2}$ and applying (8.22) obtains

$$\overline{P}_1^{-1/2} A^T \overline{P}_1 A \overline{P}_1^{-1/2} \leq \alpha I - \alpha \overline{P}_1^{-1/2} \left(\lambda^{-1} Q + \overline{P}_1 - \overline{P}_0\right) \overline{P}_1^{-1/2} \leq \alpha I,$$

i.e., $\left\| \overline{P}_1^{1/2} A \overline{P}_1^{-1/2} \right\| \leq \sqrt{\alpha}$.

For any bounded Ω, choose $\bar{c}$ such that

$$\bar{c} \geq \sup_{x\in\Omega,\ \lambda\in[\lambda_1^*,\infty)} x^T\lambda^{-1}P_1x.$$

Hence, $\Omega \subseteq \bar{S}_{\bar{c}} = \left\{x|x^T\lambda^{-1}P_1x \leq \bar{c}\right\}$.

Denote $(\bar{A}, \overline{B}, \overline{C})$ as the transformed system of (A, B, C) by nonsingular transformation $\overline{x} = \overline{P}_1^{1/2}x$, then there exists a sufficiently large $\lambda^* \geq \lambda_1^*$ such that for all $\lambda \geq \lambda^*$ and for all $x(0) \in \overline{S}_{\bar{c}}$,

$$\begin{aligned}\left\|\left(\lambda I + B^TP_1B\right)^{-1}B^TP_1Ax(0)\right\| &= \left\|\left(I + \overline{B}^T\overline{B}\right)^{-1}\overline{B}^T\bar{A}\overline{x}(0)\right\| \\ &\leq \left\|\left(I + \overline{B}^T\overline{B}\right)^{-1}\overline{B}^T\right\|\sqrt{\alpha\bar{c}} \leq \Delta\end{aligned}$$

because $\left\|\left(I + \overline{B}^T\overline{B}\right)^{-1}\overline{B}^T\right\|$ tends to be smaller when λ is increased.

Hence, for all $x(0) \in \Omega$, condition (ii) in Corollary 8.1 can be satisfied at time $k = 0$, and according to the proof of Corollary 8.3, it can also be satisfied for all $k > 0$.

In a word, if we choose $Q > 0$ by (8.19), $Q > 0$, arbitrary N and $\lambda^* \leq \lambda < \infty$, then the closed-loop system is locally exponentially stable with RoA Ω.

Corollary 8.4 ***(Semi-global stability of TSMPC)*** Suppose

(i) A has no eigenvalues outside of the unit circle;
(ii) the nonlinear equation is solved sufficiently accurately such that, in the absence of input saturation constraint, there exist suitable $\left\{\Delta = \Delta^0, b_1, b\right\}$ satisfying $b_1 > |b-1| > 0$.

Then, the conclusion in Theorem 8.2 still holds.

Proof: In the absence of input saturation constraint, determining $\left\{b_1, b\right\}$ for given Δ (or given $\left\{b_1, b\right\}$, determining Δ) is independent of the controller parameter $\left\{\lambda, P_N, N, Q\right\}$. When there is input saturation, still choose $\Delta = \Delta^0$. Then, the following two cases may happen:

Case 1: $b_1 > |b-1| > 0$ when $\lambda = \lambda_0$. Decide the parameters as in the proof of Theorem 8.2, except that $\lambda_0^* \geq \lambda_0$.

Case 2: $|b-1| \geq b_1 > 0$ when $\lambda = \lambda_0$. Apparently, the reason lies in that the control action is over-clipped by the saturation constraint. By the same reason as in the proof of Theorem 8.2 and by (8.8), we know that, for any bounded Ω, there exists $\lambda_2^* \geq \lambda_0$ such that for all $\lambda \geq \lambda_2^*$ and all $x(k) \in \Omega$, $\hat{u}(k)$ does not violate the saturation constraint. This procedure is equivalent to decreasing Δ and redetermining $\left\{b_1, b\right\}$ such that $b_1 > |b-1| > 0$.

In a word, if RoA has not been satisfactory with $\lambda = \lambda_0$, then it can be satisfied by choosing $\max\left\{\lambda^*, \lambda_2^*\right\} \leq \lambda < \infty$ and suitable $\left\{P_N, N, Q\right\}$.

Although a method is implicitly presented in the proofs of Theorem 8.2 and Corollary 8.4 for tuning the controller parameters, a large λ tends to be achieved. Then, when the desired

RoA Ω is large, the obtained controller will be very conservative. Actually, we do not have to choose λ as in Theorem 8.2 and Corollary 8.4. Moreover, a series of λ can be chosen and the following controller Algorithm 8.1 can be applied.

Algorithm 8.1 (The λ-switching algorithm of TSMPC)

- Off-line, complete the following steps 1–3.
  - Step 1. Choose the adequate b_1, b and obtain the largest possible Δ, or, choose the adequate Δ and obtain the smallest possible $|b-1|$ and largest possible b_1. Calculate $\beta > 0$.
  - Step 2. Choose Q, N, P_N as in the proof of Theorem 8.2.
  - Step 3. Gradually increase λ, until (8.17) is satisfied at $\lambda = \underline{\lambda}$. Increase λ to obtain $\lambda^M > \cdots > \lambda^2 > \lambda^1 \geq \underline{\lambda}$. The parameter λ^i corresponds to the controller Con_i and RoA S^i (S^i calculated by Corollary 8.3, $i \in \{1, 2, \ldots, M\}$). The inclusion condition $S^1 \subset S^2 \subset \cdots \subset S^M$ holds. S^M should contain the desired RoA Ω.
- On-line, at each time k,
  - A) if $x(k) \in S^1$, then Con_1;
  - B) if $x(k) \in S^i$ and $x(k) \notin S^{i-1}$, then Con_i, $i \in \{2, 3, \ldots, M\}$.

8.3.2 System Matrix Having Eigenvalues Outside of Unit Circle

In this case, semi-global stabilization cannot be implemented in a simple manner as in the above case. However, a set of ellipsoidal RoAs $i \in \{1, 2, \ldots, M\}$ can be achieved via a set of controllers with respect to different parameter sets $\left\{\lambda, P_N, N, Q\right\}^i$ and RoA S^i. In the following we give the algorithm.

Three cases may happen in Algorithm 8.2.

- Case 1: A single S^i is found to satisfy $S^i \supseteq \Omega$;
- Case 2: A set of S^i ($i \in \{1, 2, \ldots, M\}$ and $M > 1$) are found satisfying $\bigcup_{i=1}^{M} S^i \supseteq \Omega$;
- Case 3: S^i satisfying $\bigcup_{i=1}^{M} S^i \supseteq \Omega$ cannot be found with a finite number M (in the real applications, M is prescribed to be not larger than an M_0).

For Case 2, the controller switching Algorithm 8.3 can be applied.

For Case 3, we can take one of the following strategies.

(i) Decrease the equation solution error and redetermine $\left\{\Delta, b_1, b\right\}$.

(ii) When the state lies outside of $\bigcup_{i=1}^{M_0} S^i$, adopt the nonlinear separation method (optimizing $\tilde{v}(k|k)$ considering the constraint on the intermediate variable). First, transform the saturation constraint on u into the constraint on v. For a complex nonlinearity, obtaining the constraint on v could be very difficult, and it is even possible that nonlinear constraint is encountered. If f is decentralized, as in Assumption 8.5, then it is easy to obtain the constraint on the intermediate variable.

(iii) When the state lies outside of $\bigcup_{i=1}^{M_0} S^i$, substitute $v(k+i|k)$ in (8.5) by $u(k+i|k)$ and apply the pure nonlinear MPC to calculate the control action (i.e., adopt MPC based on the nonlinear prediction model and nonlinear optimization).

Algorithm 8.2 (The method for parameter search in TSMPC)

Step 1. Refer to Step 1 of Algorithm 8.1. Set $S = \{0\}, i = 1$.

Step 2. Select $\{P_N, N, Q\}$ (by changing them alternatively).

Step 3. Determine $\{S_c, \lambda, P_N, N, Q\}$ via the following steps.

Step 3.1. Check if (8.17) is satisfied. If it is not, tune λ to satisfy (8.17).

Step 3.2. Check if $Q - P_0 + P_1 > 0$ is satisfied. If it is, go to Step 3.3; otherwise, tune $\{P_N, N, Q\}$ to satisfy it and go to Step 3.1.

Step 3.3. Determine P_1 and c by $\left\|(\lambda I + B^T P_1 B)^{-1} B^T P_1 A P_1^{-1/2}\right\| \sqrt{c} = \Delta$. Then, RoA for the real system will include the level set $S_c = \{x | x^T P_1 x \le c\}$.

Step 4. Set $\{\lambda, P_N, N, Q\}^i = \{\lambda, P_N, N, Q\}$, $S^i = S_c$ and $S = S \bigcup S^i$.

Step 5. Check whether or not S contains the desired RoA Ω. If it is, go to Step 6; otherwise, set $i = i + 1$ and go to Step 2.

Step 6. Set $M = i$ and STOP.

Algorithm 8.3 (Switching algorithm of TSMPC)

Off-line, apply Algorithm 8.2 to choose a set of ellipsoidal RoAs $S^1, S^2, \ldots, S^M$ satisfying $\bigcup_{i=1}^{M} S^i \supseteq \Omega$. Arrange S^i in a proper way that results in$S^{(1)}, S^{(2)}, \ldots, S^{(M)}$ with corresponding controllers $\mathrm{Con}_{(i)}, i \in \{1, 2, \ldots, M\}$. It is unnecessary that$j \in \{1, 2, \ldots, M-1\}$ for any $S^{(j)} \subseteq S^{(j+1)}$.

On-line, at each time k, if $x(k) \in S^{(i)}$ and $x(k) \notin S^{(l)}$ for all $l < i$, then choose $\mathrm{Con}_{(i)}$, $i \in \{2, 3, \ldots, M\}$.

8.3.3 Numerical Example

First consider the case that A has no eigenvalue outside of the unit circle. The linear subsystem is $A = \begin{bmatrix} 1 & 0 \\ 1 & 1 \end{bmatrix}$, $B = \begin{bmatrix} 1 \\ 0 \end{bmatrix}$. The invertible static nonlinearity is $f(\vartheta) = 4/3\vartheta + 4/9\vartheta \mathrm{sign}\{\vartheta\} \sin(40\vartheta)$, and the input constraint is $|u| \le 1$. A simple solution $\hat{u} = 3/4v^L$ is applied to the algebraic equation. The resultant h is shown in Figure 8.1. Choose $b_1 = 2/3, b_2 = 4/3$ as in Figure 8.1, then $\beta = 1/3$. Choose $\Delta = f(1)/b_1 = 2.4968$.

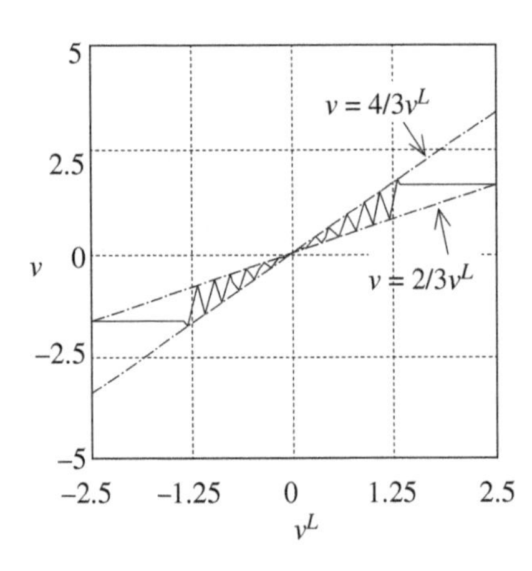

Figure 8.1 Curve h.

The initial state is $x(0) = [10, -33]^T$. Choose $N = 4$, $Q = 0.1I$, $P_N = 0.11I + A^T P_N A - A^T P_N B(\lambda + B^T P_N B)^{-1} B^T P_N A$.

Choose $\lambda = 0.225, 0.75, 2, 10, 50$, then RoAs determined by Corollary 8.3 are

$$S_c^1 = \left\{ x | x^T \begin{bmatrix} 0.6419 & 0.2967 \\ 0.2967 & 0.3187 \end{bmatrix} x \leq 1.1456 \right\},$$

$$S_c^2 = \left\{ x | x^T \begin{bmatrix} 1.1826 & 0.4461 \\ 0.4461 & 0.3760 \end{bmatrix} x \leq 3.5625 \right\},$$

$$S_c^3 = \left\{ x | x^T \begin{bmatrix} 2.1079 & 0.6547 \\ 0.6547 & 0.4319 \end{bmatrix} x \leq 9.9877 \right\},$$

$$S_c^4 = \left\{ x | x^T \begin{bmatrix} 5.9794 & 1.3043 \\ 1.3043 & 0.5806 \end{bmatrix} x \leq 62.817 \right\},$$

$$S_c^5 = \left\{ x | x^T \begin{bmatrix} 18.145 & 2.7133 \\ 2.7133 & 0.8117 \end{bmatrix} x \leq 429.51 \right\}.$$

Figure 8.2 depicts S_c^1, S_c^2, S_c^3, S_c^4, and S_c^5 from inside to outside. $x(0)$ lies in S_c^5.

The rule of Algorithm 8.1 in the simulation is as follows:

- if $x(k) \in S_c^1$, then $\lambda = 0.225$;
- else if $x(k) \in S_c^2$, then $\lambda = 0.75$;
- else if $x(k) \in S_c^3$, then $\lambda = 2$;
- else if $x(k) \in S_c^4$, then $\lambda = 10$;
- else $\lambda = 50$.

The simulation result is shown in Figure 8.2. The line with "o" is the state trajectory with Algorithm 8.1, while that with "*," with $\lambda = 50$. With Algorithm 8.1, the trajectory is

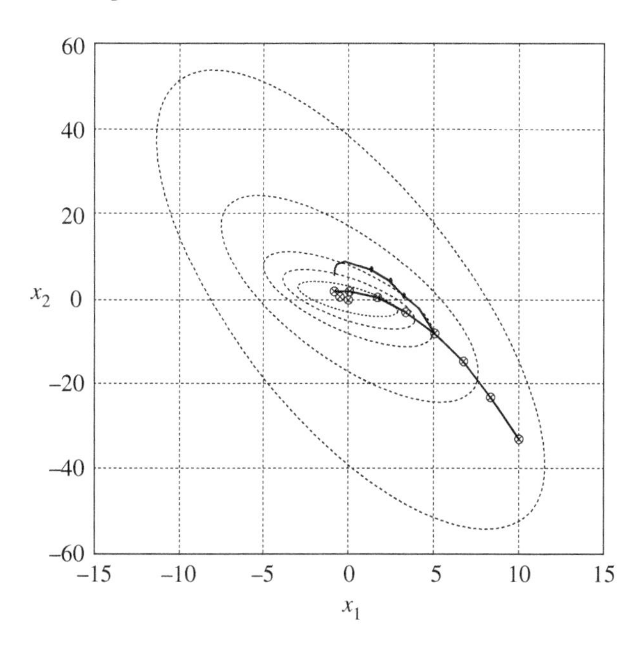

Figure 8.2 The closed-loop state trajectories when A has no eigenvalue outside of the unit circle.

very close to the origin after 15 simulation samples, but when $\lambda = 50$ is always adopted, the trajectory has just reached to the boundary of S_c^2 after 15 simulation samples.

Further, consider the case that A has eigenvalues outside of the unit circle. The linear subsystem is $A = \begin{bmatrix} 1.2 & 0 \\ 1 & 1.2 \end{bmatrix}$ and $B = \begin{bmatrix} 1 \\ 0 \end{bmatrix}$. The nonlinearities, the solution of equation and the corresponding $\{b_1, b_2, \Delta\}$ are the same as above. We obtain three ellipsoidal regions of attraction $S^1,\ S^2,\ S^3$, whose parameter sets are

$$\{\lambda, P_N, Q, N\}^1 = \left\{8.0,\ \begin{bmatrix} 1 & 0 \\ 0 & 1 \end{bmatrix},\ \begin{bmatrix} 0.01 & 0 \\ 0 & 1.01 \end{bmatrix},\ 12\right\},$$

$$\{\lambda, P_N, Q, N\}^2 = \left\{2.5,\ \begin{bmatrix} 1 & 0 \\ 0 & 1 \end{bmatrix},\ \begin{bmatrix} 0.9 & 0 \\ 0 & 0.1 \end{bmatrix},\ 4\right\},$$

$$\{\lambda, P_N, Q, N\}^3 = \left\{1.3,\ \begin{bmatrix} 3.8011 & 1.2256 \\ 1.2256 & 0.9410 \end{bmatrix},\ \begin{bmatrix} 1.01 & 0 \\ 0 & 0.01 \end{bmatrix},\ 4\right\}.$$

Arrange S^1, S^2, S^3 and the parameter sets in the following:

$$S^{(1)} = S^3,\ S^{(2)} = S^2,\ S^{(3)} = S^1,$$

$$\{\lambda, P_N, Q, N\}^{(1)} = \{\lambda, P_N, Q, N\}^3,\ \{\lambda, P_N, Q, N\}^{(2)} = \{\lambda, P_N, Q, N\}^2,$$

$$\{\lambda, P_N, Q, N\}^{(3)} = \{\lambda, P_N, Q, N\}^1.$$

Choose two sets of initial state as $x(0) = [-3.18, 4]^T$ and $x(0) = [3.18, -4]^T$, satisfying $x(0) \in S^{(3)}$. With Algorithm 8.3 applied, the resultant state trajectories are shown in Figure 8.3.

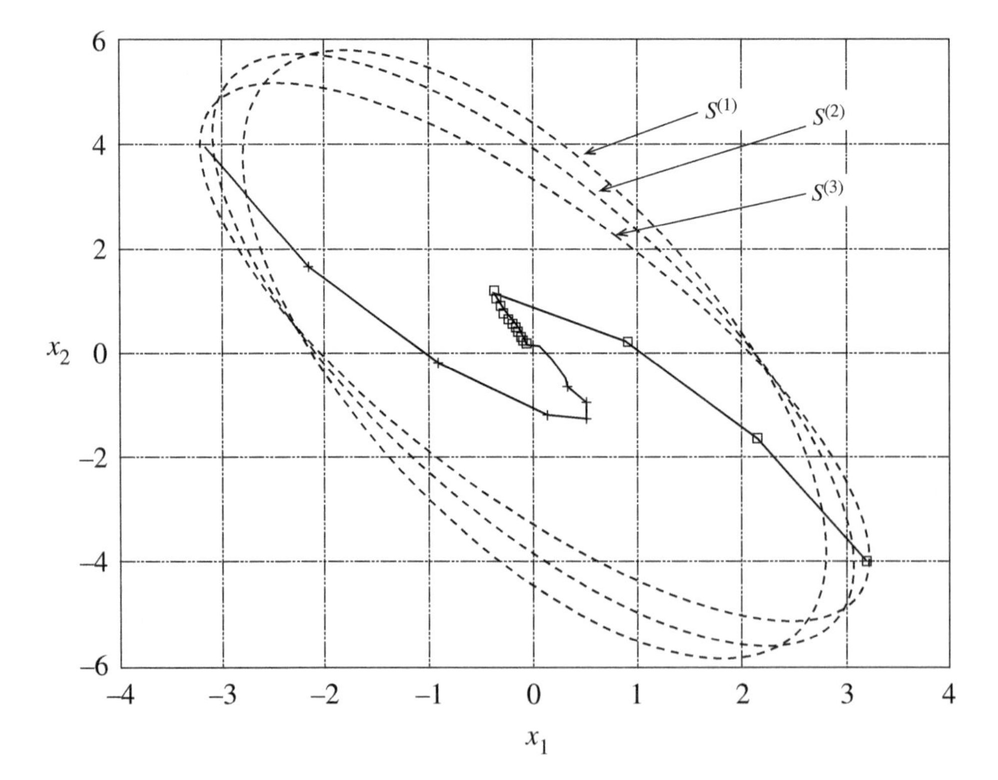

Figure 8.3 The closed-loop state trajectories when A has an eigenvalue outside of the unit circle.

8.4 Two-Step Output Feedback Model Predictive Control

Consider the system model (8.1), with notations the same as above. In two-step output feedback model predictive control (TSOFMPC), suppose (A, B, C) is completely controllable and observable. Moreover, suppose f is not an exact model of the nonlinearity, the true nonlinearity is f_0 and it is possible that $f \neq f_0$. The first step in TSOFMPC only considers the linear subsystem. The state estimation is $\hat{x}$ and the prediction model is

$$\hat{x}(k+i+1|k) = A\hat{x}(k+i|k) + Bv^L(k+i|k),\ i \geq 0. \tag{8.23}$$

Define the objective function as

$$J(N, \hat{x}(k|k)) = \|\hat{x}(k+N|k)\|_{P_N}^2 + \sum_{j=0}^{N-1} \left[\|\hat{x}(k+j|k)\|_Q^2 + \left\|v^L(k+j|k)\right\|_R^2\right], \tag{8.24}$$

where Q, R, and P_N are the same as in TSMPC. Riccati iteration (8.7) is adopted to obtain the predictive control law

$$v^L(k) \triangleq K\hat{x}(k) = -\left(R + B^T P_1 B\right)^{-1} B^T P_1 A\hat{x}(k). \tag{8.25}$$

In addition, (8.23) is consistent with the following equation:

$$\hat{x}(k+1|k) = (A - LC)\hat{x}(k|k-1) + Bv^L(k) + Ly(k), \tag{8.26}$$

where L is the observer gain matrix, being defined as

$$L = A\Sigma_1 C^T \left(R_o + C\Sigma_1 C^T\right)^{-1}, \tag{8.27}$$

where Σ_1 can be iterated from

$$\Sigma_j = Q_o + A\Sigma_{j+1}A^T - A\Sigma_{j+1}C^T\left(R_o + C\Sigma_{j+1}C^T\right)^{-1}C\Sigma_{j+1}A^T,\ j < N_o, \tag{8.28}$$

where R_o, Q_o, Σ_{N_o}, and N_o are taken as tunable parameters. If $\{R_o, Q_o, \Sigma_{N_o}, N_o\}$ are fixed, introducing Σ_{N_o} and N_o make little sense because we can let $\Sigma_{j+1} = \Sigma_j = \Sigma$ in (8.28) directly (amounting to $N_o = \infty$).

The second step in TSOFMPC is the same as in TSMPC. Hence, the actual intermediate variable is

$$v(k) = h(v^L(k)) = f_0\left(\text{sat}\left\{u(k)\right\}\right) = f_0 \circ \text{sat} \circ g(v^L(k)). \tag{8.29}$$

Equation (8.29) is the control law of TSOFMPC in terms of the intermediate variable.

By (8.1) and (8.26), denoting $e(k) = x(k) - \hat{x}(k|k-1)$, we can obtain the following closed-loop system:

$$\begin{cases} x(k+1) = (A+BK)x(k) - BKe(k) + B\left[h(v^L(k)) - v^L(k)\right] \\ e(k+1) = (A-LC)e(k) + B\left[h(v^L(k)) - v^L(k)\right] \end{cases} \tag{8.30}$$

When $h = \tilde{1}$, the nonlinear item in (8.30) will disappear, and the studied problem will become linear. However, because of, e.g., the desaturation, the error encountered in solving equation, the modeling error of nonlinearity, generally $h = \tilde{1}$ cannot hold.

Lemma 8.1 Suppose X and Y are matrices, while s and t are vectors, all with the appropriate dimensions, then

$$2s^T XYt \leq \gamma s^T XX^T s + 1/\gamma t^T Y^T Yt,\ \forall \gamma > 0. \tag{8.31}$$

In the following, we take $R = \lambda I$.

Theorem 8.3 ***(Stability of TSOFMPC)*** For system represented by (8.1), TSOFMPC (8.25)–(8.26) is adopted. Suppose there exist positive scalars γ_1 and γ_2 such that the system design satisfies

(i) $Q > P_0 - P_1$;
(ii) $(1+1/\gamma_2)\left(-Q_o + \Sigma_0 - \Sigma_1 - LR_oL^T\right) + 1/\gamma_2\hat{\Sigma}$
$< -\lambda^{-1}\left(1+1/\gamma_1\right)A^TP_1B\left(\lambda I + B^TP_1B\right)^{-1} B^T\ P_1A$;
(iii) $(A-LC)^T\hat{\Sigma}(A-LC) - \hat{\Sigma} = (A-LC)\Sigma_1(A-LC)^T - \Sigma_1$;
(iv) for all $\left[x(0)^T, e(0)^T\right] \in \Omega \subset \mathbb{R}^{2n}$ and all $k \geq 0$, $-\lambda h(v^L(k))^T h(v^L(k)) + \left[h(v^L(k)) - v^L(k)\right]^T$ $\left[\left(1+\gamma_1\right)\left(\lambda I + B^TP_1B\right) + \left(1+\gamma_2\right)\lambda B^T\hat{\Sigma}B\right]\left[h(v^L(k)) - v^L(k)\right] \leq 0$.

Then, the equilibrium $\{x=0, e=0\}$ of the closed-loop system is exponentially stable and RoA is Ω.

Proof: Choose a quadratic function as $V(k) = x(k)^TP_1x(k) + \lambda e(k)^T\hat{\Sigma}e(k)$. Applying (8.30), we derive the following (omitting the time (k)):

$$\begin{aligned}
&V(k+1) - V(k)\\
&\quad = \|\,(A+BK)x + B\left[h(v^L) - v^L\right]\|_{P_1}^2\\
&\qquad - 2e^TK^TB^TP_1\left\{(A+BK)x + B\left[h(v^L) - v^L\right]\right\}\\
&\qquad + e^TK^TB^TP_1BKe + \lambda\|\,(A-LC)e + B\left[h(v^L) - v^L\right]\|_{\hat{\Sigma}}^2 - \|x\|_{P_1}^2 - \lambda\|e\|_{\hat{\Sigma}}^2.
\end{aligned}$$

Noting that $P_0 = Q + A^TP_1A - A^TP_1B(\lambda I + B^TP_1B)^{-1}B^TP_1A = Q + (A+BK)^TP_1(A+BK) + \lambda K^TK$ and $(A+BK)^TP_1B = -\lambda K^T$, it yields

$$\begin{aligned}
&V(k+1) - V(k)\\
&\quad = \|x\|_{-Q+P_0}^2 - \lambda x^TK^TKx - 2\lambda x^TK^T\left[h(v^L) - v^L\right] + \|h(v^L) - v^L\|_{B^TP_1B}^2\\
&\qquad - 2e^TK^TB^TP_1\left\{(A+BK)x + B\left[h(v^L) - v^L\right]\right\}\\
&\qquad + e^TK^TB^TP_1BKe + \lambda\|\,(A-LC)e + B\left[h(v^L) - v^L\right]\|_{\hat{\Sigma}}^2 - \lambda\|e\|_{\hat{\Sigma}}^2 - \|x\|_{P_1}^2.
\end{aligned}$$

Define $v^x(k) \triangleq Kx(k)$ and $v^e(k) \triangleq Ke(k)$. Using $(A+BK)^TP_1B = -\lambda K^T$, it yields

$$\begin{aligned}
&V(k+1) - V(k)\\
&\quad = \|x\|_{-Q+P_0-P_1}^2 - \lambda\|v^x\|^2 - 2\lambda(v^x)^T\left[h(v^L) - v^L\right] + \|h(v^L) - v^L\|_{B^T(P_1+\lambda\hat{\Sigma})B}^2\\
&\qquad + 2\lambda(v^e)^Tv^x - 2(v^e)^TB^TP_1B\left[h(v^L) - v^L\right] + (v^e)^TB^TP_1Bv^e\\
&\qquad + \lambda e^T\left[(A-LC)^T\hat{\Sigma}(A-LC) - \hat{\Sigma}\right]e + 2\lambda e^T(A-LC)^T\hat{\Sigma}B\left[h(v^L) - v^L\right].
\end{aligned}$$

Using $v^x(k) = v^L(k) + v^e(k)$, it yields

$$\begin{aligned}
&V(k+1) - V(k)\\
&\quad = \|x\|_{-Q+P_0-P_1}^2 - \lambda\|h(v^L)\|^2 + \|h(v^L) - v^L\|_{\lambda I + B^TP_1B + \lambda B^T\hat{\Sigma}B}^2\\
&\qquad - 2(v^e)^T\left(\lambda I + B^TP_1B\right)\left[h(v^L) - v^L\right] + (v^e)^T\left(\lambda I + B^TP_1B\right)v^e\\
&\qquad + \lambda e^T\left[(A-LC)^T\hat{\Sigma}(A-LC) - \hat{\Sigma}\right]e + 2\lambda e^T(A-LC)^T\ \hat{\Sigma}B\left[h(v^L) - v^L\right].
\end{aligned}$$

By applying Lemma 8.1 twice, and $\Sigma_0 = Q_o + A\Sigma_1 A^T - A\Sigma_1 C^T (R_o + C\Sigma_1 C^T)^{-1} C\Sigma_1 A^T = Q_o + (A-LC)\Sigma_1 (A-LC)^T + LR_o L^T$, we obtain

$$
\begin{aligned}
&V(k+1) - V(k)\\
&\le x^T\left(-Q + P_0 - P_1\right)x - \lambda h(v^L)^T h(v^L)\\
&\quad + \left[h(v^L) - v^L\right]^T \left[\left(1+\gamma_1\right)\left(\lambda I + B^T P_1 B\right) + \left(1+\gamma_2\right)\lambda B^T \hat{\Sigma} B\right]\left[h(v^L) - v^L\right]\\
&\quad + \left(1 + 1/\gamma_1\right)(v^e)^T\left(\lambda I + B^T P_1 B\right)v^e\\
&\quad + \lambda e^T\left[\left(1 + 1/\gamma_2\right)\left((A-LC)^T\hat{\Sigma}(A-LC) - \hat{\Sigma}\right) + \frac{1}{\gamma_2}\hat{\Sigma}\right]e\\
&= x^T\left(-Q + P_0 - P_1\right)x - \lambda h(v^L)^T h(v^L)\\
&\quad + \left[h(v^L) - v^L\right]^T \left[\left(1+\gamma_1\right)\left(\lambda I + B^T P_1 B\right) + \left(1+\gamma_2\right)\lambda B^T \hat{\Sigma} B\right]\left[h(v^L) - v^L\right]\\
&\quad + \left(1 + 1/\gamma_1\right)(v^e)^T\left(\lambda I + B^T P_1 B\right)v^e\\
&\quad + \lambda e^T\left[\left(1 + 1/\gamma_2\right)\left((A-LC)\Sigma_1(A-LC)^T - \Sigma_1\right) + \frac{1}{\gamma_2}\hat{\Sigma}\right]e\\
&= x^T\left(-Q + P_0 - P_1\right)x - \lambda h(v^L)^T h(v^L)\\
&\quad + \left[h(v^L) - v^L\right]^T \left[\left(1+\gamma_1\right)\left(\lambda I + B^T P_1 B\right) + \left(1+\gamma_2\right)\lambda B^T \hat{\Sigma} B\right]\left[h(v^L) - v^L\right]\\
&\quad + \left(1 + 1/\gamma_1\right)(v^e)^T\left(\lambda I + B^T P_1 B\right)v^e\\
&\quad + \left(1 + 1/\gamma_2\right)\lambda e^T\left(-Q_o + \Sigma_0 - \Sigma_1 - LR_o L^T\right)e + \frac{1}{\gamma_2}\lambda e^T \hat{\Sigma} e.
\end{aligned}
$$

With the conditions (i)–(iv) satisfied, $V(k+1) - V(k) < 0$ for all $\left[x(k)^T, e(k)^T\right] \neq 0$. Hence, $V(k)$ is Lyapunov function that proves exponential stability.

The conditions (i)–(iii) in Theorem 8.3 are the requirements imposed on R, Q, P_N, N, R_o, Q_o, Σ_{N_o}, and N_o, and (iv), on h. In general, decreasing the equation solution error, γ_1 and γ_2 will be beneficial for satisfying (iv). When $h = \tilde{1}$, we can take $\gamma_1 = \infty$, then (i)–(iii) are stability conditions of the linear system. According to the structure of (8.30), however, it is easily known that when $h = \tilde{1}$, asymptotical stability are guaranteed by (i) and $Q_o > \Sigma_0 - \Sigma_1$; this stability does not require $V(k+1) - V(k) \le 0$. According to the proof of Theorem 8.3, for $V(k+1) - V(k) \le 0$, there is one more monotonic condition $\lambda \|h(v^L)\|^2 \ge (v^e)^T(\lambda I + B^T P_1 B)v^e$. This monotonic requirement reflects in the conditions (ii)–(iv) of Theorem 8.3, so the condition (ii) is stricter than $Q_o > \Sigma_0 - \Sigma_1$.

If (i)–(iii) are satisfied but $h \neq \tilde{1}$, then we can obtain more sensible conditions under (iv). For this reason we adopt Assumptions 8.4 and 8.5.

Corollary 8.5 ***(Stability of TSOFMPC)*** For systems represented by (8.1), TSOFMPC (8.25) and (8.26) is adopted, where h satisfies (8.11). Suppose

(1) whenever $[x(0), e(0)] \in \Omega \subset \mathbb{R}^{2n}$, it holds $\left\|v^L(k)\right\| \le \Delta$ for all $k \ge 0$;
(2) there exist positive scalars γ_1 and γ_2 such that the system design satisfies (i)–(iii) in Theorem 8.3 and
(iv) $-\lambda\left[b_1^2 - \left(1+\gamma_1\right)(b-1)^2\right] + (b-1)^2\sigma_{\max}\left(\left(1+\gamma_1\right)B^T P_1 B + \left(1+\gamma_2\right)\lambda B^T \hat{\Sigma} B\right) \le 0.$

Then, the equilibrium $\{x = 0, e = 0\}$ of the closed-loop system is exponentially stable with RoA Ω.

Remark 8.3 We can substitute (iv) in Corollary 8.5 by the following more conservative condition:

$$-\lambda\left[b_1^2 - (1+\gamma_1)(b-1)^2 - (1+\gamma_2)(b-1)^2\ \sigma_{\max}\left(B^T\hat{\Sigma}B\right)\right] + (1+\gamma_1)(b-1)^2\sigma_{\max}\left(B^T\ P_1B\right) \le 0.$$

The aforementioned condition will be useful in the following.

For RoA in Theorem 8.3 and Corollary 8.5, we have the following result.

Theorem 8.4 ***(RoA of TSOFMPC)*** For systems represented by (8.1), TSOFMPC (8.25)–(8.26) is adopted where h satisfies (8.11). Suppose there exist positive scalars Δ, γ_1 and γ_2 such that the conditions (i)–(iv) in Corollary 8.5 are satisfied. Then, RoA for the closed-loop system is not smaller than

$$S_c = \left\{(x,e) \in \mathbb{R}^{2n} | x^T P_1 x + \lambda e^T \hat{\Sigma} e \le c\right\}, \tag{8.32}$$

where

$$c = (\Delta/d)^2,\ d = \left\|(\lambda I + B^T P_1 B)^{-1} B^T P_1 A \left[P_1^{-1/2},\ -\lambda^{-1/2}\hat{\Sigma}^{-1/2}\right]\right\|. \tag{8.33}$$

Proof: Having satisfied the conditions (i)–(iv) in Corollary 8.5, we need only to verify that for all $[x(0),\ e(0)] \in S_c$, it holds $\|v^L(k)\| \le \Delta$ for all $k \ge 0$.

We adopt two nonsingular transformations: $\bar{x} = P_1^{1/2}x$ and $\bar{e} = \lambda^{1/2}\hat{\Sigma}^{1/2}e$. Then for all $[x(0), e(0)] \in S_c$, it holds $\left\|\left[\bar{x}(0)^T\ \bar{e}(0)^T\right]\right\| \le \sqrt{c}$ and

$$\begin{aligned}\|v^L(0)\| &= \left\|\left(\lambda I + B^T P_1 B\right)^{-1}\ B^T\ P_1 A\,[x(0) - e(0)]\right\| \\ &\le \left\|\left[\left(\lambda I + B^T P_1 B\right)^{-1} B^T P_1 A P_1^{-1/2}\ -\left(\lambda I + B^T P_1 B\right)^{-1} B^T P_1 A \lambda^{-1/2}\hat{\Sigma}^{-1/2}\right]\right\| \\ &\quad\times \left\|\left[\bar{x}(0)^T\ \bar{e}(0)^T\right]\right\| \le \Delta.\end{aligned} \tag{8.34}$$

Thus, all the conditions in Corollary 8.5 are satisfied at time $k = 0$ if $[x(0), e(0)] \in S_c$. According to the proof of Theorem 8.3, $[x(1), e(1)] \in S_c$ if $[x(0),\ e(0)] \in S_c$. Therefore, $\|v^L(1)\| \le \Delta$ for all $[x(0), e(0)] \in S_c$, which shows that all the conditions in Corollary 8.5 are satisfied at time $k = 1$. By analogy, we can conclude that as long as $[x(0), e(0)] \in S_c$, it holds $\|v^L(k)\| \le \Delta$ for all $k \ge 0$. Thus, S_c is RoA.

By applying Theorem 8.4, we can tune the control parameters so as to satisfy the conditions (i)–(iv) in Corollary 8.5 and obtain the desired RoA. The following Algorithm 8.4 may serve as a guideline.

Certainly, this does not mean that any desired RoA can be obtained for any system. However, if A has all its eigenvalues inside or on the circle, we have the following conclusion.

Algorithm 8.4 (Parameter tuning guideline for achieving the desired RoA Ω)

Step 1. Define the accuracy of the equation solution. Choose the initial Δ. Determine b_1 and b.
Step 2. Choose $\left\{R_o, Q_o, \Sigma_{N_o}, N_o\right\}$ rendering a convergent observer.
Step 3. Choose $\{\lambda, Q, P_N, N\}$ (mainly Q, P_N, N) satisfying (i).
Step 4. Choose $\{\gamma_1, \gamma_2, \lambda, Q, P_N, N\}$ (mainly $\gamma_1, \gamma_2, \lambda$) satisfying (ii)–(iv). If they cannot be satisfied, then go to one of Steps 1–3 (depending on the actual situation).
Step 5. Check whether or not (i)–(iv) are all satisfied. If they are not, go to Step 3; otherwise, decrease γ_1 and γ_2 (maintaining satisfaction of (ii)) and increase Δ (b_1 is decreased accordingly, maintaining satisfaction of (iv)).
Step 6. Calculate c using (8.33). If $S_c \supseteq \Omega$, STOP; otherwise, turn to Step 1.

Theorem 8.5 ***(Semi-global stability of TSOFMPC)*** For systems represented by (8.1), TSOFMPC (8.25)–(8.26) is adopted where h satisfies (8.11). Suppose A has all its eigenvalues inside or on the circle, and there exist Δ and γ_1 such that

$$b_1^2 - (1+\gamma_1)(b-1)^2 > 0 \tag{8.35}$$

in the absence of input saturation constraint. Then, for any bounded set $\Omega \subset \mathbb{R}^{2n}$, the controller and the observer parameters can be adjusted to make the closed-loop system possess RoA not smaller than Ω.

Proof: In the absence of saturation, b_1 and b are determined independently of the controller parameters. Denote the parameters $\{\gamma_1, \Delta\}$ that make (8.35) hold as $\{\gamma_1^0, \Delta^0\}$. When there is saturation, still choose $\gamma_1 = \gamma_1^0$ and $\Delta = \Delta^0$. Then, the following two cases may occur.

Case 1: (8.35) holds when $\lambda = \lambda_0$. Decide the parameters in the following.

(A) Choose

$$P_N = Q + A^T P_N A - A^T P_N B\left(\lambda I + B^T P_N B\right)^{-1} B^T P_N A.$$

Then $P_0 - P_1 = 0$. Furthermore, choose $Q > 0$, then $Q > P_0 - P_1$ and condition (i) in Corollary 8.5 will be satisfied for all λ and N. Choose an arbitrary N.

(B) Choose $R_o = \varepsilon I$, $Q_o = \varepsilon I > 0$, where ε is a scalar. Choose

$$\Sigma_{N_o} = Q_o + A\Sigma_{N_o}A^T - A\Sigma_{N_o}C^T\left(R_o + C\Sigma_{N_o}C^T\right)^{-1}C\Sigma_{N_o}A^T$$

and an arbitrary N_o. Then, $\Sigma_0 - \Sigma_1 = 0$. Apparently, there exist $\gamma_2 > 0$ and $\xi > 0$ such that $\left(1+1/\gamma_2\right)(Q_o - \Pi) - 1/\gamma_2\hat{\Sigma} \geq \varepsilon\xi I$ for all ε. Choose a sufficiently small ε such that

$$b_1^2 - (1+\gamma_1)(b-1)^2 - (1+\gamma_2)(b-1)^2\sigma_{\max}\left(B^T\hat{\Sigma}B\right) > 0.$$

Note that, if Σ_1 is small enough, then $\hat{\Sigma}$ is small enough. At this point,

$$\begin{aligned}&\left(1+1/\gamma_2\right)\left(-Q_o + \Sigma_0 - \Sigma_1 - LR_oL^T\right) + 1/\gamma_2\hat{\Sigma}_1\\ &\quad = \left(1+1/\gamma_2\right)\left(-Q_o - LR_oL^T\right) + 1/\gamma_2\hat{\Sigma} \leq -\varepsilon\xi I.\end{aligned}$$

(C) Multiplying both sides of

$$P_1 = Q + A^T P_1 A - A^T P_1 B\left(\lambda I + B^T P_1 B\right)^{-1} B^T P_1 A$$

by λ^{-1}, then

$$\overline{P}_1 = \lambda^{-1} Q + A^T \overline{P}_1 A - A^T \overline{P}_1 B\left(I + B^T \overline{P}_1 B\right)^{-1} B^T \ \overline{P}_1 A.$$

Since A has all its eigenvalues inside or on the circle, we know that $\overline{P}_1 \to 0$ as $\lambda \to \infty$ (refer to [Lin et al., 1996]). Hence, there exists $\lambda_1 \geq \lambda_0$ such that for all $\lambda \geq \lambda_1$,

(C.1) $\lambda^{-1}\left(1 + 1/\gamma_1\right) A^T P_1 B\left(\lambda I + B^T P_1 B\right)^{-1} \ B^T P_1 A = \left(1 + 1/\gamma_1\right) A^T \overline{P}_1 B \left(I + B^T \ \overline{P}_1 B\right)^{-1} B^T \ \overline{P}_1 A < \varepsilon \xi I$, i.e., the condition (ii) in Corollary 8.5 is satisfied,

(C.2) $-\left[b_1^2 - \left(1 + \gamma_1\right)(b-1)^2 - \left(1 + \gamma_2\right)(b-1)^2 \sigma_{\max}\left(B^T \hat{\Sigma} B\right)\right] + (b-1)^2 \left(1 + \gamma_1\right) \sigma_{\max}\left(B^T \overline{P}_1 B\right) \leq 0$, i.e., the inequality in Remark 8.3 is satisfied and in turn, the condition (iv) in Corollary 8.5 is satisfied.

(D) Further, there exists $\lambda_2 \geq \lambda_1$ such that for all $\lambda \geq \lambda_2$, $\left\|\overline{P}_1^{1/2} A \overline{P}_1^{-1/2}\right\| \leq \sqrt{2}$ (refer to [Lin et al., 1996]). Now, let

$$\overline{c} = \sup_{\lambda \in [\lambda_2, \infty), (x,e) \in \Omega} \left(x^T \overline{P}_1 x + e^T \hat{\Sigma} e\right),$$

then $\Omega \subseteq \overline{S}_{\overline{c}} = \left\{(x, e) \in \mathbb{R}^{2n} | x^T \overline{P}_1 x + e^T \hat{\Sigma} e \leq \overline{c}\right\}$. Define two transformations: $\overline{x} = \overline{P}_1^{1/2} x$ and $\overline{e} = \hat{\Sigma}^{1/2} e$. Then, there exists $\lambda_3 \geq \lambda_2$ such that for all $\lambda \geq \lambda_3$ and all $\left[x(0)^T, e(0)^T\right] \in \overline{S}_{\overline{c}}$,

$$\begin{aligned}
\|v^L(0)\| &= \|-K\left[x(0) - e(0)\right]\| = \left\|[-K, \ K]\left[x(0)^T, e(0)^T\right]^T\right\| \\
&= \left\|\left[-K\overline{P}_1^{-1/2}, \ K\hat{\Sigma}^{-1/2}\right]\left[\overline{x}(0)^T, \overline{e}(0)^T\right]^T\right\| \\
&\leq \left\|\left[-K\overline{P}_1^{-1/2}, \ K\hat{\Sigma}^{-1/2}\right]\right\| \left\|\left[\overline{x}(0)^T, \overline{e}(0)^T\right]\right\| \\
&\leq \left\|\sqrt{2}(I + B^T \overline{P}_1 B)^{-1} B^T \ \overline{P}_1^{1/2}, -(I + B^T \ \overline{P}_1 B)^{-1} B^T \ \overline{P}_1 A \hat{\Sigma}^{-1/2}\right\| \sqrt{\overline{c}} \leq \Delta.
\end{aligned}$$

Hence, for all $\left[x(0)^T, \ e(0)^T\right] \in \Omega$, the conditions in Corollary 8.5 can be satisfied at time $k = 0$, and by the proof of Theorem 8.4, they are also satisfied for $\forall k > 0$.

Through the aforesaid decision procedure, the designed controller will have the desired RoA.

Case 2: (8.35) does not hold when $\lambda = \lambda_0$, which apparently is due to the fact that the control action is clipped too much by the saturation constraint. For the same reason as in Case 1 (C), by (8.25), we know that for any bounded Ω, there exists a sufficiently large $\lambda_4 \geq \lambda_0$ such that for all $\lambda \geq \lambda_4$ and all $\left[x(k)^T, e(k)^T\right] \in \Omega$, $\hat{u}(k)$ does not violate the saturation constraint. This procedure is equivalent to decreasing Δ and re-determining b_1 and b such that (8.35) holds.

In a word, if RoA is unsatisfactory with $\lambda = \lambda_0$, then it can be satisfied by choosing $\lambda \geq \max\left\{\lambda_3, \lambda_4\right\}$ and suitable $\left\{Q, P_N, N, R_o, Q_o, \Sigma_{N_o}, N_o\right\}$.

In the proof of Theorem 8.5, we have emphasized the effect of tuning λ. If A has all its eigenvalues inside or on the circle, then by properly fixing other parameters, we can tune λ to obtain an arbitrarily large bounded RoA. This is very important because many industrial processes can be represented by the stable models plus integrals. When A has no eigenvalue outside of the unit circle, but has eigenvalues on the unit circle, then the corresponding system can be critical stable or unstable; however, by slight controls this system can be stabilized.

Solving nonlinear equations can be referred to Ortega and Rheinboldt [2000].

8.5 Generalized Predictive Control: Basics

8.5.1 Output Prediction

Consider the following SISO controlled auto-regressive integral moving average (CARIMA) model:

$$A(z^{-1})y(k) = B(z^{-1})u(k-1) + \frac{\xi(k)}{\Delta}, \tag{8.36}$$

where

$$A(z^{-1}) = 1 + a_1 z^{-1} + \cdots + a_{n_a} z^{-n_a}, \ \deg A(z^{-1}) = n_a,$$
$$B(z^{-1}) = b_0 + b_1 z^{-1} + \cdots + b_{n_b} z^{-n_b}, \ \deg B(z^{-1}) = n_b.$$

z^{-1} is the backward shift operator, i.e., $z^{-1}y(k) = y(k-1)$, $z^{-1}u(k) = u(k-1)$; $\Delta = 1 - z^{-1}$ is the difference operator; $\{\xi(k)\}$ is the white noise sequence with zero mean value. For the systems with q samples time delay, $b_0 \sim b_{q-1} = 0$ when $n_b \geq q$.

In order to deduce the prediction $y(k+j|k)$ via (8.36), let us first introduce Diophantine equation

$$1 = E_j(z^{-1})A(z^{-1})\Delta + z^{-j}F_j(z^{-1}), \tag{8.37}$$

where $E_j(z^{-1})$, $F_j(z^{-1})$ are polynomials uniquely determined by $A(z^{-1})$ and prediction length j,

$$E_j(z^{-1}) = e_{j,0} + e_{j,1}z^{-1} + \cdots + e_{j,j-1}z^{-(j-1)},$$
$$F_j(z^{-1}) = f_{j,0} + f_{j,1}z^{-1} + \cdots + f_{j,n_a}z^{-n_a}.$$

Multiplying (8.36) by $E_j(z^{-1})\Delta z^j$, and utilizing (8.37), we can write out the output prediction at time $k+j$,

$$y(k+j|k) = E_j(z^{-1})B(z^{-1})\Delta u(k+j-1|k) + F_j(z^{-1})y(k) + E_j(z^{-1})\xi(k+j). \tag{8.38}$$

Since, at time k, the future noises $\xi(k+i)$, $i \in \{1, \ldots, j\}$ are unknown, the most suitable predicted value of $y(k+j)$ can be represented by

$$\bar{y}(k+j|k) = E_j(z^{-1})B(z^{-1})\Delta u(k+j-1|k) + F_j(z^{-1})y(k). \tag{8.39}$$

In (8.39), denote $G_j(z^{-1}) = E_j(z^{-1})B(z^{-1})$. Combining with (8.37) yields

$$G_j(z^{-1}) = \frac{B(z^{-1})}{A(z^{-1})\Delta}[1 - z^{-j}F_j(z^{-1})]. \tag{8.40}$$

Let us introduce another Diophantine equation

$$G_j(z^{-1}) = E_j(z^{-1})B(z^{-1}) = \tilde{G}_j(z^{-1}) + z^{-(j-1)}H_j(z^{-1}),$$

where

$$\tilde{G}_j(z^{-1}) = g_{j,0} + g_{j,1}z^{-1} + \cdots + g_{j,j-1}z^{-(j-1)},$$
$$H_j(z^{-1}) = h_{j,1}z^{-1} + h_{j,2}z^{-2} + \cdots + h_{j,n_b}z^{-n_b}.$$

Then, applying (8.38) and (8.39) yields

$$\bar{y}(k+j|k) = \tilde{G}_j(z^{-1})\Delta u(k+j-1|k) + H_j(z^{-1})\Delta u(k) + F_j(z^{-1})y(k), \tag{8.41}$$

$$y(k+j|k) = \bar{y}(k+j|k) + E_j(z^{-1})\xi(k+j). \tag{8.42}$$

In order to predict the future output by applying (8.38) or (8.39), one has to first know $E_j(z^{-1})$, $F_j(z^{-1})$. Ref. Clarke et al. [1987] gives an iterative algorithm for calculating $E_j(z^{-1})$, $F_j(z^{-1})$. First, according to (8.37),

$$1 = E_j(z^{-1})A(z^{-1})\Delta + z^{-j}F_j(z^{-1}),$$
$$1 = E_{j+1}(z^{-1})A(z^{-1})\Delta + z^{-(j+1)}F_{j+1}(z^{-1}).$$

Taking subtractions on both sides of the aforementioned two equations yields

$$A(z^{-1})\Delta[E_{j+1}(z^{-1}) - E_j(z^{-1})] + z^{-j}[z^{-1}F_{j+1}(z^{-1}) - F_j(z^{-1})] = 0.$$

Denote

$$\tilde{A}(z^{-1}) = A(z^{-1})\Delta = 1 + \tilde{a}_1 z^{-1} + \cdots + \tilde{a}_{n_a}z^{-n_a} + \tilde{a}_{n_a+1}z^{-(n_a+1)}$$
$$= 1 + (a_1 - 1)z^{-1} + \cdots + (a_{n_a} - a_{n_a-1})z^{-n_a} - a_{n_a}z^{-(n_a+1)},$$
$$E_{j+1}(z^{-1}) - E_j(z^{-1}) = \tilde{E}(z^{-1}) + e_{j+1,j}z^{-j}.$$

Then,

$$\tilde{A}(z^{-1})\tilde{E}(z^{-1}) + z^{-j}[z^{-1}F_{j+1}(z^{-1}) - F_j(z^{-1}) + \tilde{A}(z^{-1})e_{j+1,j}] = 0. \tag{8.43}$$

Since the coefficient for the first item in $\tilde{A}(z^{-1})$ is 1, it is easy to obtain that the necessary condition for consistently satisfying (8.43) is

$$\tilde{E}(z^{-1}) = 0. \tag{8.44}$$

Further, the necessary and sufficient condition for consistently satisfying (8.43) is that (8.44) and the following equation holds:

$$F_{j+1}(z^{-1}) = z[F_j(z^{-1}) - \tilde{A}(z^{-1})e_{j+1,j}]. \tag{8.45}$$

Comparing the items in the same order on both sides of (8.45), we obtain

$$e_{j+1,j} = f_{j,0},$$
$$f_{j+1,i} = f_{j,i+1} - \tilde{a}_{i+1}e_{j+1,j} = f_{j,i+1} - \tilde{a}_{i+1}f_{j,0},\ i \in \{0,\ldots,n_a-1\},$$
$$f_{j+1,n_a} = -\tilde{a}_{n_a+1}e_{j+1,j} = -\tilde{a}_{n_a+1}f_{j,0}.$$

This formulation for deducing the coefficients of $F_j(z^{-1})$ can be written in the vector form

$$f_{j+1} = \tilde{A}f_j,$$

where

$$f_{j+1} = [f_{j+1,0}, \cdots, f_{j+1,n_a}]^T,$$
$$f_j = [f_{j,0}, \cdots, f_{j,n_a}]^T,$$
$$\tilde{A} = \begin{bmatrix} 1-a_1 & 1 & 0 & \cdots & 0 \\ a_1 - a_2 & 0 & 1 & \ddots & 0 \\ \vdots & \vdots & \ddots & \ddots & 0 \\ a_{n_a-1} - a_{n_a} & 0 & \cdots & 0 & 1 \\ a_{n_a} & 0 & \cdots & 0 & 0 \end{bmatrix}.$$

Moreover, the iteration formula for coefficients of $E_j(z^{-1})$ is

$$E_{j+1}(z^{-1}) = E_j(z^{-1}) + e_{j+1,j}z^{-j} = E_j(z^{-1}) + f_{j,0}z^{-j}.$$

When $j = 1$, (8.37) is

$$1 = E_1(z^{-1})\tilde{A}(z^{-1}) + z^{-1}F_1(z^{-1}).$$

Hence, we should choose $E_1(z^{-1}) = 1, F_1(z^{-1}) = z[1 - \tilde{A}(z^{-1})]$ as the initial values of $E_j(z^{-1})$, $F_j(z^{-1})$. Thus, $E_{j+1}(z^{-1})$ and $F_{j+1}(z^{-1})$ can be iteratively calculated by

$$\left.\begin{aligned} f_{j+1} &= \tilde{A}f_j,\ f_0 = [1, 0, \dots, 0]^T, \\ E_{j+1}(z^{-1}) &= E_j(z^{-1}) + f_{j,0}z^{-j},\ E_0 = 0. \end{aligned}\right\} \tag{8.46}$$

From (8.44), it is shown that $e_{j,i}$, $i < j$ is not related with j. Hence, denote $e_i \triangleq e_{j,i}$, $i < j$.

Consider the second Diophantine equation. From (8.40) we know that the first j items of $G_j(z^{-1})$ have no relation with j; the first j coefficients of $G_j(z^{-1})$ are the unit impulse response values, which are denoted as $g_1, \dots, g_j$. Thus,

$$G_j(z^{-1}) = E_j(z^{-1})B(z^{-1}) = g_1 + g_2z^{-1} + \cdots + g_jz^{-(j-1)} + z^{-(j-1)}H_j(z^{-1}).$$

Therefore, $g_{j,i} = g_{i+1}$ for $i < j$.

8.5.2 Receding Horizon Optimization

In generalized predictive control (GPC), the cost function at time k has the following form:

$$\min J(k) = \mathrm{E}\left\{ \sum_{j=N_1}^{N_2} [y(k+j|k) - y_s(k+j)]^2 + \sum_{j=1}^{N_u} \lambda(j)\Delta u(k+j-1|k)^2 \right\}, \tag{8.47}$$

where $\mathrm{E}\{\cdot\}$ represents the mathematical expectation; $y_s(k+j)$ is the desired value of the output; N_1 and N_2 are the starting and ending instant for the prediction horizon; N_u is the control horizon, i.e., the control moves after N_u steps keep invariant, $u(k+j-1|k) = u(k+N_u-1|k)$ for all $j > N_u$; $\lambda(j)$ is the control weight coefficient. For simplicity, λ is taken as constant.

In the cost function (8.47), N_1 should be larger than the number of delayed intervals, N_2 should be as large as the plant dynamics is sufficiently represented (e.g., as large as the effect of the current control increment is included). Furthermore, the choice of N_1, N_2, N_u can be referred to, e.g, Clarke et al. [1987]. Since the multi-step prediction and the optimization are adopted, even when the delay is not estimated correctly, or the delay is changed, one can still achieve reasonable control performance from the overall optimization. This is the important reason why GPC has robustness with respect to the modeling inaccuracy. Although there is the difference brought by the stochastic system, the above cost function is similar to that in DMC, except that the weight coefficients Q_i before N_1 are taken as zeros.

Applying the prediction model (8.39) yields

$$\begin{aligned}
\bar{y}(k+1|k) &= G_1(z^{-1})\Delta u(k) + F_1(z^{-1})y(k) = g_{1,0}\Delta u(k|k) + f_1(k)\\
\bar{y}(k+2|k) &= G_2(z^{-1})\Delta u(k+1|k) + F_2(z^{-1})y(k)\\
&= g_{2,0}\Delta u(k+1|k) + g_{2,1}\Delta u(k|k) + f_2(k)\\
&\vdots\\
\bar{y}(k+N|k) &= G_N(z^{-1})\Delta u(k+N-1|k) + F_N(z^{-1})y(k)\\
&= g_{N,0}\Delta u(k+N-1|k) + \cdots + g_{N,N-N_u}\Delta u(k+N_u-1|k)\\
&\quad + \cdots + g_{N,N-1}\Delta u(k|k) + f_N(k)\\
&= g_{N,N-N_u}\Delta u(k+N_u-1|k) + \cdots + g_{N,N-1}\Delta u(k) + f_N(k),
\end{aligned}$$

where

$$\left.\begin{aligned}
f_1(k) &= [G_1(z^{-1}) - g_{1,0}]\Delta u(k) + F_1(z^{-1})y(k)\\
f_2(k) &= z[G_2(z^{-1}) - z^{-1}g_{2,1} - g_{2,0}]\Delta u(k) + F_2(z^{-1})y(k)\\
&\vdots\\
f_N(k) &= z^{N-1}[G_N(z^{-1}) - z^{-(N-1)}g_{N,N-1} - \cdots - g_{N,0}]\Delta u(k) + F_N(z^{-1})y(k)
\end{aligned}\right\} \tag{8.48}$$

can be calculated by applying $\{y(\tau), \tau \leq k\}$ and $\{u(\tau), \tau < k\}$ which are known at time k.

Denote

$$\begin{aligned}
\vec{y}(k|k) &= [\bar{y}(k+N_1|k), \dots, \bar{y}(k+N_2|k)]^T,\\
\Delta\tilde{u}(k|k) &= [\Delta u(k|k), \dots, \Delta u(k+N_u-1|k)]^T,\\
\overleftarrow{f}(k) &= [f_{N_1}(k), \dots, f_{N_2}(k)]^T.
\end{aligned}$$

Note that $g_{j,i} = g_{i+1}$ $(i < j)$ is the step response coefficient. Then

$$\vec{y}(k|k) = G\Delta\tilde{u}(k|k) + \overleftarrow{f}(k), \tag{8.49}$$

where $G = \begin{bmatrix} g_{N_1} & g_{N_1-1} & \cdots & g_{N_1-N_u+1}\\ g_{N_1+1} & g_{N_1} & \cdots & g_{N_1-N_u+2}\\ \vdots & \vdots & \ddots & \vdots\\ g_{N_2} & g_{N_2-1} & \cdots & g_{N_2-N_u+1} \end{bmatrix}$ with $g_j = 0$ for all $j \leq 0$.

By using $\bar{y}(k+j|k)$ to substitute $y(k+j|k)$ in (8.47), the cost function can be written in the vector form, i.e.,

$$J(k) = [\vec{y}(k|k) - \vec{\omega}(k)]^T[\vec{y}(k|k) - \vec{\omega}(k)] + \lambda\Delta\tilde{u}(k|k)^T\Delta\tilde{u}(k|k),$$

where $\vec{\omega}(k) = [y_s(k+N_1), \dots, y_s(k+N_2)]^T$. Thus, when $\lambda I + G^T G$ is nonsingular, the optimal solution to cost function (8.47) is obtained as

$$\Delta\tilde{u}(k|k) = (\lambda I + G^T G)^{-1}G^T[\vec{\omega}(k) - \overleftarrow{f}(k)]. \tag{8.50}$$

The real-time optimal control moves is given by (assuming $\Delta u(k) = \Delta u(k|k)$ and $u(k) = u(k|k)$)

$$u(k) = u(k-1) + d^T[\vec{\omega}(k) - \overleftarrow{f}(k)], \tag{8.51}$$

where d^T is the first row of $(\lambda I + G^T G)^{-1} G^T$.

One can further utilize(8.42) to write the output prediction in the following vector form:

$$\tilde{y}(k|k) = G\Delta\tilde{u}(k|k) + F(z^{-1})y(k) + H(z^{-1})\Delta u(k) + \tilde{\varepsilon}(k),$$

where

$$\begin{aligned}
\tilde{y}(k|k) &= [y(k+N_1|k), \dots, y(k+N_2|k)]^T, \\
F(z^{-1}) &= [F_{N_1}(z^{-1}), \dots, F_{N_2}(z^{-1})]^T, \\
H(z^{-1}) &= [H_{N_1}(z^{-1}), \dots, H_{N_2}(z^{-1})]^T, \\
\tilde{\varepsilon}(k) &= [E_{N_1}(z^{-1})\xi(k+N_1), \dots, E_{N_2}(z^{-1})\xi(k+N_2)]^T.
\end{aligned}$$

Thus, the cost function is written as the vector form

$$J(k) = \mathrm{E}\left\{[\tilde{y}(k|k) - \vec{\omega}(k)]^T[\tilde{y}(k|k) - \vec{\omega}(k)] + \lambda\Delta\tilde{u}(k|k)^T\Delta\tilde{u}(k|k)\right\}.$$

Thus, when $\lambda I + G^T G$ is nonsingular, the optimal control law is

$$\Delta\tilde{u}(k|k) = (\lambda I + G^T G)^{-1} G^T \left[\vec{\omega}(k) - F(z^{-1})y(k) - H(z^{-1})\Delta u(k)\right].$$

Since the mathematical expectation is adopted, $\tilde{\varepsilon}(k)$ does not appear in the above control law. The real-time optimal control move is given by

$$u(k) = u(k-1) + d^T \left[\vec{\omega}(k) - F(z^{-1})y(k) - H(z^{-1})\Delta u(k)\right]. \tag{8.52}$$

Define

$$\begin{aligned}
\Delta\,\overleftarrow{u}(k) &= \left[\Delta u(k-1), \Delta u(k-2), \dots, \Delta u(k-n_b)\right]^T, \\
\overleftarrow{y}(k) &= \left[y(k), y(k-1), \dots, y(k-n_a)\right]^T.
\end{aligned}$$

The future output prediction $\vec{y}(k|k) = \left[\bar{y}(k+N_1|k), \bar{y}(k+N_1+1|k), \dots, \bar{y}(k+N_2|k)\right]^T$ can be denoted as

$$\vec{y}(k|k) = G\Delta\tilde{u}(k|k) + H\Delta\,\overleftarrow{u}(k) + F\,\overleftarrow{y}(k), \tag{8.53}$$

where

$$H = \begin{bmatrix} h_{N_1,1} & h_{N_1,2} & \cdots & h_{N_1,n_b} \\ h_{N_1+1,1} & h_{N_1+1,2} & \cdots & h_{N_1+1,n_b} \\ \vdots & \vdots & \ddots & \vdots \\ h_{N_2,1} & h_{N_2,2} & \cdots & h_{N_2,n_b} \end{bmatrix},$$

$$F = \begin{bmatrix} f_{N_1,0} & f_{N_1,1} & \cdots & f_{N_1,n_a} \\ f_{N_1+1,0} & f_{N_1+1,1} & \cdots & f_{N_1+1,n_a} \\ \vdots & \vdots & \ddots & \vdots \\ f_{N_2,0} & f_{N_2,1} & \cdots & f_{N_2,n_a} \end{bmatrix},$$

i.e., when $\lambda I + G^T G$ is nonsingular, the optimal control law is

$$\Delta u(k) = d^T \left[\vec{\omega}(k) - H\Delta\,\overleftarrow{u}(k) - F\,\overleftarrow{y}(k)\right]. \tag{8.54}$$

Theorem 8.6 ***(Optimal solution to the cost function)*** Suppose $\vec{\omega}(k) = 0$, then the optimal solution to the cost function applying GPC is

$$J^*(k) = \overleftarrow{f}(k)^T \left[I - G(\lambda I + G^T G)^{-1} G^T\right] \overleftarrow{f}(k), \tag{8.55}$$

$$J^*(k) = \lambda \overleftarrow{f}(k)^T (\lambda I + GG^T)^{-1} \overleftarrow{f}(k),\ \lambda \neq 0, \tag{8.56}$$

where $\overleftarrow{f}(k) = H\Delta\,\overleftarrow{u}(k) + F\,\overleftarrow{y}(k)$.

Proof: Substituting (8.49) and (8.50) into the cost function, we obtain (8.55). Apply the following matrix inverse formula:

$$(Q + MTS)^{-1} = Q^{-1} - Q^{-1}M\left(SQ^{-1}M + T^{-1}\right)^{-1} SQ^{-1}, \tag{8.57}$$

where $Q,\ M,\ T,\ S$ are arbitrary inverse matrices. Then, (8.56) is obtained from (8.55).

8.5.3 Dead-Beat Property of Generalized Predictive Control

Consider the model in (8.36), $C(z^{-1}) = 1$, which is transformed to

$$\tilde{A}(z^{-1})y(k) = \tilde{B}(z^{-1})\Delta u(k) + \xi(k), \tag{8.58}$$

where $\tilde{A}(z^{-1}) = 1 + \tilde{a}_1 z^{-1} + \cdots + \tilde{a}_{n_A} z^{-n_A}$, $\tilde{B}(z^{-1}) = \tilde{b}_1 z^{-1} + \tilde{b}_2 z^{-2} + \cdots + \tilde{b}_{n_B} z^{-n_B}$, $n_A = n_a + 1$, $n_B = n_b + 1$. Assume that $\tilde{a}_{n_A} \neq 0$ and $\tilde{b}_{n_B} \neq 0$. Assume that $\left(\tilde{A}(z^{-1}), \tilde{B}(z^{-1})\right)$ is the irreducible pair. Take $\vec{\omega} = [\omega, \omega, \dots, \omega]^T$.

In order to transform GPC problem into LQR in the receding horizon manner, without considering $\xi(k)$, transform (8.58) into the following state space model (observable canonical form, minimal realization):

$$x(k+1) = Ax(k) + B\Delta u(k),\ y(k) = Cx(k), \tag{8.59}$$

where $x \in \mathbb{R}^n$, $n = \max\left\{n_A, n_B\right\}$, $A = \begin{bmatrix} -\tilde{\alpha}^T & -\tilde{a}_n \\ I_{n-1} & 0 \end{bmatrix}$, $B = [1\ 0\ \cdots\ 0]^T$, $C = \left[\tilde{b}_1\ \tilde{b}_2\ \cdots\ \tilde{b}_n\right]$, and I_{n-1} is an $n-1$-order identity matrix, $\tilde{\alpha}^T = \left[\tilde{a}_1\ \tilde{a}_2\ \cdots\ \tilde{a}_{n-1}\right]$. When $i > n_A$, $\tilde{a}_i = 0$; when $i > n_B$, $\tilde{b}_i = 0$; when $n_A < n_B$, A is a singular matrix.

Since we discuss the stability, considering $\omega = 0$ does not lose the generality. Take

$$Q_i = \begin{cases} C^T C, & N_1 \leq i \leq N_2 \\ 0, & i < N_1 \end{cases},\quad \lambda_j = \begin{cases} \lambda, & 1 \leq j \leq N_u \\ \infty, & j > N_u \end{cases}$$

Then, the cost function (8.47) can be equivalently transformed into the following cost function of LQR:

$$J(k) = x(k+N_2)^T C^T Cx(k+N_2) + \sum_{i=0}^{N_2-1} \left[x(k+i)^T Q_i x(k+i) + \lambda_{i+1}\Delta u(k+i)^2\right]. \tag{8.60}$$

The LQR control law is

$$\Delta u(k) = -\left(\lambda + B^T P_1 B\right)^{-1} B^T P_1 A x(k). \tag{8.61}$$

Hence, it regards GPC as LQR, which is called the LQR law of GPC. P_1 can be obtained by the following Riccati iteration:

$$\begin{aligned} P_i &= Q_i + A^T P_{i+1} A - A^T P_{i+1} B\left(\lambda_{i+1} + B^T P_{i+1} B\right)^{-1} B^T P_{i+1} A, \\ i &= N_2 - 1, \ldots, 2, 1, \ P_{N_2} = C^T C. \end{aligned} \tag{8.62}$$

The control law (8.61) and the regular GPC control law (8.51) are equivalent in terms of stability (see [Kwon et al., 1992]).

If A is singular, then taking a nonsingular linear transformation to (8.59), we can obtain

$$\bar{x}(k+1) = \bar{A}\bar{x}(k) + \bar{B}\Delta u(k), \ y(k) = \bar{C}\bar{x}(k)$$

where $\bar{A} = \begin{bmatrix} A_0 & 0 \\ 0 & A_1 \end{bmatrix}$, $\bar{B} = \begin{bmatrix} B_0 \\ B_1 \end{bmatrix}$, $\bar{C} = \left[C_0, \ C_1\right]$, A_0 is invertible, $A_1 = \begin{bmatrix} 0 & 0 \\ I_{p-1} & 0 \end{bmatrix} \in \mathbb{R}^{p\times p}$, $C_1 = \left[0 \ \cdots \ 0 \ 1\right]$, p is the number of zero eigenvalues of A, I_{p-1} is the $p-1$-ordered identity matrix.

Lemma 8.2 When $n_A \geq n_B$, choose $N_1 \geq N_u$, $N_2 - N_1 \geq n - 1$. Then,

1) for $\lambda \geq 0$, P_{N_u} is nonsingular;
2) for $\lambda > 0$, GPC control law (8.61) can be transformed into

$$\Delta u(k) = -B^T\left(A^T\right)^{N_u-1}\left[\lambda P_{N_u}^{-1} + \sum_{h=0}^{N_u-1} A^h B B^T \left(A^T\right)^h\right]^{-1} A^{N_u} x(k). \tag{8.63}$$

Lemma 8.3 When $n_A \leq n_B$, choose $N_u \geq N_1$, $N_2 - N_u \geq n - 1$. Then,

1) for $\lambda \geq 0$, P_{N_1} is nonsingular;
2) for $\lambda > 0$, GPC control law (8.61) can be transformed into

$$\Delta u(k) = -B^T\left(A^T\right)^{N_1-1}\left[\lambda P_{N_1}^{-1} + \sum_{h=0}^{N_1-1} A^h B B^T \left(A^T\right)^h\right]^{-1} A^{N_1} x(k). \tag{8.64}$$

Lemma 8.4 When $n_A \leq n_B$, let $p = n_B - n_A$ and $N^p = \min\{N_1 - p, N_u\}$, $N_1 \geq N_u$, $N_2 - N_1 \geq n - p - 1$, $N_2 - N_u \geq n - 1$. Then,

1) for $\lambda \geq 0$, $P_{N^p} = \begin{bmatrix} P_{0,N^p} & 0 \\ 0 & 0 \end{bmatrix}$, where $P_{0,N^p} \in \mathbb{R}^{(n-p)\times(n-p)}$ is nonsingular;
2) for $\lambda > 0$, GPC control law (8.61) can be transformed into

$$\Delta u(k) = -\left[B_0^T\left(A_0^T\right)^{N^p-1}\left[\lambda P_{0,N^p}^{-1} + \sum_{h=0}^{N^p-1} A_0^h B_0 B_0^T \left(A_0^T\right)^h\right]^{-1} A_0^{N^p}, \ 0\right]\bar{x}(k). \tag{8.65}$$

Lemma 8.5 When $n_A \geq n_B$, let $q = n_A - n_B$, $N^q = \min\{N_1 + q, N_u\}$ and choose $N_u \geq N_1, N_2 - N_u \geq n - q - 1, N_2 - N_1 \geq n - 1$. Then,

1) for $\lambda \geq 0$, if P_{N_1} is taken as the initial value to calculate $P^*_{N^q}$ via

$$P^*_i = A^T P^*_{i+1} A - A^T P^*_{i+1} B\left(\lambda + B^T P^*_{i+1} B\right)^{-1} B^T P^*_{i+1} A,$$
$$P^*_{N_1} = P_{N_1},\ i \in \{N_1, N_1 + 1, \ldots, N^q - 1\},$$

then $P^*_{N^q}$ is nonsingular;

2) for $\lambda > 0$, GPC control law (8.61) can be transformed into

$$\Delta u(k) = -B\left(A^T\right)^{N^q-1}\left[\lambda P^{*-1}_{N^q} + \sum_{h=0}^{N^q-1} A^h BB^T \left(A^T\right)^h\right]^{-1} A^{N^q} x(k). \tag{8.66}$$

Theorem 8.7 There exists a sufficiently small λ_0 such that, for any $0 < \lambda < \lambda_0$, the closed-loop system of GPC is stable if the following condition is satisfied:

$$N_u \geq n_A,\ N_1 \geq n_B,\ N_2 - N_u \geq n_B - 1,\ N_2 - N_1 \geq n_A - 1. \tag{8.67}$$

Theorem 8.8 Suppose $\xi(k) = 0$. GPC is a dead beat controller if either of the following two conditions is satisfied:

1) $\lambda = 0, N_u = n_A, N_1 \geq n_B, N_2 - N_1 \geq n_A - 1$;
2) $\lambda = 0, N_u \geq n_A, N_1 = n_B, N_2 - N_u \geq n_B - 1$.

Stability was not guaranteed in the routine GPC. This has been overcome since the 1990s with new versions of GPC. One idea is that stability of the closed-loop system could be guaranteed if, in the last part of the prediction horizon, the future outputs are constrained at the desired sp and the prediction horizon is properly selected. The obtained predictive control is the predictive control with terminal equality constraint, or stabilizing input/output receding horizon control (SIORHC; see [Mosca and Zhang, 1992]) or constrained receding horizon predictive control (CRHPC; see [Clarke and Scattolini, 1991]).

Consider the model (8.58), with $\xi(k) = 0$. At sampling time k the objective function of GPC with terminal equality constraint is

$$J = \sum_{i=N_0}^{N_1-1} q_i y(k+i|k)^2 + \sum_{j=1}^{N_u} \lambda_j \Delta u^2(k+j-1|k), \tag{8.68}$$

$$\text{s.t. } y(k+l|k) = 0,\ l \in \{N_1, \ldots, N_2\}, \tag{8.69}$$

$$\Delta u(k+l-1|k) = 0,\ l \in \{N_u + 1, \ldots, N_2\}, \tag{8.70}$$

where $q_i \geq 0$ and $\lambda_j \geq 0$ are the weight coefficients, N_0, N_1 and N_1, N_2 are the starting and end points of the prediction horizon and constraint horizon respectively, and N_u is the control horizon.

Theorem 8.9 The closed-loop system of GPC with terminal equality constraint is dead-beat stable, if either of the following two conditions is satisfied:

$$\text{(i) } N_u = n_A,\ N_1 \geq n_B,\ N_2 - N_1 \geq n_A - 1;$$
$$\text{(ii) } N_u \geq n_A,\ N_1 = n_B,\ N_2 - N_u \geq n_B - 1. \tag{8.71}$$

8.5.4 On-line Identification and Feedback Correction

In GPC, the modeling coefficients are on-line estimated continuously based on the real-time input/output data, and the control law is modified correspondingly.

Let us write (8.36) as $A(z^{-1})\Delta y(k) = B(z^{-1})\Delta u(k-1) + \xi(k)$. Then $\Delta y(k) = -A_1(z^{-1})\Delta y(k) + B(z^{-1})\Delta u(k-1) + \xi(k)$, where $A_1(z^{-1}) = A(z^{-1}) - 1$. Denote the modeling parameters and data as the vector forms,

$$\theta = [a_1 \cdots a_{n_a}\ b_0 \cdots b_{n_b}]^T,$$
$$\varphi(k) = [-\Delta y(k-1) \cdots - \Delta y(k-n_a)\ \Delta u(k-1) \cdots \Delta u(k-n_b-1)]^T$$

which is written as $\Delta y(k) = \varphi(k)^T\theta + \xi(k)$.

Here, we can utilize the iterative LS with fading memory to estimate the parameters, i.e.,

$$\left.\begin{aligned} \hat{\theta}(k) &= \hat{\theta}(k-1) + K(k)[\Delta y(k) - \varphi(k)^T\hat{\theta}(k-1)] \\ K(k) &= P(k-1)\varphi(k)[\varphi(k)^T P(k-1)\varphi(k) + \mu]^{-1} \\ P(k) &= \frac{1}{\mu}[I - K(k)\varphi(k)^T]P(k-1) \end{aligned}\right\}, \tag{8.72}$$

where $0 < \mu < 1$ is the forgetting factor usually chosen as $0.95 < \mu < 1$; $K(k)$ is the weight factor; $P(k)$ is the positive-definite covariance matrix. In the startup of the controller, it needs to set the initial values of the parameter vector θ and covariance matrix P. Usually, we can set $\hat{\theta}(-1) = 0$, $P(-1) = \alpha^2 I$ where α is a sufficiently large positive scalar. At each control step, first setup the data vector, and then calculate $K(k)$, $\hat{\theta}(k)$ and $P(k)$ by applying (8.72).

After the parameters in $A(z^{-1})$, $B(z^{-1})$ are obtained by identification, d^T and $\overleftarrow{f}(k)$ can be recalculated and the optimal control move can be computed; see Algorithm 8.5.

Algorithm 8.5 (Adaptive GPC)

The on-line implementation of GPC falls into the following steps.

Step 1. Based on the newly obtained input/output data, use the iterative formula (8.72) to estimate the modeling parameters, so as to obtain $A(z^{-1})$, $B(z^{-1})$.

Step 2. Based on the obtained $A(z^{-1})$, iteratively calculate $E_j(z^{-1})$, $F_j(z^{-1})$ according to $A(z^{-1})$.

Step 3. Based on $B(z^{-1})$, $E_j(z^{-1})$, $F_j(z^{-1})$, calculate the elements g_i of G, and calculate $f_i(k)$ according to (8.49).

Step 4. Recompute d^T, and calculate $u(k)$ according to (8.52). Implement $u(k)$ to the plant.

Step 4 in Algorithm 8.5 involves the inversion of a N_u dimensional matrix and, hence, the on-line computational burden should be considered in selecting N_u.

8.6 Two-Step Generalized Predictive Control

The Hammerstein model is composed of a static nonlinear model followed by a dynamic linear submodel. Static nonlinearity is

$$v(k) = f(u(k)),\ f(0) = 0, \tag{8.73}$$

where u is the input, v the intermediate variable; in the literature, f is usually called invertible nonlinearity. The linear part adopts the controlled auto-regressive moving average (CARMA) model, i.e.,

$$a(z^{-1})y(k) = b(z^{-1})v(k-1), \tag{8.74}$$

where y is the output, $a_{n_a} \neq 0$, $b_{n_b} \neq 0$, $\{a, b\}$ is irreducible. Since the exact $v(k)$ is often unknown, (8.74) has the following form:

$$a(z^{-1})y^L(k) = b(z^{-1})v^L(k-1), \tag{8.75}$$

where $v^L(k)$ is the ideal value of $v(k)$, and $y^L(k)$ corresponds to $v^L(k)$.

8.6.1 Unconstrained Algorithm

First, utilize (8.75) for designing the linear generalized predictive control (LGPC) such that the desired $v^L(k)$ is obtained. Adopt the following cost function:

$$J(k) = \sum_{i=N_1}^{N_2} \left[y^L(k+i|k) - y_s(k+i)\right]^2 + \sum_{j=1}^{N_u} \lambda(\Delta v^L)^2(k+j-1|k). \tag{8.76}$$

Usually $y_s(k+i) = \omega$ for all $i > 0$. Thus, the control law of LGPC is

$$\Delta v^L(k) = d^T(\vec{\omega} - \overleftarrow{f}), \tag{8.77}$$

where $\vec{\omega} = [\omega, \omega, \dots, \omega]^T$, $\overleftarrow{f}$ is a vector composed of the past intermediate variable, the past output and the current output. u is revised as v, while y used by $\overleftarrow{f}$ is not revised as y^L.

Then, use

$$v^L(k) = v^L(k-1) + \Delta v^L(k) \tag{8.78}$$

to calculate $u(k)$ which is applied to the true plant, i.e., solve the following equation with $u(k)$:

$$f(u(k)) - v^L(k) = 0, \tag{8.79}$$

with the solution (not necessarily exact) denoted as

$$u(k) = g\left(v^L(k)\right). \tag{8.80}$$

When the aforementioned method was first proposed, it was called nonlinear generalized predictive control (NLGPC) (see [Zhu et al., 1991]).

8.6.2 Algorithm with Input Saturation

The input saturation constraint is usually inevitable in the real applications. Now, suppose the control move is clipped by the saturation constraint $|u| \leq U$, where U is a positive scalar. After $\Delta v^L(k)$ is obtained by applying (8.77), solve the equation

$$f(\hat{u}(k)) - v^L(k) = 0 \tag{8.81}$$

to decide $\hat{u}(k)$, with the solution (not necessarily exact) denoted as

$$\hat{u}(k) = \hat{f}^{-1}(v^L(k)). \tag{8.82}$$

Then, the desaturation is invoked to obtain the actual control move $u(k) = \text{sat}\{\hat{u}(k)\}$, where $\text{sat}\{s\} = \text{sign}\{s\} \min\{|s|, U\}$, being denoted as (8.80).

The above control strategy is called TSGPC-I (type-I two-step GPC).

In order to handle the input saturation, one can also transform the input saturation constraint to the constraint on the intermediate variable. Then, another TSGPC strategy is obtained. First, use the constraint on u, i.e., $|u| \le U$, to determine the constraint on v^L, i.e., $v_{\min} \le v^L \le v_{\max}$. After $\Delta v^L(k)$ is obtained by applying (8.77), let

$$\hat{v}(k) = \begin{cases} v_{\min}, & v^L(k) \le v^L_{\min} \\ v^L(k), & v_{\min} < v^L(k) < v_{\max}. \\ v_{\max}, & v^L(k) \ge v_{\max} \end{cases} \tag{8.83}$$

Then, solve the nonlinear algebraic equation

$$f(u(k)) - \hat{v}(k) = 0, \tag{8.84}$$

and let the solution $u(k)$ satisfy saturation constraint, being denoted as

$$u(k) = \hat{g}(\hat{v}(k)), \tag{8.85}$$

which can also be denoted as (8.80). This control strategy is called type-II TSGPC (TSGPC-II).

Remark 8.4 After the constraint on the intermediate variable is obtained, we can design the nonlinear separation generalized predictive control (NSGPC). In NSGPC, solving $\Delta v^L(k)$ no longer adopts (8.77), but is through the following optimization problem:

$$\begin{aligned} &\min_{\Delta v^L(k|k),\cdots,\Delta v^L(k+N_u-1|k)} J(k) \\ &= \sum_{i=N_1}^{N_2} \left[y^L(k+i|k) - y_s(k+i)\right]^2 + \sum_{j=1}^{N_u} \lambda(\Delta v^L)^2(k+j-1|k), \end{aligned} \tag{8.86}$$

$$\begin{aligned} \text{s.t. } & \Delta v^L(k+l|k) = 0, \ l \ge N_u, \\ & v_{\min} \le v^L(k+j-1|k) \le v_{\max}, \ j \in \{1,\ldots,N_u\}. \end{aligned} \tag{8.87}$$

The remaining computation and notations are the same as NLGPC. We can easily find the difference between TSGPC and NSGPC.

As addressed aforesaid, NLGPC, TSGPC-I and TSGPC-II are all called TSGPC. Now, suppose the true plant is "static nonlinearity + dynamic linear model" and the nonlinear submodel is $v(k) = f_0(u(k))$. We call the procedure determining $u(k)$ via $v^L(k)$ as nonlinear inversion. An ideal inversion will achieve $f_0 \circ g = 1$, i.e.,

$$v(k) = f_0(g(v^L(k))) = v^L(k). \tag{8.88}$$

If $f_0 \ne f$ or $f \ne g^{-1}$, then it is difficult to achieve $f_0 = g^{-1}$. In fact, it is usually impossible to achieve $f_0 = g^{-1}$. If there is no input saturation, theoretically, finding $u(k)$ via $v^L(k)$ is determined by the magnitude of $v^L(k)$ and the formulation of f. It is well-known that, even for the monotonic function $v = f(u)$, its inversion function $u = f^{-1}(v)$ does not necessarily exist for all the possible values of v. In the real applications, because of computational time and computational accuracy, the algebraic equation may not be exactly solved. Hence, in general, the approximate solution to the algebraic equation is adopted. When there is input saturation, the effect of desaturation may incur $v(k) \neq v^L(k)$. In summary, due to, e.g., the inaccuracy in equation solving, the desaturation, and modeling error, the $v^L(k)$ obtained through the linear model may not be implemented, and what is implemented is $v(k)$.

We will analyze the closed-loop stability of TSGPC in the sequel.

8.6.3 Stability Results Based on Popov's Theorem

Since the reserved nonlinear item in the closed-loop system is $f_0 \circ g$, the inaccuracy in the nonlinear submodel and the nonlinearity of the real actuator can also be incorporated into $f_0 \circ g$. Hence, stability results of TSGPC are also the robustness results.

Lemma 8.6 ***(Popov's stability Theorem)*** Suppose $G(z)$ in the Figure 8.4 is stable and $0 \leq \varphi(\vartheta)\vartheta \leq K_\varphi \vartheta^2$. Then, the closed-loop system is stable if $\frac{1}{K_\varphi} + \mathrm{Re}\{G(z)\} > 0$, for all $|z| = 1$.

$\mathrm{Re}\{\cdot\}$ refers to the real part of a complex and $|z|$ is mode of the complex number z. Applying Lemma 8.6, we can obtain the following stability result of TSGPC.

Theorem 8.10 ***(TSGPC's stability)*** Suppose the linear submodel applied by TSGPC is accurate and there exist two constants $k_1, k_2 > 0$ such that

1) the roots of $a(1 + d^T H)\Delta + (1 + k_1)z^{-1}d^T Fb = 0$ are all located in the unit circle;
2) for any $|z| = 1$,

$$\frac{1}{k_2 - k_1} + \mathrm{Re}\left\{\frac{z^{-1}d^T Fb}{a(1 + d^T H)\Delta + (1 + k_1)z^{-1}d^T Fb}\right\} > 0. \tag{8.89}$$

Then, the closed-loop system of TSGPC is stable if the following is satisfied:

$$k_1\vartheta^2 \leq (f_0 \circ g - 1)(\vartheta)\vartheta \leq k_2\vartheta^2. \tag{8.90}$$

Remark 8.5 For the given $\{\lambda, N_1, N_2, N_u\}$, we may find multiple sets of $\{k_0, k_3\}$, such that for all $k_1 \in \{k_0, k_3\}$, the roots of $a(1 + d^T H)\Delta + (1 + k_1)z^{-1}d^T Fb = 0$ are all located in the unit circle. In this way, $[k_1, k_2] \subseteq [k_0, k_3]$ satisfying the conditions (i) and (ii) in Theorem 8.10 may be innumerable. Suppose that the nonlinear item in the true system satisfies

$$k_1^0\vartheta^2 \leq (f_0 \circ g - 1)(\vartheta)\vartheta \leq k_2^0\vartheta^2, \tag{8.91}$$

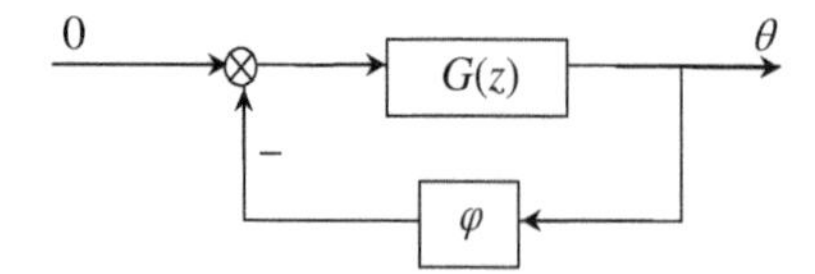

Figure 8.4 The static nonlinear feedback form.

where $k_1^0, k_2^0 > 0$ are constants, Then, Theorem 8.10 means that, if any set of $\{k_1, k_2\}$ satisfies

$$[k_1, k_2] \supseteq [k_1^0, k_2^0], \tag{8.92}$$

the corresponding system is stable.

In fact, with (8.91) known, verifying stability can directly apply the following conclusion.

Corollary 8.6 ***(TSGPC's stability)*** Suppose the linear submodel applied by TSGPC is accurate and the nonlinear item satisfies (8.91). Then, under the following two conditions the closed-loop system of TSGPC will be stable:

(i) all the roots of $a(1 + d^T H)\Delta + (1 + k_1^0)z^{-1}d^T Fb = 0$ are located in the unit circle;
(ii) for any $|z| = 1$,

$$\frac{1}{k_2^0 - k_1^0} + \mathrm{Re}\left\{\frac{z^{-1}d^T Fb}{a(1 + d^T H)\Delta + (1 + k_1^0)z^{-1}d^T Fb}\right\} > 0. \tag{8.93}$$

Theorem 8.10 and Corollary 8.6 can also be applied to the design of the controller parameters $\{\lambda, N_1, N_2, N_u\}$ in order to stabilize the system. In the following we discuss two cases in the form of Algorithms 8.6 and 8.7.

Algorithm 8.6 (Given $\{k_1^0, k_2^0\}$, design the controller parameters $\{\lambda, N_1, N_2, N_u\}$ to stabilize the closed-loop system of TSGPC)

Step 1. Search $\{\lambda, N_1, N_2, N_u\}$ by variable alternation method (their permissible ranges are determined before variable alternation; "variable alternation" comes from mathematical programming). If the search is finished, then terminate the whole algorithm, otherwise choose one set of $\{\lambda, N_1, N_2, N_u\}$ and determine $a(1 + d^T H)\Delta + z^{-1}d^T Fb$.

Step 2. Apply Jury's criterion to examine whether or not all roots of $a(1 + d^T H)\Delta + (1 + k_1^0)z^{-1}d^T Fb = 0$ are located in the unit circle. If it is not, then go to Step 1.

Step 3. Transform $-z^{-1}d^T Fb/[a(1 + d^T H)\Delta + (1 + k_1^0)z^{-1}d^T Fb]$ into irreducible form, denoted as $G(k_1^0, z)$.

Step 4. Substitute $z = \sigma + \sqrt{1 - \sigma^2}i$ into $G(k_1^0, z)$ to obtain $\mathrm{Re}\{G(k_1^0, z)\} = G_R(k_1^0, \sigma)$.

Step 5. Let $M = \max_{\sigma\in[-1,1]} G_R(k_1^0, \sigma)$. If $k_2^0 \le k_1^0 + \frac{1}{M}$, then terminate; otherwise, go to Step 1.

If the open-loop system has no eigenvalues outside of the unit circle, generally Algorithm 8.6 can obtain satisfactory $\{\lambda, N_1, N_2, N_u\}$. Otherwise, satisfactory $\{\lambda, N_1, N_2, N_u\}$ may not be found for all given $\{k_1^0, k_2^0\}$; in this case, one can restrict the degree of desaturation, i.e., try to increase k_1^0. Algorithm 8.7 can be used to determine a smallest k_1^0.

In the above, given $\{k_1^0, k_2^0\}$, we have described the algorithms for determining the controller parameters. In the following we briefly illustrate how to decide $\{k_1^0, k_2^0\}$ so as to

Algorithm 8.7 (Given the desired $\{k_1^0, k_2^0\}$, determine the controller parameters $\{\lambda, N_1, N_2, N_u\}$ such that $\{k_{10}^0, k_2^0\}$ satisfies stability requirements and $k_{10}^0 - k_1^0$ is minimized)

Step 1. Let $k_{10}^{0,\text{old}} = k_2^0$.
Step 2. Perform the same as Step 1 in Algorithm 8.6.
Step 3. Utilize root locus or Jury's criterion to decide $\{k_0, k_3\}$ such that $[k_0, k_3] \supset [k_{10}^{0,\text{old}}, k_2^0]$ and all roots of $a(1 + d^T H)\Delta + (1 + k_1)z^{-1}d^T Fb = 0$ are located in the unit circle, for all $k_1 \in [k_0, k_3]$. If such $\{k_0, k_3\}$ does not exist, then go to Step 2.
Step 4. Search k_{10}^0 in the range $k_{10}^0 \in \left[\max\{k_0, k_1^0\}, k_{10}^{0,\text{old}}\right]$ by increasing it gradually. If the search is finished, then go to Step 2; otherwise, transform $-z^{-1}d^T Fb/[a(1 + d^T H)\Delta + (1 + k_{10}^0)z^{-1}d^T Fb]$ into an irreducible form, being denoted as $G(k_{10}^0, z)$.
Step 5. Substitute $z = \sigma + \sqrt{1 - \sigma^2}i$ into $G(k_{10}^0, z)$ to obtain $\text{Re}\{G(k_{10}^0, z)\} = G_R(k_{10}^0, \sigma)$.
Step 6. Let $M = \max_{\sigma \in [-1,1]} G_R(k_{10}^0, \sigma)$. If $k_2^0 \le k_{10}^0 + \frac{1}{M}$ and $k_{10}^0 \le k_{10}^{0,\text{old}}$, then take $k_{10}^{0,\text{old}} = k_{10}^0$, denote $\{\lambda, N_1, N_2, N_u\}^* = \{\lambda, N_1, N_2, N_u\}$, and go to Step 2. Otherwise, go to Step 4.
Step 7. On finishing the search, let $k_{10}^0 = k_{10}^{0,\text{old}}$ and $\{\lambda, N_1, N_2, N_u\} = \{\lambda, N_1, N_2, N_u\}^*$.

bring Theorem 8.10 and Corollary 8.6 into play. We know that $f_0 \circ g \neq 1$ may be due to the following reasons:

- desaturation effect;
- solution error of nonlinear algebraic equation, including the case where an approximate solution is given since no accurate real-valued solution exists;
- inaccuracy in modeling of the nonlinearity;
- execution error of the actuator in a true system.

Suppose TSGPC-II is adopted. Then $f_0 \circ g$ is shown in Figure 8.5. Further, suppose

- no error exists in solving nonlinear equation;
- $k_{0,1}f(\vartheta)\vartheta \le f_0(\vartheta)\vartheta \le k_{0,2}f(\vartheta)\vartheta$ for all $v_{\min} \le v \le v_{\max}$;
- the desaturation level satisfies $k_{s,1}\vartheta^2 \le \text{sat}(\vartheta)\vartheta \le \vartheta^2$,

then

- $f_0 \circ g = f_0 \circ \hat{g} \circ \text{sat}$;

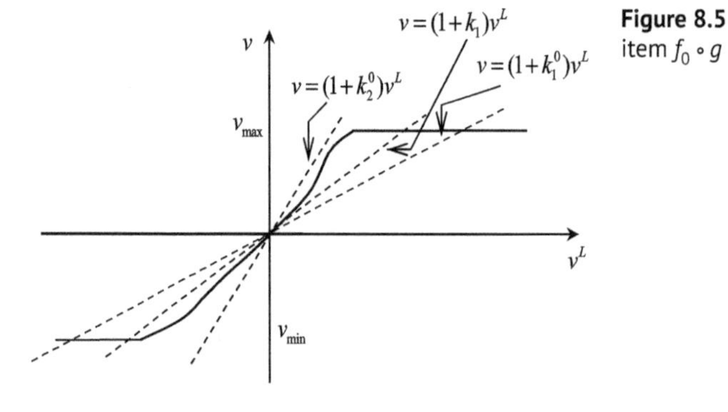

Figure 8.5 The sketch map of nonlinear item $f_0 \circ g$

- $k_{0,1}\text{sat}(\vartheta)\vartheta \le f_0 \circ g(\vartheta)\vartheta \le k_{0,2}\text{sat}(\vartheta)\vartheta$;
- $k_{0,1}k_{s,1}\vartheta^2 \le f_0 \circ g(\vartheta)\vartheta \le k_{0,2}\vartheta^2$,

and finally, $k_1^0 = k_{0,1}k_{s,1} - 1$ and $k_2^0 = k_{0,2} - 1$. Since $(1 - k_{s,1})$ is the desaturation level, $k_{s,1}$ can be called the unsaturation level.

8.7 Region of Attraction for Two-Step Generalized Predictive Control

For a fixed $k_{s,1}$, if all the conditions in Corollary 8.6 are satisfied, then the closed-loop system is stable. However, when TSGPC is applied for input-saturated systems, $k_{s,1}$ will change along with the level of desaturation. In the above, the issue that $\{k_1^0, k_2^0\}$ changes with $k_{s,1}$ is not handled. This issue is directly involved with RoA for the closed-loop system, which needs discussion by the state space equation.

8.7.1 State Space Description

Transform (8.75) into the following state-space model:

$$x^L(k+1) = Ax^L(k) + B\Delta v^L(k),\ y^L(k) = Cx^L(k), \tag{8.94}$$

where $x^L \in \mathbb{R}^n$. For $0 < i \le N_2$ and $0 < j \le N_2$, take

$$q_i = \begin{cases} 1, & N_1 \le i \le N_2 \\ 0, & i < N_1 \end{cases},\ \lambda_j = \begin{cases} \lambda, & 1 \le j \le N_u \\ \infty, & j > N_u \end{cases}. \tag{8.95}$$

Moreover, take a vector L such that $CL = 1$ (since $C \ne 0$, such an L exists but is not unique). Then, the cost function (8.76) of LGPC can be equivalently transformed into the following cost function of LQR (refer to [Kwon and Byun, 1989]):

$$\begin{aligned} J(k) = {} & \|x^L(k+N_2|k) - Ly_s(k+N_2)\|^2_{C^T q_{N_2} C} \\ & + \sum_{i=0}^{N_2-1} \left\{ \|x^L(k+i|k) - Ly_s(k+i)\|^2_{C^T q_i C} + \lambda_{i+1}(\Delta v^L)^2(k+i|k) \right\}, \end{aligned} \tag{8.96}$$

where $x^L(k|k) = x(k)$, and LQR law is

$$\Delta v^L(k) = -\left(\lambda + B^T P_1 B\right)^{-1} B^T \left[P_1 A x(k) + r(k+1)\right], \tag{8.97}$$

where P_1 can be obtained by the following Riccati iteration:

$$P_i = q_i C^T C + A^T P_{i+1} A - A^T P_{i+1} B\left(\lambda_{i+1} + B^T P_{i+1} B\right)^{-1} B^T P_{i+1} A,\ P_{N_2} = C^T C, \tag{8.98}$$

and $r(k+1)$ can be calculated by

$$r(k+1) = -\sum_{i=N_1}^{N_2} \Psi^T(i,1) C^T y_s(k+i), \tag{8.99}$$

$$\begin{aligned} & \Psi(1,1) = I, \\ & \Psi(j,1) = \prod_{i=1}^{j-1} \left[A - B\left(\lambda_{i+1} + B^T P_{i+1} B\right)^{-1} B^T P_{i+1} A\right],\ \forall j > 1. \end{aligned} \tag{8.100}$$

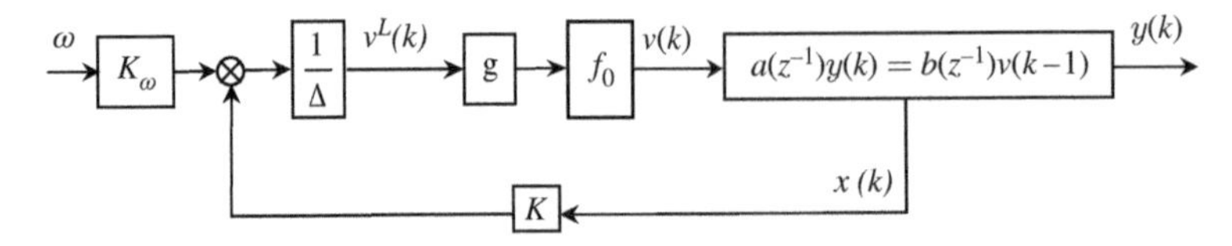

Figure 8.6 The block diagram of TSGPC

Denote (8.97) as

$$\Delta v^L(k) = Kx^L(k) + K_r r(k+1) = \begin{bmatrix} K & K_r \end{bmatrix} \begin{bmatrix} x(k)^T & r(k+1)^T \end{bmatrix}^T. \tag{8.101}$$

Take $y_s(k+i) = \omega$ for all $i > 0$. Then,

$$v^L(k) = v^L(k-1) + Kx(k) + K_\omega y_s(k+1), \tag{8.102}$$

where $K_\omega = -K_r \sum_{i=N_1}^{N_2} \Psi^T(i,1)C^T$. Figure 8.6 shows the equivalent block diagram of TSGPC.

8.7.2 Stability with Region of Attraction

When (8.91) is satisfied, let $\delta \in \text{Co}\left\{\delta_1, \delta_2\right\} = \text{Co}\left[k_1^0 + 1,\ k_2^0 + 1\right]$, i.e., $\delta = \xi\delta_1 + (1-\xi)\delta_2$, where ξ is any value satisfying $0 \le \xi \le 1$. If we use δ to replace $f_0 \circ g$, then since δ is a scalar, it can move in the block diagram. Hence, Figure 8.6 is transformed into Figure 8.7. It is easy to know that if the uncertain system in Figure 8.7 is robustly stable, then the closed-loop system of the original TSGPC is stable.

Next, we deduce the extended state space model of the system in Figure 8.7. First,

$$\Delta v(k) = \delta Kx(k) + \delta K_\omega y_s(k+1). \tag{8.103}$$

Hence,

$$x(k+1) = (A + \delta BK)x(k) + \delta BK_\omega y_s(k+1). \tag{8.104}$$

Since

$$y_s(k+2) = y_s(k+1), \tag{8.105}$$

there is

$$\begin{bmatrix} v^L(k) \\ x(k+1) \\ y_s(k+2) \end{bmatrix} = \begin{bmatrix} 1 & K & K_\omega \\ 0 & A + \delta BK & \delta BK_\omega \\ 0 & 0 & 1 \end{bmatrix} \begin{bmatrix} v^L(k-1) \\ x(k) \\ y_s(k+1) \end{bmatrix}. \tag{8.106}$$

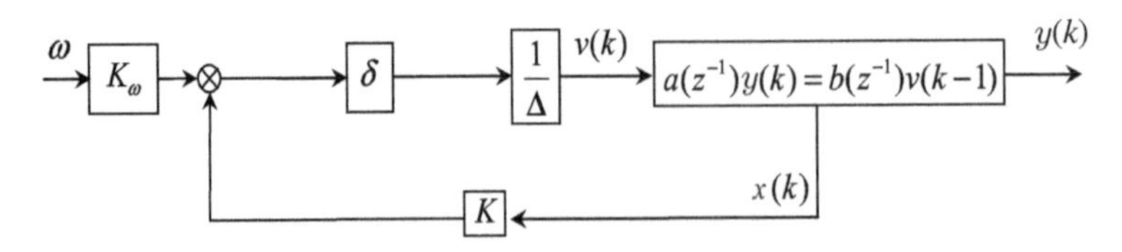

Figure 8.7 The uncertain system representation of TSGPC

Denote (8.106) as

$$x^{\mathrm{E}}(k+1) = \Phi(\delta)x^{\mathrm{E}}(k), \tag{8.107}$$

and call $x^{\mathrm{E}} \in \mathbb{R}^{n+2}$ as the extended state.

Both Theorem 8.10 and Corollary 8.6 are not related with RoA. RoA Ω of TSGPC with respect to the equilibrium point (u_e, y_e) is specially defined as the set of initial extended state $x^{\mathrm{E}}(0)$ that satisfies the following conditions:

$$\forall x^{\mathrm{E}}(0) \in \Omega \subset \mathbb{R}^{n+2}, \ \lim_{k\to\infty} u(k) = u_e, \ \lim_{k\to\infty} y(k) = y_e. \tag{8.108}$$

For the given $v(-1)$ and ω, RoA Ω_x of TSGPC with respect to the equilibrium point (u_e, y_e) is specially defined as the set of initial extended state $x(0)$ that satisfies the following conditions:

$$\forall x(0) \in \Omega_x \subset \mathbb{R}^{n}, \ \lim_{k\to\infty} u(k) = u_e, \ \lim_{k\to\infty} y(k) = y_e. \tag{8.109}$$

According to the aforementioned descriptions and Corollary 8.6, we can easily obtain the following result.

Theorem 8.11 ***(TSGPC's stability)*** Suppose the linear submodel is the same as the true system, and

1) for all $x^{\mathrm{E}}(0) \in \Omega$ and all $k > 0$, the unsaturation level $k_{s,1}$ is such that $f_0 \circ g$ satisfies (8.91);
2) the roots of $a(1 + d^T H)\Delta + (1 + k_1^0)z^{-1}d^T Fb = 0$ are all located inside of the unit circle;
3) (8.93) is satisfied.

Then, the equilibrium point (u_e, y_e) of TSGPC is stable with RoA Ω.

8.7.3 Computation of Region of Attraction

Denote $\Phi_1 = \begin{bmatrix} 1 & K & K_\omega \end{bmatrix}$. In (8.106),

$$\begin{aligned}\Phi(\delta) &\in \mathrm{Co}\left\{\Phi^{(1)}, \Phi^{(2)}\right\} \\ &= \mathrm{Co}\left\{\begin{bmatrix} 1 & K & K_\omega \\ 0 & A + \delta_1 BK & \delta_1 BK_\omega \\ 0 & 0 & 1 \end{bmatrix}, \begin{bmatrix} 1 & K & K_\omega \\ 0 & A + \delta_2 BK & \delta_2 BK_\omega \\ 0 & 0 & 1 \end{bmatrix}\right\}.\end{aligned} \tag{8.110}$$

Suppose all the conditions in Corollary 8.6 are satisfied, then we can adopt Algorithm 8.8 to calculate RoA.

RoA calculated by Algorithm 8.8 is also called the "maximal output admissible set" of the following system:

$$\begin{aligned}&x^{\mathrm{E}}(k+1) = \Phi(\delta)x^{\mathrm{E}}(k), \ v^L(k) = \Phi_1 x^{\mathrm{E}}(k), \\ &v_{\min}/k_{s,1} \le v^L(k) \le v_{\max}/k_{s,1} \ (\text{or } v^L_{\min} \le v^L(k) \le v^L_{\max}).\end{aligned}$$

For the maximal output admissible set, one can refer to, e.g., Gilbert and Tan [1991]; note that, here, the "output" refers to the output $v^L(k)$ of the above system, rather than the output y of the system (8.74); "admissible" refers to satisfaction of constraints. In Algorithm 8.8, the iterative method is adopted: define S_0 as the zero-step admissible set, then S_1 is 1-step

Algorithm 8.8 (The theoretical method for calculating RoA)

Step 1. Decide $k_{s,1}$ that satisfies all the conditions in Corollary 8.6 (if TSGPC-II is adopted, then $k_{s,1} = (k_1^0 + 1)/k_{0,1}$). Let

$$\begin{aligned} S_0 &= \{\theta \in \mathbb{R}^{n+2} | \Phi_1 \theta \leq v_{\max}/k_{s,1},\ \Phi_1 \theta \geq v_{\min}/k_{s,1}\} \\ &= \{\theta \in \mathbb{R}^{n+2} | F^{(0)}\theta \leq g^{(0)}\}, \end{aligned} \tag{8.111}$$

where $g^{(0)} = \begin{bmatrix} v_{\max}/k_{s,1} \\ -v_{\min}/k_{s,1} \end{bmatrix}$, $F^{(0)} = \begin{bmatrix} \Phi_1 \\ -\Phi_1 \end{bmatrix}$. Let $j = 1$. In this step, if the extremums $v^L_{\min}$ and $v^L_{\max}$ of v^L are given, then we can let

$$\begin{aligned} S_0 &= \{\theta \in \mathbb{R}^{n+2} | \Phi_1 \theta \leq v^L_{\max},\ \Phi_1 \theta \geq v^L_{\min}\} \\ &= \{\theta \in \mathbb{R}^{n+2} | F^{(0)}\theta \leq g^{(0)}\}. \end{aligned} \tag{8.112}$$

Step 2. Let

$$N_j = \{\theta \in \mathbb{R}^{n+2} | F^{(j-1)}\Phi^{(l)}\theta \leq g^{(j-1)},\ l = 1, 2\}, \tag{8.113}$$

and

$$S_j = S_{j-1} \bigcap N_j = \{\theta \in \mathbb{R}^{n+2} | F^{(j)}\theta \leq g^{(j)}\}. \tag{8.114}$$

Step 3. If $S_j = S_{j-1}$, then let $S = S_{j-1}$ and STOP; otherwise, let $j = j + 1$ and turn to Step 2.

admissible set, ..., S_j is j-step admissible set; the satisfaction of constraints means that they are always satisfied irrespective of how many times the sets have iterated.

Definition 8.2 If there exists $d > 0$ such that $S_d = S_{d+1}$, then S is finite determined and $S = S_d$. $d^* = \min\{d | S_d = S_{d+1}\}$ is the determinedness index (or, the output admissibility index).

Since the judgment of $S_j = S_{j-1}$ can be transformed into the optimization problem, Algorithm 8.8 can be transformed into Algorithm 8.9.

Remark 8.6 $J^*_{i,l} \leq 0$ indicates that, when (8.116) is satisfied, $F^{(j-1)}\Phi^{(l)}\theta \leq g^{(j-1)}$ is also satisfied. In S_j calculated by (8.113) and (8.114), the redundant inequalities can be removed by the similar optimizations.

In real applications, it may not be possible to find a finite number of inequalities to precisely express RoA S, i.e., d^* is not a finite value. It may also happen that d^* is finite but is very large, so that the convergence of the Algorithms 8.8 and 8.9 is very slow. In order to speed up the convergence, or when the algorithms do not converge, to approximate RoA, one can introduce $\varepsilon > 0$ and apply Algorithm 8.10. Denote $\tilde{1} = [1, 1, \ldots, 1]^T$ and, in (8.113), let

$$N_j = \{\theta \in \mathbb{R}^{n+2} | F^{(j-1)}\Phi^{(l)}\theta \leq g^{(j-1)} - \varepsilon\tilde{1},\ l = 1, 2\}. \tag{8.117}$$

Algorithm 8.9 (The iterative algorithm for calculating RoA)

Step 1. Decide $k_{s,1}$ satisfying all the conditions in Corollary 8.6. Calculate S_0 according to (8.111) or (8.112). Take $j = 1$.

Step 2. Solve the following optimization problem:

$$\max_{\theta} J_{i,l}(\theta) = \left(F^{(j-1)}\Phi^{(l)}\theta - g^{(j-1)}\right)_i,\ i \in \{1, \dots, n_j\},\ l \in \{1, 2\}, \tag{8.115}$$

such that the following constraint is satisfied:

$$F^{(j-1)}\theta - g^{(j-1)} \leq 0, \tag{8.116}$$

where n_j is the number of rows in $F^{(j-1)}$ and $(\cdot)_i$ denotes the ith row. Let $J_{i,l}^*$ be the optimum of $J_{i,l}(\theta)$. If

$$J_{i,l}^* \leq 0,\ \forall l \in \{1, 2\},\ \forall i \in \{1, \dots, n_j\},$$

then STOP and take $d^* = j - 1$; otherwise, continue.

Step 3. Calculate N_j via (8.113), and S_j via (8.114). Let $j = j + 1$ and turn to Step 2.

Algorithm 8.10 (The ε-iteration algorithm for calculating RoA)

All the details are the same as Algorithm 8.9 except that N_j is calculated by (8.117).

8.7.4 Numerical Example

The linear part of the system is $y(k) - 2y(k-1) = v(k-1)$. Take $N_1 = 1$, $N_2 = N_u = 2$, $\lambda = 10$, then $k_0 = 0.044$, $k_3 = 1.8449$, and when $k_1 \in [k_0, k_3]$, the condition (i) of Theorem 8.10 is satisfied. Take $k_1 = 0.287$, then the largest k_2 satisfying condition (ii) of Theorem 8.10 is $k_2 = 1.8314$; this is the set of $\{k_1, k_2\}$ satisfying $[k_1, k_2] \subseteq [k_0, k_3]$ such that $k_2 - k_1$ is maximized.

Take Hammerstein nonlinearity as $f_0(\theta) = 2.3f(\theta) + 0.5\sin f(\theta)$, $f(\theta) = \text{sign}\{\theta\}\theta \sin\left(\frac{\pi}{4}\theta\right)$. The input constraint is $|u| \leq 2$. Let the solution to the algebraic equation be utterly accurate. Then by utilizing the expression of f, it is known that $|\hat{v}| \leq 2$. Let the level

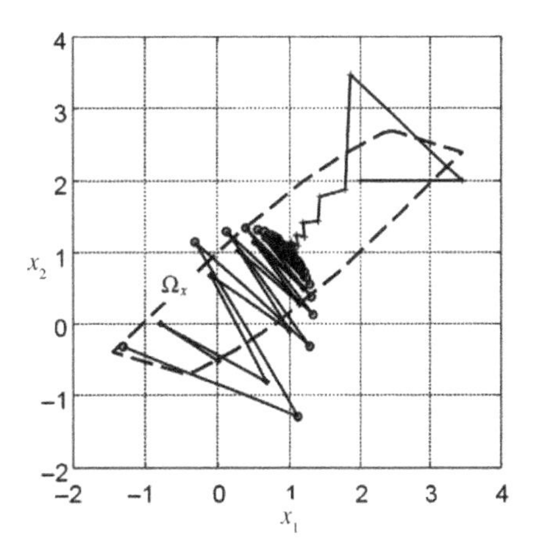

Figure 8.8 RoA and closed-loop state trajectory of TSGPC

of desaturation satisfy $k_{s,1} = 3/4$, then according to the above description, it is known that $1.35\theta^2 \leq f_0 \circ g(\theta)\theta \leq 2.8\theta^2$, i.e, $k_1^0 = 0.35$, $k_2^0 = 1.8$.

Under the above parameterizations, by applying Corollary 8.6 it is known that the system can be stabilized within a certain region of the initial extended state.

Take the two system states as $x_1(k) = y(k)$ and $x_2(k) = y(k-1)$. In Figure 8.8, the region in the dotted line is RoA Ω_x for $v(-1) = 0$ and $\omega = 1$, which is calculated according to Algorithm 8.9. Take the three sets of initial values as

- $y(-1) = 2, y(0) = 2, v^L(-1) = 0$;
- $y(-1) = -1.3, y(0) = -0.3, v^L(-1) = 0$;
- $y(-1) = 0, y(0) = -0.5, v^L(-1) = 0$.

The sp value is $\omega = 1$. According to Theorem 8.11, the system should be stable. In Figure 8.8, the state trajectories shown with solid lines indicate closed-loop stability.

Notice that, Ω_x is the projection of a cross-section of Ω on the $x_1 - x_2$ plane, which is not an invariant set.

9

Heuristic Model Predictive Control for LPV Model

When it is involved with parametric uncertainties and nonlinearities, if we want to inherit/integrate the analysis methods of linear model and to reduce the computation, we can often adopt LPV (the linear polytopic varying, or linear parameter varying) model, i.e.,

$$x_{k+1} = A_k x_k + B_k u_k, \ [A_k|B_k] \in \Omega, \tag{9.1}$$

where $x \in \mathbb{R}^n$, $u \in \mathbb{R}^m$; Ω is defined as the following polytope (convex hull):

$$\Omega = \mathrm{Co}\left\{[A_1|B_1], \ [A_2|B_2], \ \dots, \ [A_L|B_L]\right\},$$

i.e., there exist L nonnegative combing coefficients $\omega_{l,k}$, $l \in \{1, \dots, L\}$ such that

$$\sum_{l=1}^{L} \omega_{l,k} = 1, \ \left[A_k|B_k\right] = \sum_{l=1}^{L} \omega_{l,k} \left[A_l|B_l\right], \tag{9.2}$$

where $\left[A_l|B_l\right]$ are the prespecified vertices of the polytope. Usually, we assume that $\omega_{l,k}$ is unknown, unmeasurable, and not directly computable.

In practice, in general there is the output equation. Hence, extend (9.1) as

$$\begin{cases} x_{k+1} = A_k x_k + B_k u_k \\ y_k = C_k x_k \end{cases}, \ \left[A_k|B_k|C_k\right] \in \Omega, \tag{9.3}$$

where Ω is defined as the following polytope:

$$\Omega = \mathrm{Co}\left\{[A_1|B_1|C_1], \ [A_2|B_2|C_2], \ \dots, \ [A_L|B_L|C_L]\right\},$$

i.e., there exist L nonnegative coefficients $\omega_{l,k}$, $l \in \{1, \dots, L\}$ such that

$$\sum_{l=1}^{L} \omega_{l,k} = 1, \ \left[A_k|B_k|C_k\right] = \sum_{l=1}^{L} \omega_{l,k} \left[A_l|B_l|C_l\right]. \tag{9.4}$$

$\left[A_l|B_l|C_l\right]$ are called the vertices of the polytope, which are known/prespecified. Usually, we assume that the value of $\omega_{l,k}$ is unknown, unmeasurable, and not directly computable.

When it involves state estimation, we also adopt the following extended form of (9.3):

$$\begin{cases} x_{k+1} = A_k x_k + B_k u_k + B_{w,k} w_k \\ y_k = C_k x_k + C_{w,k} w_k \\ z_k = C_k x_k + C_{w,k} w_k \\ z'_k = C'_k x_k + C'_{w,k} w_k \end{cases}, \tag{9.5}$$

Model Predictive Control, First Edition. Baocang Ding and Yuanqing Yang.

where $z_k \in \mathbb{R}^{n_z}$ (see [Ding et al., 2014, Ding and Pan, 2016a]) and $z'_k \in \mathbb{R}^{n_{z'}}$ (see [Ding et al., 2015, Ding and Pan, 2016b]) are constrained signal and penalized signal, respectively. The value of the disturbance w is unknown, persistent (existing for all $k \geqslant 0$), and norm-bounded.

Assumption 9.1 For all $k \geqslant 0$, it holds $\|w_k\| \leqslant 1$.

Assumption 9.2 $[A|B|B_w|C|C_w|C|C_w|C'|C'_w]_k \in \Omega := \mathrm{Co}\{[A_l|B_l|B_{w,l}|C_l|C_{w,l}|C_l|C_{w,l}, |C'_l|C'_{w,l}]|l = 1, \dots, L]\}$, i.e., there exist nonnegative coefficients $\omega_{l,k}$, $l \in \{1, \dots, L\}$ such that $\sum_{l=1}^{L} \omega_{l,k} = 1$ and

$$[A|B|B_w|C|C_w|C|C_w|C'|C'_w]_k = \sum_{l=1}^{L} \omega_{l,k}[A_l|B_l|B_{w,l}|C_l|C_{w,l}|C_l|C_{w,l}|C'_l|C'_{w,l}].$$

Since $B_{w,k}$, $C_{w,k}$ are shaping matrices, Assumption 9.1 is suitable for any 2-norm-bounded disturbance. If the combining coefficients $\omega_{l,k}$ are exactly known at the current time k, but its future predictions $\omega_{l,i|k}$ are unknown for all $i > 0$, then we call (9.5) the quasi-LPV model.

In this chapter, we introduce two types of approaches; both can be solved by quadratic programming.

9.1 A Heuristic Approach Based on Open-Loop Optimization

Suppose that the state x_k is measurable, i.e., the value of x_k is exactly measured at each time k. We often consider the following constraints on inputs and states:

$$-\underline{u} \leqslant u_k \leqslant \overline{u}, -\underline{\psi} \leqslant \Psi x_{k+1} \leqslant \overline{\psi},\ \forall k \geqslant 0. \tag{9.6}$$

This equation means that the input and some linear combination of state should not exceed a given range. Herein, define $\underline{u} := [\underline{u}_1, \underline{u}_2, \dots, \underline{u}_m]$, $\overline{u} := [\overline{u}_1, \overline{u}_2, \dots, \overline{u}_m]$; $\underline{\psi} := [\underline{\psi}_1, \underline{\psi}_2, \dots, \underline{\psi}_q]$, $\overline{\psi} := [\overline{\psi}_1, \overline{\psi}_2, \dots, \overline{\psi}_q]$; $\underline{u}_i > 0, \overline{u}_i > 0$, $i \in \{1, \dots, m\}$; $\underline{\psi}_j > 0, \overline{\psi}_j > 0$, $j \in \{1, \dots, q\}$; $\Psi \in \mathbb{R}^{q \times n}$.

Define the vertex control move (see [Ding, 2010b])

$$u_{k|k}, u^{l_0}_{k+1|k}, \dots, u^{l_{N-2}\cdots l_0}_{k+N-1|k},\quad l_j \in \{1, \dots, L\},\ j \in \{0, \dots, N-2\},$$

where N is the control horizon. Then we obtain

$$\begin{aligned} x^{l_0}_{k+1|k} &= A_{l_0} x_k + B_{l_0} u_{k|k}, \\ x^{l_i \cdots l_0}_{k+i+1|k} &= A_{l_i} x^{l_{i-1}\cdots l_0}_{k+i|k} + B_{l_i} u^{l_{i-1}\cdots l_0}_{k+i|k}, \\ & i \in \{1, \dots, N-1\},\ l_j \in \{1, \dots L\},\ j \in \{0, \dots, N-1\}. \end{aligned}$$

We call $x^{l_{i-1}\cdots l_0}_{k+i|k}$, $i \in \{1, \dots, N\}$ the vertex state predictions. Define

$$\begin{aligned} u_{k+i|k} &= \sum_{l_0 \cdots l_{i-1}=1}^{L} \left(\left(\prod_{h=0}^{i-1} \omega_{l_h,k+h} \right) u^{l_{i-1}\cdots l_0}_{k+i|k} \right), \\ & \sum_{l_0 \cdots l_{i-1}=1}^{L} \left(\prod_{h=0}^{i-1} \omega_{l_h,k+h} \right) = 1,\ i \in \{1, \dots, N-1\}. \end{aligned} \tag{9.7}$$

The control input $u_{k+i|k}$ defined in (9.7) belongs to the polytope. In other words, $u_{k+i|k}$ defined in (9.7) is parameter-dependent, i.e., dependent on the parameters $\prod_{h=0}^{i-1} \omega_{l_h,k+h}$. According to (9.1) and (9.7), making predictions on the future state, yields

$$
\begin{aligned}
x_{k+1|k} &= A_k x_{k|k} + B_k u_{k|k} \\
&= \sum_{l_0=1}^{L} \omega_{l_0,k} \left[A_{l_0} x_k + B_{l_0} u_{k|k} \right] \\
&= \sum_{l_0=1}^{L} \omega_{l_0,k} x_{k+1|k}^{l_0}, \\
x_{k+2|k} &= A_{k+1} x_{k+1|k} + B_{k+1} u_{k+1|k} \\
&= \sum_{l_1=1}^{L} \omega_{l_1,k+1} \left[A_{l_1} x_{k+1|k} + B_{l_1} u_{k+1|k} \right] \\
&= \sum_{l_1=1}^{L} \omega_{l_1,k+1} \left[A_{l_1} \sum_{l_0=1}^{L} \omega_{l_0,k} x_{k+1|k}^{l_0} + B_{l_1} \sum_{l_0=1}^{L} \omega_{l_0,k} u_{k+1|k}^{l_0} \right] \\
&= \sum_{l_1=1}^{L} \sum_{l_0=1}^{L} \omega_{l_1,k+1} \omega_{l_0,k} x_{k+2|k}^{l_1 l_0}, \\
&\vdots
\end{aligned}
$$

Hence, $x_{k+i|k}$ belongs to the polytope. Writing it in vector form, yields

$$
\begin{gathered}
\begin{bmatrix} x_{k+1|k} \\ x_{k+2|k} \\ \vdots \\ x_{k+N|k} \end{bmatrix} = \sum_{l_0 \cdots l_{N-1}=1}^{L} \left(\prod_{h=0}^{N-1} \omega_{l_h,k+h} \begin{bmatrix} x_{k+1|k}^{l_0} \\ x_{k+2|k}^{l_1 l_0} \\ \vdots \\ x_{k+N|k}^{l_{N-1} \cdots l_1 l_0} \end{bmatrix} \right), \\
\sum_{l_0 \cdots l_{i-1}=1}^{L} \left(\prod_{h=0}^{i-1} \omega_{l_h,k+h} \right) = 1, \ i \in \{1, \ldots, N\},
\end{gathered}
\tag{9.8}
$$

where

$$
\begin{bmatrix} x_{k+1|k}^{l_0} \\ x_{k+2|k}^{l_1 l_0} \\ \vdots \\ x_{k+N|k}^{l_{N-1} \cdots l_1 l_0} \end{bmatrix} = \begin{bmatrix} A_{l_0} \\ A_{l_1} A_{l_0} \\ \vdots \\ \prod_{i=0}^{N-1} A_{l_{N-1-i}} \end{bmatrix} x_k
+ \begin{bmatrix} B_{l_0} & 0 & \cdots & 0 \\ A_{l_1} B_{l_0} & B_{l_1} & \ddots & \vdots \\ \vdots & \vdots & \ddots & 0 \\ \prod_{i=0}^{N-2} A_{l_{N-1-i}} B_{l_0} & \prod_{i=0}^{N-3} A_{l_{N-1-i}} B_{l_1} & \cdots & B_{l_{N-1}} \end{bmatrix} \begin{bmatrix} u_{k|k} \\ u_{k+1|k}^{l_0} \\ \vdots \\ u_{k+N-1|k}^{l_{N-2} \cdots l_1 l_0} \end{bmatrix}. \tag{9.9}
$$

Now, let us define the quadratic positive-definite function of the vertices $[x^{l_0}_{k+1|k}; x^{l_1 l_0}_{k+2|k}; \vdots; x^{l_{N-1}\cdots l_1 l_0}_{k+N|k}]$ and $[u_{k|k}; u^{l_0}_{k+1|k}; \vdots; u^{l_{N-2}\cdots l_1 l_0}_{k+N-1|k}]$ as

$$
\begin{aligned}
\hat{J}^N_{0,k} =& \sum_{l_0=1}^{L} \|Cx^{l_0}_{k+1|k} - y_{ss}\|^2_{\mathcal{Q}_{1,l_0}} + \|u_{k|k} - u_{ss}\|^2_{\mathcal{R}_0} \\
&+ \sum_{l_1=1}^{L}\sum_{l_0=1}^{L} \|Cx^{l_1 l_0}_{k+2|k} - y_{ss}\|^2_{\mathcal{Q}_{2,l_1,l_0}} + \|u^{l_0}_{k+1|k} - u_{ss}\|^2_{\mathcal{R}_{1,l_0}} \\
&+ \cdots \\
&+ \sum_{l_{N-1}=1}^{L} \cdots \sum_{l_1=1}^{L}\sum_{l_0=1}^{L} \|Cx^{l_{N-1}\cdots l_1 l_0}_{k+N|k} - y_{ss}\|^2_{\mathcal{Q}_{N,l_{N-1}\cdots l_1 l_0}} \\
&+ \sum_{l_{N-2}=1}^{L} \cdots \sum_{l_1=1}^{L}\sum_{l_0=1}^{L} \|u^{l_{N-2}\cdots l_1 l_0}_{k+N-1|k} - u_{ss}\|^2_{\mathcal{R}_{N-1,l_{N-2}\cdots l_1 l_0}},
\end{aligned}
$$

where $\mathcal{Q}_{1,l_0}, \mathcal{R}_0, \mathcal{Q}_{2,l_1,l_0}, \mathcal{R}_{1,l_0}, \dots, \mathcal{Q}_{N,l_{N-1}\cdots l_1 l_0}, \mathcal{R}_{N-1,l_{N-2}\cdots l_1 l_0}$ are nonnegative weight matrices, y_{ss} is ss (sp) of $y = Cx$, and u_{ss} is ss (sp) of u. Note that ss is unnecessarily the equilibrium.

At each time k, take $[x^{l_0}_{k+1|k}; x^{l_1 l_0}_{k+2|k}; \vdots; x^{l_{N-1}\cdots l_1 l_0}_{k+N|k}]$ and $[u_{k|k}; u^{l_0}_{k+1|k}; \vdots; u^{l_{N-2}\cdots l_1 l_0}_{k+N-1|k}]$ as the decision variables, and solve the following quadratic program:

$$
\min \hat{J}^N_{0,k}, \quad \text{s.t. (9.9), (9.11), (9.12).} \tag{9.10}
$$

$$
-\underline{u} \leqslant u_{k|k} \leqslant \underline{u}, -\underline{u} \leqslant u^{l_{i-1}\cdots l_1 l_0}_{k+i|k} \leqslant \underline{u}, \;\; i \in \{1, \dots, N-1\},\; l_{i-1} \in \{1, \dots, L\}, \tag{9.11}
$$

$$
-\underline{\psi}^s \leqslant \Psi^{\mathrm{d}} \begin{bmatrix} x^{l_0}_{k+1|k} \\ x^{l_1 l_0}_{k+2|k} \\ \vdots \\ x^{l_{N-1}\cdots l_1 l_0}_{k+N|k} \end{bmatrix} \leqslant \overline{\psi}^s, \;\; l_j \in \{1, \dots, L\},\; j \in \{0, \dots, N-1\}. \tag{9.12}
$$

Then, send $u_{k|k}$ to the true system. Ψ^{d} is a block-diagonal matrix with Ψ being the diagonal block. We call the method based on (9.10) the heuristic ol MPC. The vertex control moves, the vertex state predictions, and the above performance index $\hat{J}^N_{0,k}$ (taking a quadratic function for each vertex of state and input) have been adopted in Wang and Rawlings [2004].

The heuristic ol MPC has the following characteristics:

(1) It is computationally less expensive than MPC with stability ingredients.
(2) Its stability cannot be theoretically proved.
(3) Assuming that all the weight matrices $\mathcal{Q}_\bullet$ and $\mathcal{R}_\bullet$ are positive-definite, when $y_{ss} \neq 0$ and $u_{ss} \neq 0$, even if the closed-loop system is asymptotically stable, it is difficult to achieve $\hat{J}^N_{0,\infty} = 0$, i.e., difficult to guarantee offset-free.

Even if the system reaches the steady state, it is hard to make $Cx^{l_{i-1}\cdots l_1 l_0}_{\infty|\infty} = y_{ss} \neq 0$ for all $i \in \{1, 2, \dots, N\}$ and all $l_{i-1} \cdots l_1 l_0$. If there is some $Cx^{l_{i-1}\cdots l_1 l_0}_{k+i|k} \neq y_{ss}$, then the optimization of

$\hat{J}^N_{0,k}$ will be balanced among the quadratic terms; therefore, $\hat{J}^N_{0,\infty} \neq 0$ leads to $Cx_{ss} \neq y_{ss}$, i.e., induces offset.

When we discuss on the removal of offset, there is a delicate question: where do y_{ss} and u_{ss} come from? Apparently, y_{ss} has to be related to u_{ss}. Since we cannot know the exact value of $[A_k|B_k]$, usually we cannot determine the relationship by

$$\begin{cases} x_{ss} = A_{ss}x_{ss} + B_{ss}u_{ss} \\ y_{ss} = Cx_{ss} \end{cases}. \tag{9.13}$$

But one exception does exist: when $k \to \infty$, $[A_k|B_k]$ converges to the fixed value $[A_{ss}|B_{ss}]$. For this exception, in (9.11), y_{ss} and u_{ss} have to satisfy (9.13). A more reliable and general viewpoint is that $\{y_{ss}, u_{ss}, x_{ss}\}$ satisfy some steady-state nonlinear equation, i.e.,

$$f(x_{ss} + x_{eq}, u_{ss} + u_{eq}) = 0, \tag{9.14}$$

$$g(y_{ss} + y_{eq}, u_{ss} + u_{eq}) = 0. \tag{9.15}$$

If (9.14) and (9.15) are true, then there is a strategy that can remove the offset. With x_{ss} and u_{ss} obtained via (9.14), calculate

$$\begin{bmatrix} x_{ss}^{l_0} \\ x_{ss}^{l_1 l_0} \\ \vdots \\ x_{ss}^{l_{N-1}\cdots l_1 l_0} \end{bmatrix} = \begin{bmatrix} A_{l_0} \\ A_{l_1}A_{l_0} \\ \vdots \\ \prod_{i=0}^{N-1} A_{l_{N-1-i}} \end{bmatrix} x_{ss} + \begin{bmatrix} B_{l_0} & 0 & \cdots & 0 \\ A_{l_1}B_{l_0} & B_{l_1} & \ddots & \vdots \\ \vdots & \vdots & \ddots & 0 \\ \prod_{i=0}^{N-2} A_{l_{N-1-i}} B_{l_0} & \prod_{i=0}^{N-3} A_{l_{N-1-i}} B_{l_1} & \cdots & B_{l_{N-1}} \end{bmatrix} \begin{bmatrix} u_{ss} \\ u_{ss} \\ \vdots \\ u_{ss} \end{bmatrix}. \tag{9.16}$$

Then, in order to remove the offset, we can change the performance index as

$$\begin{aligned} \tilde{J}^N_{0,k} = & \sum_{l_0=1}^{L} \|x^{l_0}_{k+1|k} - x^{l_0}_{ss}\|^2_{\mathcal{Q}_{1,l_0}} + \|u_{k|k} - u_{ss}\|^2_{\mathcal{R}_0} \\ & + \sum_{l_1=1}^{L}\sum_{l_0=1}^{L} \|x^{l_1 l_0}_{k+2|k} - x^{l_1 l_0}_{ss}\|^2_{\mathcal{Q}_{2,l_1,l_0}} + \|u^{l_0}_{k+1|k} - u_{ss}\|^2_{\mathcal{R}_{1,l_0}} \\ & + \cdots \\ & + \sum_{l_{N-1}=1}^{L} \cdots \sum_{l_1=1}^{L}\sum_{l_0=1}^{L} \|x^{l_{N-1}\cdots l_1 l_0}_{k+N|k} - x^{l_{N-1}\cdots l_1 l_0}_{ss}\|^2_{\mathcal{Q}_{N,l_{N-1}\cdots l_1 l_0}} \\ & + \sum_{l_{N-2}=1}^{L} \cdots \sum_{l_1=1}^{L}\sum_{l_0=1}^{L} \|u^{l_{N-2}\cdots l_1 l_0}_{k+N-1|k} - u_{ss}\|^2_{\mathcal{R}_{N-1,l_{N-2}\cdots l_1 l_0}}. \end{aligned}$$

If the input and state have to satisfy some constraints, then $\{u_{ss}, x^{l_0}_{ss}, x^{l_1 l_0}_{ss}, \ldots, x^{l_{N-1}\cdots l_1 l_0}_{ss}\}$ should satisfy the same constraints.

In Wang and Rawlings [2004], if $\hat{J}_{0,k}^{N}$ is adopted, then the offset can still be removed. The reason is that Wang and Rawlings [2004] adopt a more delicate method for computing x_{ss} and u_{ss}. A necessary condition for "Theorem 5.2" of [Wang and Rawlings, 2004] is to adopt $\tilde{J}_{0,k}^{N}$, if it did not adopt that "more delicate method."

Remark 9.1 What are f and g in (9.14) and (9.15)? The nonlinear equations of the controlled system are represented by f and g; at least, the steady-state equation can be represented by f and g. Equation (9.1) is the dynamic equation of the system in the neighborhood of the equilibrium y_{eq} and u_{eq}.

Example 9.1 Consider the continuous stirred-tank reactor (CSTR) model. With a constant volume, CSTR for an exothermic, irreversible reaction A→B is described by

$$\dot{C}_A(t) = \frac{q}{V}(C_{Af} - C_A(t)) - k_0 \exp\left(-\frac{E/R}{T(t)}\right) C_A(t),$$
$$\dot{T}(t) = \frac{q}{V}(T_f - T(t)) + \frac{(-\Delta H)}{\rho C_p} k_0 \exp\left(-\frac{E/R}{T(t)}\right) C_A(t) + \frac{UA}{V\rho C_p}(T_c(t) - T(t)), \quad (9.17)$$

where C_A is the concentration of material A in the reactor, T is the reactor temperature, and T_c is the coolant stream temperature. V and UA denote the volume of the reactor and the rate of heat input, respectively. k_0, E, and ΔH denote the pre-exponential constant, the activation energy, and the enthalpy of the reaction, respectively. C_p and ρ stand for the heat capacity and density of the fluid in the reactor, respectively. The objective is to regulate T by manipulating T_c satisfying $328\text{K} \leqslant T_c \leqslant 348\text{K}$.

Denote the nonzero equilibrium as $\{C_A^{eq}, T^{eq}, T_c^{eq}\}$. Choose $C_A^{eq} = 0.5\,\text{mol/l}$, $T^{eq} = 350\text{K}$, $T_c^{eq} = 338\text{K}$, $340\text{K} \leqslant T \leqslant 360\text{K}$, $0 \leqslant C_A \leqslant 1\,\text{mol/l}$, $q = 100\,\text{l/min}$, $C_{Af} = 0.9\,\text{mol/l}$, $T_f = 350\text{K}$, $V = 100\text{l}$, $\rho = 1000\,\text{g/l}$, $C_p = 0.239\ \text{J/(g}\cdot\text{K)}$, $\Delta H = -2.5 \times 10^4\,\text{J/mol}$, $E/R = 8750\text{K}$, $k_0 = 3.456 \times 10^{10}\text{min}^{-1}$, $UA = 5 \times 10^4\,\text{J/(min K)}$.

Define $x = [C_A - C_A^{eq}, T - T^{eq}]^T$, $u = T_c - T_c^{eq}$. Denote the bounds on u and x as $\underline{u} \leqslant u \leqslant \overline{u}$ $(-10 \leqslant u \leqslant 10)$, $\underline{x}_1 \leqslant x_1 \leqslant \overline{x}_1$ $(-0.5 \leqslant x_1 \leqslant 0.5)$, $\underline{x}_2 \leqslant x_2 \leqslant \overline{x}_2$ $(-10 \leqslant x_2 \leqslant 10)$.

Define

$$\varphi_1(x_2) = k_0 \exp\left(-\frac{E/R}{x_2 + T^{eq}}\right),$$
$$\varphi_2(x_2) = k_0 \left[\exp\left(-\frac{E/R}{x_2 + T^{eq}}\right) - \exp\left(-\frac{E/R}{T^{eq}}\right)\right] C_A^{eq} \frac{1}{x_2},$$

$\varphi_1^0 = [\varphi_1(\underline{x}_2) + \varphi_1(\overline{x}_2)]/2$, $\varphi_2^0 = [\varphi_2(\underline{x}_2) + \varphi_2(\overline{x}_2)]/2$, $g_1(x_2) = \varphi_1(x_2) - \varphi_1^0$, $g_2(x_2) = \varphi_2(x_2) - \varphi_2^0$,

$$\omega_1 = \frac{1}{2}\frac{g_1(x_2) - g_1(\underline{x}_2)}{g_1(\overline{x}_2) - g_1(\underline{x}_2)}, \quad \omega_2 = \frac{1}{2}\frac{g_1(\overline{x}_2) - g_1(x_2)}{g_1(\overline{x}_2) - g_1(\underline{x}_2)},$$
$$\omega_3 = \frac{1}{2}\frac{g_2(x_2) - g_2(\underline{x}_2)}{g_2(\overline{x}_2) - g_2(\underline{x}_2)}, \quad \omega_4 = \frac{1}{2}\frac{g_2(\overline{x}_2) - g_2(x_2)}{g_2(\overline{x}_2) - g_2(\underline{x}_2)}.$$

Then (9.17) can be exactly represented by

$$x(t) = \tilde{A}(t)x(t) + \tilde{B}(t)u(t), \; \tilde{A}(t) = \sum_{l=1}^{4} \omega_l(t)\tilde{A}_l, \; \tilde{B}(t) = \sum_{l=1}^{4} \omega_l(t)\tilde{B}_l, \quad (9.18)$$

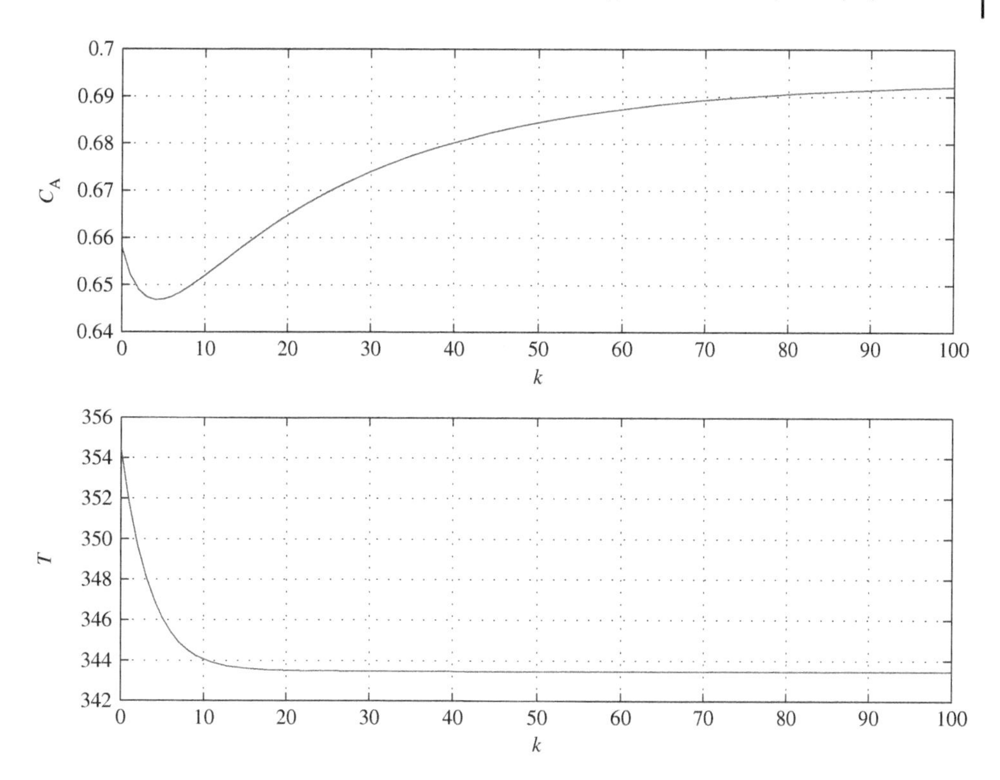

Figure 9.1 The state responses.

where

$$\tilde{A}_1 = \begin{bmatrix} -\frac{q}{V} - \varphi_1^0 - 2g_1(\overline{x}_2) & -\varphi_2^0 \\ \frac{(-\Delta H)}{\rho C_p}\varphi_1^0 + 2\frac{(-\Delta H)}{\rho C_p}g_1(\overline{x}_2) & -\frac{q}{V} - \frac{UA}{V\rho C_p} + \frac{(-\Delta H)}{\rho C_p}\varphi_2^0 \end{bmatrix},$$

$$\tilde{A}_2 = \begin{bmatrix} -\frac{q}{V} - \varphi_1^0 - 2g_1(\underline{x}_2) & -\varphi_2^0 \\ \frac{(-\Delta H)}{\rho C_p}\varphi_1^0 + 2\frac{(-\Delta H)}{\rho C_p}g_1(\underline{x}_2) & -\frac{q}{V} - \frac{UA}{V\rho C_p} + \frac{(-\Delta H)}{\rho C_p}\varphi_2^0 \end{bmatrix},$$

$$\tilde{A}_3 = \begin{bmatrix} -\frac{q}{V} - \varphi_1^0 & -\varphi_2^0 - 2g_2(\overline{x}_2) \\ \frac{(-\Delta H)}{\rho C_p}\varphi_1^0 & -\frac{q}{V} - \frac{UA}{V\rho C_p} + \frac{(-\Delta H)}{\rho C_p}\varphi_2^0 + 2\frac{(-\Delta H)}{\rho C_p}g_2(\overline{x}_2) \end{bmatrix},$$

$$\tilde{A}_4 = \begin{bmatrix} -\frac{q}{V} - \varphi_1^0 & -\varphi_2^0 - 2g_2(\underline{x}_2) \\ \frac{(-\Delta H)}{\rho C_p}\varphi_1^0 & -\frac{q}{V} - \frac{UA}{V\rho C_p} + \frac{(-\Delta H)}{\rho C_p}\varphi_2^0 + 2\frac{(-\Delta H)}{\rho C_p}g_2(\underline{x}_2) \end{bmatrix},$$

$$\tilde{B}_1 = \tilde{B}_2 = \tilde{B}_3 = \tilde{B}_4 = \begin{bmatrix} 0 \\ \frac{UA}{V\rho C_p} \end{bmatrix}.$$

By discretizing the four continuous-time LTI models (sampling period $T_s = 0.05$ min), we obtain the discrete-time LPV model. Based on (9.17), letting $\dot{C}_A = 0$ and $\dot{T} = 0$, we obtain the steady-state model $f(x_{ss}, u_{ss}) = 0$ of (9.17). Choose the input ss $T_{c,ss} = 330$, and find $C_{A,ss}$ and T_{ss} satisfying (9.14), i.e., $C_{A,ss} = 0.693$, $T_{ss} = 343.465$. Thus, we have $x_{ss} = [0.193; -6.535]$,

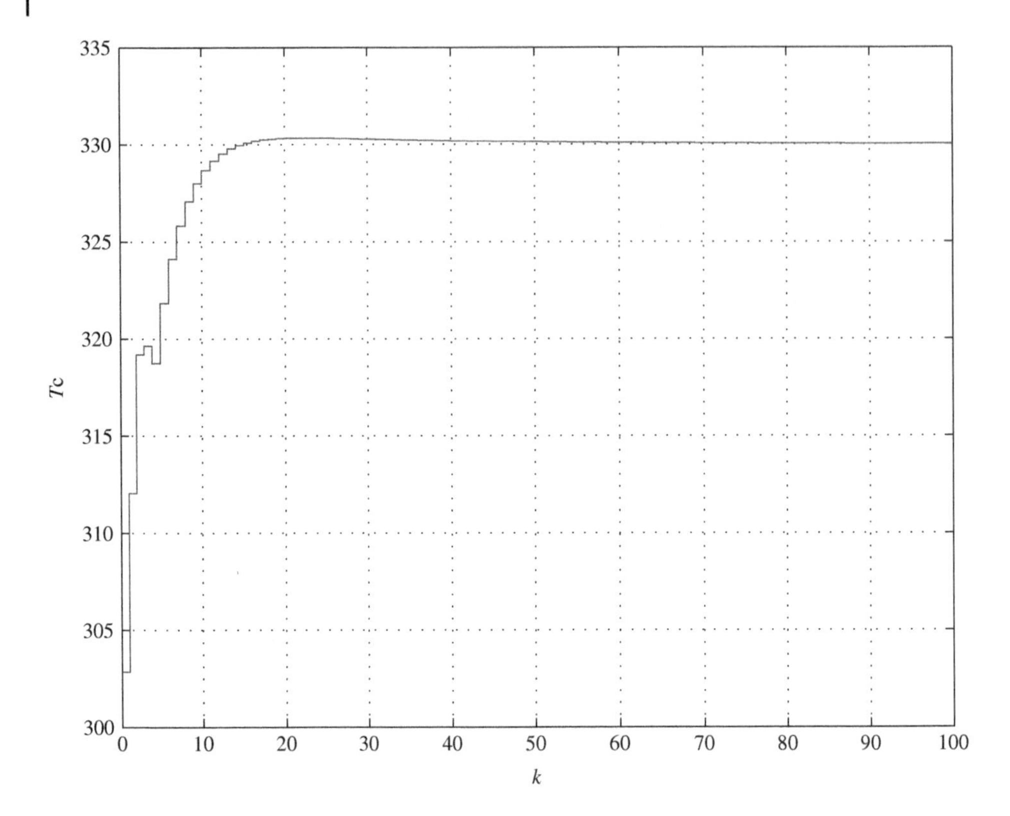

Figure 9.2 The control input.

$u_{ss} = -8$. Choose the initial state $x_0 = [0.3; 3]$, $N = 5$. Apply the control strategy (9.10), replacing $\hat{J}^N_{0,k}$ with $\tilde{J}^N_{0,k}$. The simulation results are shown in Figures 9.1 and 9.2. From the simulation results, it is shown that the values of the input and state finally converge to the steady state, while the deviations between the true steady-state values and ss are zeros. It reveals that the offset-free is achieved.

9.2 Open-Loop MPC for Unmeasurable State

Consider the model in (9.3)–(9.4). Assume that x_k is unmeasurable (all x are not measurable), and y is measurable, i.e., the value of y_k is exactly measured at each time k. We often consider the following constraints on inputs and states:

$$-\underline{u} \leqslant u_k \leqslant \overline{u}, -\underline{\psi} \leqslant \Psi x_{k+1} \leqslant \overline{\psi},\ \forall k \geqslant 0 \tag{9.19}$$

$\Psi \in \mathbb{R}^{q\times n}$. Take $\Psi = [C_1^{\mathrm{T}}, C_2^{\mathrm{T}}, \ldots, C_L^{\mathrm{T}}]^{\mathrm{T}}$ and take $\{\underline{\psi}, \overline{\psi}\}$ appropriately, then $-\underline{\psi} \leqslant \Psi x_{k+1} \leqslant \overline{\psi}$ denotes the output constraint.

Since the state cannot be measured, apply the following observer to estimate the real state:

$$\hat{x}_{k+1} = A_o\hat{x}_k + B_o u_k + L_o y_k, \tag{9.20}$$

where $\hat{x} \in \mathbb{R}^n$ is the estimated state, and $\{A_o, B_o, L_o\}$ are observer parameter matrices. Certainly, if the real-time value of $\omega_{l,k}$ is exactly known, (9.20) can be replaced by

$$\hat{x}_{k+1} = A_k \hat{x}_k + B_k u_k + L_o[y_k - C_k \hat{x}_k]. \tag{9.21}$$

Defining the vertex control move and control horizon N as in Section 9.1, yields

$$\begin{aligned}
\hat{x}^{l_0}_{k+1|k} &= A_{l_0} \hat{x}_{k|k} + B_{l_0} u_{k|k}, \\
\hat{x}^{l_i \cdots l_0}_{k+i+1|k} &= A_{l_i} \hat{x}^{l_{i-1} \cdots l_0}_{k+i|k} + B_{l_i} u^{l_{i-1} \cdots l_0}_{k+i|k}, \\
& i \in \{1, \dots, N-1\},\ l_j \in \{1, \dots, L\},\ j \in \{0, \dots, N-1\}.
\end{aligned}$$

We call $\hat{x}^{l_{i-1} \cdots l_0}_{k+i|k}$, $i \in \{1, \dots, N\}$ the vertex state predictions. Based on (9.3)–(9.4), taking predictions on the future state, yields

$$\begin{aligned}
\hat{x}_{k+1|k} &= A_k \hat{x}_{k|k} + B_k u_{k|k} \\
&= \sum_{l_0=1}^{L} \omega_{l_0,k} \left[A_{l_0} \hat{x}_{k|k} + B_{l_0} u_{k|k} \right] \\
&= \sum_{l_0=1}^{L} \omega_{l_0,k} \hat{x}^{l_0}_{k+1|k}, \\
\hat{x}_{k+2|k} &= A_{k+1} \hat{x}_{k+1|k} + B_{k+1} u_{k+1|k} \\
&= \sum_{l_1=1}^{L} \omega_{l_1,k+1} \left[A_{l_1} \hat{x}_{k+1|k} + B_{l_1} u_{k+1|k} \right] \\
&= \sum_{l_1=1}^{L} \omega_{l_1,k+1} \left[A_{l_1} \sum_{l_0=1}^{L} \omega_{l_0,k} \hat{x}^{l_0}_{k+1|k} + B_{l_1} \sum_{l_0=1}^{L} \omega_{l_0,k} u^{l_0}_{k+1|k} \right] \\
&= \sum_{l_1=1}^{L} \sum_{l_0=1}^{L} \omega_{l_1,k+1} \omega_{l_0,k} \hat{x}^{l_1 l_0}_{k+2|k} \\
& \vdots .
\end{aligned}$$

Hence, $\hat{x}_{k+i|k}$ belongs to the polytope. Writing it in vector form, yields

$$\begin{aligned}
\begin{bmatrix} \hat{x}_{k+1|k} \\ \hat{x}_{k+2|k} \\ \vdots \\ \hat{x}_{k+N|k} \end{bmatrix} &= \sum_{l_0 \cdots l_{N-1}=1}^{L} \left(\prod_{h=0}^{N-1} \omega_{l_h,k+h} \begin{bmatrix} \hat{x}^{l_0}_{k+1|k} \\ \hat{x}^{l_1 l_0}_{k+2|k} \\ \vdots \\ \hat{x}^{l_{N-1} \cdots l_1 l_0}_{k+N|k} \end{bmatrix} \right), \\
\sum_{l_0 \cdots l_{i-1}=1}^{L} \left(\prod_{h=0}^{i-1} \omega_{l_h,k+h} \right) &= 1,\ i \in \{1, \dots, N\},
\end{aligned} \tag{9.22}$$

where

$$\begin{bmatrix} \hat{x}^{l_0}_{k+1|k} \\ \hat{x}^{l_1 l_0}_{k+2|k} \\ \vdots \\ \hat{x}^{l_{N-1}\cdots l_1 l_0}_{k+N|k} \end{bmatrix} = \begin{bmatrix} A_{l_0} \\ A_{l_1}A_{l_0} \\ \vdots \\ \prod_{i=0}^{N-1} A_{l_{N-1-i}} \end{bmatrix} \hat{x}_{k|k}$$
$$+ \begin{bmatrix} B_{l_0} & 0 & \cdots & 0 \\ A_{l_1}B_{l_0} & B_{l_1} & \ddots & \vdots \\ \vdots & \vdots & \ddots & 0 \\ \prod_{i=0}^{N-2} A_{l_{N-1-i}}B_{l_0} & \prod_{i=0}^{N-3} A_{l_{N-1-i}}B_{l_1} & \cdots & B_{l_{N-1}} \end{bmatrix} \begin{bmatrix} u_{k|k} \\ u^{l_0}_{k+1|k} \\ \vdots \\ u^{l_{N-2}\cdots l_1 l_0}_{k+N-1|k} \end{bmatrix}. \tag{9.23}$$

Now, let us define the quadratic positive-definite function of the vertices $[\hat{x}^{l_0}_{k+1|k}; \hat{x}^{l_1 l_0}_{k+2|k}; \vdots; \hat{x}^{l_{N-1}\cdots l_1 l_0}_{k+N|k}]$ and $[u_{k|k}; u^{l_0}_{k+1|k}; \vdots; u^{l_{N-2}\cdots l_1 l_0}_{k+N-1|k}]$ as

$$\begin{aligned} \tilde{J}^N_{0,k} =& \sum_{l_0=1}^{L} \|\hat{x}^{l_0}_{k+1|k} - \hat{x}^{l_0}_{ss}\|^2_{\mathcal{Q}_{1,l_0}} + \|u_{k|k} - u_{ss}\|^2_{\mathcal{R}_0} \\ &+ \sum_{l_1=1}^{L}\sum_{l_0=1}^{L} \|\hat{x}^{l_1 l_0}_{k+2|k} - \hat{x}^{l_1 l_0}_{ss}\|^2_{\mathcal{Q}_{2,l_1,l_0}} + \|u^{l_0}_{k+1|k} - u_{ss}\|^2_{\mathcal{R}_{1,l_0}} \\ &+ \cdots \\ &+ \sum_{l_{N-1}=1}^{L} \cdots \sum_{l_1=1}^{L}\sum_{l_0=1}^{L} \|\hat{x}^{l_{N-1}\cdots l_1 l_0}_{k+N|k} - \hat{x}^{l_{N-1}\cdots l_1 l_0}_{ss}\|^2_{\mathcal{Q}_{N,l_{N-1}\cdots l_1 l_0}} \\ &+ \sum_{l_{N-2}=1}^{L} \cdots \sum_{l_1=1}^{L}\sum_{l_0=1}^{L} \|u^{l_{N-2}\cdots l_1 l_0}_{k+N-1|k} - u_{ss}\|^2_{\mathcal{R}_{N-1,l_{N-2}\cdots l_1 l_0}}, \end{aligned}$$

where $\mathcal{Q}_{1,l_0}$, $\mathcal{R}_0$, $\mathcal{Q}_{2,l_1,l_0}$, $\mathcal{R}_{1,l_0}$, ..., $\mathcal{Q}_{N,l_{N-1}\cdots l_1 l_0}$ and $\mathcal{R}_{N-1,l_{N-2}\cdots l_1 l_0}$ are nonnegative weight matrices, u_{ss} is ss (sp) of u, and

$$\begin{bmatrix} \hat{x}^{l_0}_{ss} \\ \hat{x}^{l_1 l_0}_{ss} \\ \vdots \\ \hat{x}^{l_{N-1}\cdots l_1 l_0}_{ss} \end{bmatrix} = \begin{bmatrix} A_{l_0} \\ A_{l_1}A_{l_0} \\ \vdots \\ \prod_{i=0}^{N-1} A_{l_{N-1-i}} \end{bmatrix} \hat{x}_{ss}$$
$$+ \begin{bmatrix} B_{l_0} & 0 & \cdots & 0 \\ A_{l_1}B_{l_0} & B_{l_1} & \ddots & \vdots \\ \vdots & \vdots & \ddots & 0 \\ \prod_{i=0}^{N-2} A_{l_{N-1-i}}B_{l_0} & \prod_{i=0}^{N-3} A_{l_{N-1-i}}B_{l_1} & \cdots & B_{l_{N-1}} \end{bmatrix} \begin{bmatrix} u_{ss} \\ u_{ss} \\ \vdots \\ u_{ss} \end{bmatrix}. \tag{9.24}$$

At each time, take $[\hat{x}^{l_0}_{k+1|k}; \hat{x}^{l_1 l_0}_{k+2|k}; \vdots; \hat{x}^{l_{N-1}\cdots l_1 l_0}_{k+N|k}]$ and $[u_{k|k}; u^{l_0}_{k+1|k}; \vdots; u^{l_{N-2}\cdots l_1 l_0}_{k+N-1|k}]$ as the decision variables, and solve the following QP:

$$\min \tilde{J}^N_{0,k}, \ \text{s.t. (9.23), (9.24), (9.26), and (9.27).} \tag{9.25}$$

$$-\underline{u} \leqslant u_{k|k} \leqslant \bar{u}, -\underline{u} \leqslant u^{l_{i-1}\cdots l_1 l_0}_{k+i|k} \leqslant \bar{u}, \ i \in \{1,\ldots,N-1\}, \ l_{i-1} \in \{1,\ldots,L\}, \tag{9.26}$$

$$-\underline{\psi}^s \leqslant \Psi^d \begin{bmatrix} \hat{x}^{l_0}_{k+1|k} \\ \hat{x}^{l_1 l_0}_{k+2|k} \\ \vdots \\ \hat{x}^{l_{N-1}\cdots l_1 l_0}_{k+N|k} \end{bmatrix} \leqslant \overline{\psi}^s, \ l_j \in \{1,\ldots,L\}, \ j \in \{0,\ldots,N-1\}. \tag{9.27}$$

Then, send $u_{k|k}$ to the true system.

The approach of (9.25) is called heuristic ol output-feedback MPC, and x of heuristic ol MPC in Section 9.1 is replaced with $\hat{x}$. Apparently, the change is so small. The heuristic ol output-feedback MPC has the following characteristics:

(1) It is computationally less expensive than those output-feedback MPC with stability ingredients.
(2) Its stability cannot be theoretically proved.
(3) A more reliable and general viewpoint is that $\hat{x}_{ss}$ and u_{ss} satisfy certain steady-state nonlinear equation $g(\hat{x}_{ss} + x_{eq}, u_{ss} + u_{eq}) = 0$, or satisfy $\hat{x}_{ss} = A_o\hat{x}_{ss} + B_o u_{ss} + L_o y_{ss}$.

Aiming at (3), there is still room for improvement, which is not mentioned in Section 9.1. Wang and Rawlings [2004] give an idea of how to calculate ss using the polytopic model, which can be applied to solve $\{x_{ss}, u_{ss}, d_l\}_k$ satisfying the following constraints:

$$x_{ss,k} = A_l x_{ss,k} + B_l u_{ss,k} + d_{l,k}, \ l \in \{1,2,\ldots,L\}, \tag{9.28}$$

$$-\underline{u} \leqslant u_{ss,k} \leqslant \bar{u}, \tag{9.29}$$

$$-\underline{\psi} \leqslant \Psi x_{ss,k} \leqslant \overline{\psi}. \tag{9.30}$$

Add $\sum_{l=1}^{L} \|d_{l,k}\|^2$ to the performance index of solving $\{x_{ss}, u_{ss}, d_l\}_k$, so that the magnitude of $d_{l,k}$ is minimized to some extent. In order to achieve the offset-free, the dynamic prediction of the state must be consistent with its steady-state prediction. Since (9.28)–(9.30) is applied, (9.3) and (9.4) should be rewritten as

$$\begin{aligned} x_{k+1} &= A_k x_k + B_k u_k + d_k, \\ d_{l,k+1} &= d_{l,k}, \ l \in \{1,2,\ldots,L\}, \\ y_k &= C_k x_k, \ \ [A_k | B_k | C_k] \in \Omega, \end{aligned} \tag{9.31}$$

where $d_k = \sum_{l=1}^{L} \omega_{l,k} d_{l,k}$. Obviously, (9.28)–(9.31) are independent of the measurability of state.

Based on (9.31), the state prediction equation (9.23) should be modified as

$$\begin{bmatrix} \hat{x}^{l_0}_{k+1|k} \\ \hat{x}^{l_1 l_0}_{k+2|k} \\ \vdots \\ \hat{x}^{l_{N-1}\cdots l_1 l_0}_{k+N|k} \end{bmatrix} = \begin{bmatrix} A_{l_0} \\ A_{l_1}A_{l_0} \\ \vdots \\ \prod_{i=0}^{N-1} A_{l_{N-1-i}} \end{bmatrix} \hat{x}_{k|k}$$
$$+ \begin{bmatrix} B_{l_0} & 0 & \cdots & 0 \\ A_{l_1}B_{l_0} & B_{l_1} & \ddots & \vdots \\ \vdots & \vdots & \ddots & 0 \\ \prod_{i=0}^{N-2} A_{l_{N-1-i}} B_{l_0} & \prod_{i=0}^{N-3} A_{l_{N-1-i}} B_{l_1} & \cdots & B_{l_{N-1}} \end{bmatrix} \begin{bmatrix} u_{k|k} \\ u^{l_0}_{k+1|k} \\ \vdots \\ u^{l_{N-2}\cdots l_1 l_0}_{k+N-1|k} \end{bmatrix}$$
$$+ \begin{bmatrix} I & 0 & \cdots & 0 \\ A_{l_1} & I & \ddots & \vdots \\ \vdots & \vdots & \ddots & 0 \\ \prod_{i=0}^{N-2} A_{l_{N-1-i}} & \prod_{i=0}^{N-3} A_{l_{N-1-i}} & \cdots & I \end{bmatrix} \begin{bmatrix} d_{l_0,k} \\ d_{l_1,k} \\ \vdots \\ d_{l_{N-1},k} \end{bmatrix}. \tag{9.32}$$

The real-time state prediction still takes (9.20).

Let us redefine the quadratic positive-definite function of the vertices $[\hat{x}^{l_0}_{k+1|k}; \hat{x}^{l_1 l_0}_{k+2|k}; \vdots; \hat{x}^{l_{N-1}\cdots l_1 l_0}_{k+N|k}]$ and $[u_{k|k}; u^{l_0}_{k+1|k}; \vdots; u^{l_{N-2}\cdots l_1 l_0}_{k+N-1|k}]$ as

$$\begin{aligned} \breve{J}^N_{0,k} = & \sum_{l_0=1}^{L} \|\hat{x}^{l_0}_{k+1|k} - \check{x}_{ss,k}\|^2_{\mathcal{Q}_{1,l_0}} + \|u_{k|k} - u_{ss}\|^2_{\mathcal{R}_0} \\ & + \sum_{l_1=1}^{L}\sum_{l_0=1}^{L} \|\hat{x}^{l_1 l_0}_{k+2|k} - \check{x}_{ss,k}\|^2_{\mathcal{Q}_{2,l_1,l_0}} + \|u^{l_0}_{k+1|k} - u_{ss}\|^2_{\mathcal{R}_{1,l_0}} \\ & + \cdots \\ & + \sum_{l_{N-1}=1}^{L} \cdots \sum_{l_1=1}^{L}\sum_{l_0=1}^{L} \|\hat{x}^{l_{N-1}\cdots l_1 l_0}_{k+N|k} - \check{x}_{ss,k}\|^2_{\mathcal{Q}_{N,l_{N-1}\cdots l_1 l_0}} \\ & + \sum_{l_{N-2}=1}^{L} \cdots \sum_{l_1=1}^{L}\sum_{l_0=1}^{L} \|u^{l_{N-2}\cdots l_1 l_0}_{k+N-1|k} - u_{ss}\|^2_{\mathcal{R}_{N-1,l_{N-2}\cdots l_1 l_0}}, \end{aligned}$$

where x_{ss} satisfies (consistent with (9.28))

$$\begin{bmatrix} \check{x}_{ss,k} \\ \check{x}_{ss,k} \\ \vdots \\ \check{x}_{ss,k} \end{bmatrix} = \begin{bmatrix} A_{l_0} \\ A_{l_1}A_{l_0} \\ \vdots \\ \prod_{i=0}^{N-1} A_{l_{N-1-i}} \end{bmatrix} \check{x}_{ss,k}$$

$$+\begin{bmatrix} B_{l_0} & 0 & \cdots & 0 \\ A_{l_1}B_{l_0} & B_{l_1} & \ddots & \vdots \\ \vdots & \vdots & \ddots & 0 \\ \prod_{i=0}^{N-2} A_{l_{N-1-i}}B_{l_0} & \prod_{i=0}^{N-3} A_{l_{N-1-i}}B_{l_1} & \cdots & B_{l_{N-1}} \end{bmatrix}\begin{bmatrix} u_{ss,k} \\ u_{ss,k} \\ \vdots \\ u_{ss,k} \end{bmatrix}$$
$$+\begin{bmatrix} I & 0 & \cdots & 0 \\ A_{l_1} & I & \ddots & \vdots \\ \vdots & \vdots & \ddots & 0 \\ \prod_{i=0}^{N-2} A_{l_{N-1-i}} & \prod_{i=0}^{N-3} A_{l_{N-1-i}} & \cdots & I \end{bmatrix}\begin{bmatrix} d_{l_0,k} \\ d_{l_1,k} \\ \vdots \\ d_{l_{N-1},k} \end{bmatrix}.$$

At each time, instead of solving (9.25), we solve the following QP:

$$\min \check{J}_{0,k}^{N}, \text{ s.t. (9.32), (9.26), and (9.27),} \tag{9.33}$$

and send $u_{k|k}$ to the real system.

Which one is better, (9.25) or (9.33)? Neither. The former is suitable when x_{ss} and u_{ss} satisfy certain steady-state nonlinear equation $g(x_{ss}+x_{eq}, u_{ss}+u_{eq})=0$. The latter is suitable when x_{ss} and u_{ss} satisfy (9.28).

Note that $d_{l,k}$, appearing in (9.31), change the linear polytopic model $x_{k+1}=A_kx_k+B_ku_k$. This implies that the linear polytopic model $x_{k+1}=A_kx_k+B_ku_k$ is not suitable for the offset-free problem in the tracking control, but suitable in the regulation case. Defining $\bar{z}=x-x_{ss}$ and $v=u-u_{ss}$ for (9.31) (not for (9.3)–(9.4)), yields

$$\bar{z}_{k+1}=A_k\bar{z}_k+B_k\bar{v}_k,\ y_k=C_k\bar{z}_k+C_kx_{ss,k},\quad [A_k|B_k|C_k]\in\Omega. \tag{9.34}$$

$d_{l,k}$ is not in (9.34). If $C_k=C$ in y_k is time-invariant, then the "tail" in the expression of y_k can be removed. However, as mentioned in Section 9.1, if $x_{ss,k}$ is time-variant, similarly as the industrial hierarchical MPC, then the constraint bounds of $\bar{z}$ and $\bar{v}$ are time-variant.

Example 9.2 Consider CSTR model

$$\begin{aligned} \dot{x}_1 &= -x_1-\left(3.6\times10^{10}\exp\left(-\frac{8750}{x_2+350}\right)\right)(2x_1+1)+0.5,\\ \dot{x}_2 &= 10^{10}\exp\left(-\frac{8750}{x_2+350}\right)(361.506x_1+180.753)\\ &\quad -3.0921x_2-25.103+2.0921u. \end{aligned} \tag{9.35}$$

The parameters of the linear polytopic model are

$$A_1=\begin{bmatrix}0.8227 & -0.0017\\ 6.1234 & 0.9367\end{bmatrix}, A_2=\begin{bmatrix}0.9654 & -0.0018\\ -0.6759 & 0.9433\end{bmatrix},$$
$$A_3=\begin{bmatrix}0.8895 & -0.0029\\ 2.9447 & 0.9968\end{bmatrix}\mathrm{T}, A_4=\begin{bmatrix}0.8930 & -0.0006\\ 2.7738 & 0.8864\end{bmatrix}\mathrm{T},$$
$$B_1=\begin{bmatrix}-0.0001 & 0.1014\end{bmatrix}^{\mathrm{T}}, B_2=\begin{bmatrix}-0.0001 & 0.1016\end{bmatrix}^{\mathrm{T}},$$
$$B_3=\begin{bmatrix}-0.0002 & 0.1045\end{bmatrix}\mathrm{T}, B_4=\begin{bmatrix}-0.000034 & 0.0986\end{bmatrix}\mathrm{T},$$
$$C_1=C_2=C_3=C_4=\begin{bmatrix}0 & 1\end{bmatrix}.$$

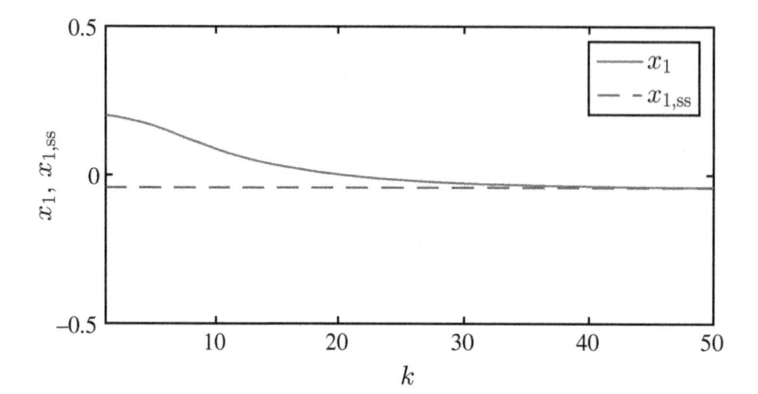

Figure 9.3 Response of state $x_1(k)$.

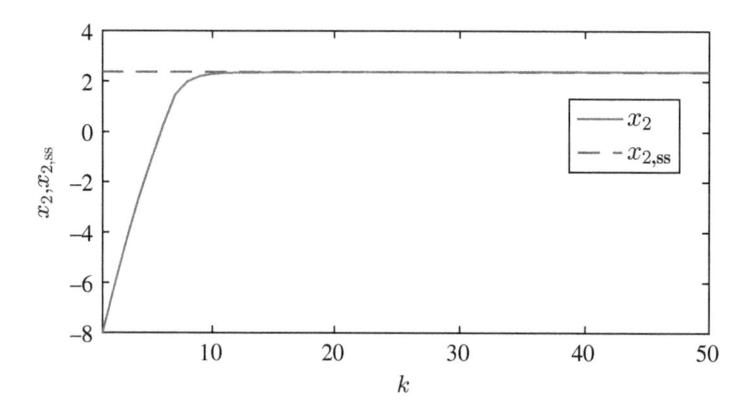

Figure 9.4 Response of state $x_2(k)$.

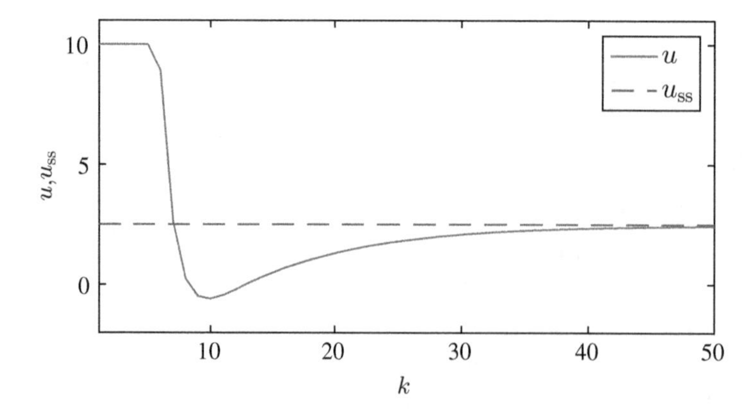

Figure 9.5 Control input signal $u(k)$.

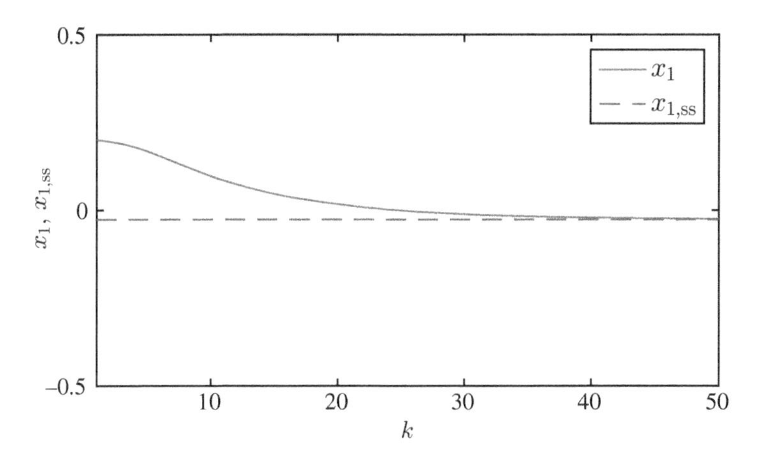

Figure 9.6 Response of state $x_1(k)$.

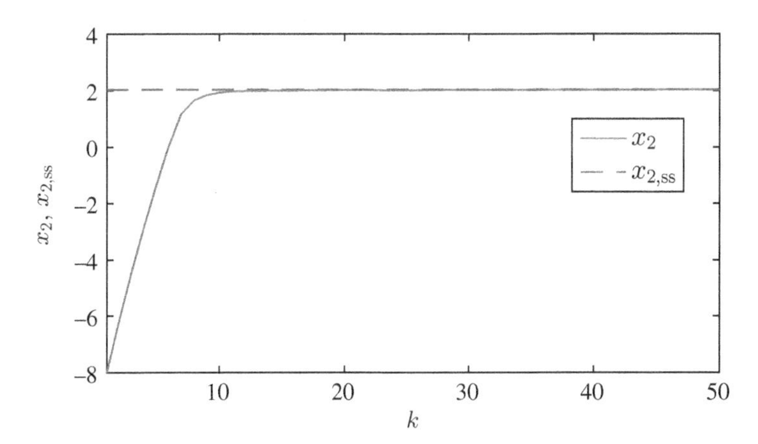

Figure 9.7 Response of state $x_2(k)$.

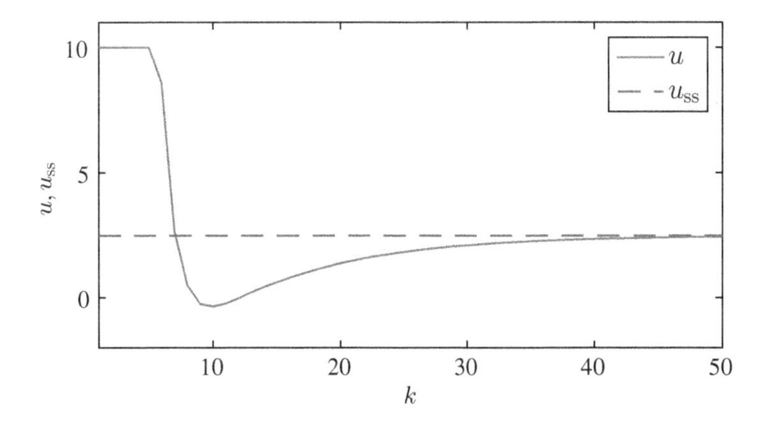

Figure 9.8 Control input signal $u(k)$.

The coefficients are $\omega_1 = \frac{1}{2}\frac{\varphi_1(x_2)-\varphi_1(-\bar{x}_2)}{\varphi_1(\bar{x}_2)-\varphi_1(-\bar{x}_2)}$, $\omega_2 = \frac{1}{2}\frac{\varphi_1(\bar{x}_2)-\varphi_1(x_2)}{\varphi_1(\bar{x}_2)-\varphi_1(-\bar{x}_2)}$, $\omega_3 = \frac{1}{2}\frac{\varphi_2(x_2)-\varphi_2(-\bar{x}_2)}{\varphi_2(\bar{x}_2)-\varphi_2(-\bar{x}_2)}$, $\omega_4 = \frac{1}{2}\frac{\varphi_2(\bar{x}_2)-\varphi_2(x_2)}{\varphi_2(\bar{x}_2)-\varphi_2(-\bar{x}_2)}$, where $\varphi_1(x_2) = 7.2\times 10^{10}\exp\left(-\frac{8750}{x_2+350}\right)$, $\varphi_2(x_2) = 3.6\times 10^{10}\left[\exp\left(-\frac{8750}{x_2+350}\right) - \exp\left(-\frac{8750}{350}\right)\right]\frac{1}{x_2}$. The input and output constraints are $|u| \leq 10$ and $|y| \leq 10$, and the state constraint $|x_1| \leqslant 0.5$.

Take initial state $x_0 = [0.2, -8]^{\mathrm{T}}$, $y_0 = 3.3$, and ET $u_t = 2.5$, $x_t = [-0.04, 2.37]^{\mathrm{T}}$, and $L_0 = [1, 1]^{\mathrm{T}}$, $N = 3$, $\mathscr{Q} = 10$, $\mathscr{R} = 1$. By applying (9.25) and the steady-state nonlinear model in SSTC, the simulation results are shown in Figures 9.3–9.5. The control move and system state track the reachable ET well. The model consistency between the upper and lower layers guarantees the offset-free.

By applying (9.33), the linear model and introducing $d_{l,k}$, the simulation results are shown in Figures 9.6–9.8. The control move and the system state can track ss of the upper layer, which also guarantees the offset-free.

10

Robust Model Predictive Control

Notations in this chapter:

ε_M: the ellipsoid associated with the positive-definite matrix M, i.e., $\varepsilon_M = \{\xi | \xi^T M \xi \leqslant 1\}$;
Co$\mathcal{S}$: an element belonging to Co$\mathcal{S}$ means that it is a convex combination of the elements in the polytope $\mathcal{S}$, with the scalar combing coefficients being nonnegative and summing as 1;
$\star$: this symbol induces a symmetric structure in any square matrix;
$*$: a value with superscript $*$ means that it is the solution of the optimization problem.

10.1 A Cornerstone Method

In 1996, a robust MPC approach appeared in Kothare et al. [1996]. This work gives us "KBM formula" which has great value.

10.1.1 KBM Formula

In this chapter, we define Kothare–Balakrishnan–Morari (KBM) formula as

$$\begin{bmatrix} 1 & x_{k|k}^T \\ x_{k|k} & Q_k \end{bmatrix} \geq 0, \tag{10.1}$$

$$\begin{bmatrix} Q_k & \star & \star & \star \\ A_l Q_k + B_l Y_k & Q_k & \star & \star \\ \mathscr{Q}^{1/2} Q_k & 0 & \gamma_k I & \star \\ \mathscr{R}^{1/2} Y_k & 0 & 0 & \gamma_k I \end{bmatrix} \geq 0, \ l \in \{1, \dots, L\}, \tag{10.2}$$

$$\begin{bmatrix} Q_k & Y_{j,k}^T \\ Y_{j,k} & \bar{u}_j^2 \end{bmatrix} \geq 0, \ j \in \{1, \dots, m\}, \tag{10.3}$$

$$\begin{bmatrix} Q_k & \star \\ \Psi_s (A_l Q_k + B_l Y_k) & \overline{\psi}_s^2 \end{bmatrix} \geq 0, \ l \in \{1, 2, \dots, L\}, \ s \in \{1, 2, \dots, q\}. \tag{10.4}$$

Model Predictive Control, First Edition. Baocang Ding and Yuanqing Yang.

With $\{\gamma_k, Q_k, Y_k\}$ being the unknown variables, (10.1)–(10.4) are linear matrix inequalities (LMIs).

Let us explain the symbols and connotations of KBM formula.

(1) $x_{k|k}$ is the recognition of x_k at time k. By assuming that the state is measurable, it directly yields $x_{k|k} = x_k$. Equation (10.1) means that $x_{k|k}^T Q_k^{-1} x_{k|k} \leqslant 1$. Since Q_k is positive-definite matrix, (10.1) means that $x_{k|k}$ lies in the ellipsoid sized by Q_k; this ellipsoid is denoted as $\mathcal{E}_{Q_k^{-1}}$. Can any eigenvalue of Q_k be zero? Yes, from (10.1) to (10.4), this can happen, but this can hardly be met in real applications.
(2) The original intention of (10.1) is more control-oriented: adopting a quadratic performance index $J_{0,k}^{\infty} = \sum_{i=0}^{\infty} \left[\|x_{k+i|k}\|_{\mathcal{Q}}^2 + \|u_{k+i|k}\|_{\mathcal{R}}^2\right]$. Such a quadratic performance index is the same as the famous LQR. Hence, it is more traditional to define the quadratic function $V(x) = x^T P_k x$, $P_k > 0$. Like LQR, by appropriate treatment, it yields $J_{0,k}^{\infty} \leqslant V(x_{k|k})$, i.e., the upper bound of the infinite-horizon quadratic performance cost is $V(x_{k|k})$. This result is not as accurate as $J_{0,k}^{\infty} = V(x_{k|k})$ for the nominal model $x_{k+1} = Ax_k + Bu_k$. Let $V(x_{k|k}) \leqslant \gamma_k$ and $P_k = \gamma_k Q_k^{-1}$, then we obtain $x_{k|k} \in \mathcal{E}_{Q_k^{-1}}$.
(3) Equation (10.2) guarantees

$$V(x_{k+i+1|k}) - V(x_{k+i|k}) \leqslant -\left[\|x_{k+i|k}\|_{\mathcal{Q}}^2 + \|u_{k+i|k}\|_{\mathcal{R}}^2\right]. \tag{10.5}$$

For LQR of nominal model, basically we can say that (10.2) is equivalent to (10.5). Certainly, for LQR of nominal model, it is unnecessary to adopt (10.2). Equation (10.2) is the result via treating Lyapunov inequality; while for LQR of nominal model, one adopts the Lyapunov equation (equality). What are the benefits to guarantee (10.5) so as to obtain the Lyapunov inequality? The benefits are twofold, i.e., both guaranteeing stability and handling optimality. It is ingenious, but conservative and robust, to handle optimality in this manner; via (10.5), it is shown that $J_{0,k}^{\infty} \leqslant x_{k|k}^T P_k x_{k|k}$ holds for all modeling uncertainty.
(4) Due to the above treatment of optimality and definition $P_k = \gamma_k Q_k^{-1}$, γ_k (the key variable embodying performance index, to be optimized, and to be the Lyapunov function) does not appear in (10.1) but appear in (10.2); this is an excellent turnover.
(5) With this successful turnover, $\mathcal{E}_{Q_k^{-1}}$ becomes the invariant set of $x_{k+i|k}$, $i \geq 0$. That is, the power of (10.1) and (10.2) is to keep $x_{k+i|k}$, $i \geq 0$ inside of $\mathcal{E}_{Q_k^{-1}}$. Since the range of $x_{k+i|k}$ is restrained, the treatment of constraints on $x_{k+i|k}$ will be more promising. KBM formula guarantees the constraints for the infinite-horizon (in comparison, Section 9.1 only handles the constraints within the horizon N). Equations (10.3) and (10.4) guarantee that, when $u_{j,k+i|k} = F_{j,k} x_{k+i|k}$ for $i \geq 0$ and $F_j = Y_j Q^{-1}$ for $j \in \{1, 2, \ldots, m\}$,

$$-\bar{u} \leqslant u_{k+i|k} \leqslant \bar{u},\ -\overline{\psi} \leqslant \Psi x_{k+i+1|k} \leqslant \overline{\psi},\ \forall i \geq 0. \tag{10.6}$$

F_j is the jth row of F, Y_j is the jth row of Y, and Ψ_s is the sth row of Ψ. The upper and lower bounds of (10.6) are the same value, otherwise KBM formula becomes more inappropriate.
(6) If there is no such turnover, then KBM formula becomes

$$\begin{bmatrix} \beta_k & \beta_k x_{k|k}^T \\ \beta_k x_{k|k} & X_k \end{bmatrix} \geq 0, \tag{10.7}$$

$$\begin{bmatrix} X_k & \star & \star & \star \\ A_l X_k + B_l Y_k & X_k & \star & \star \\ \mathcal{Q}^{1/2} X_k & 0 & I & \star \\ \mathcal{R}^{1/2} Y_k & 0 & 0 & I \end{bmatrix} \geq 0, \; l \in \{1, \dots, L\}, \tag{10.8}$$

$$\begin{bmatrix} X_k & Y_{j,k}^T \\ Y_{j,k} & \beta_k \bar{u}_j^2 \end{bmatrix} \geq 0, \; j \in \{1, \dots, m\}, \tag{10.9}$$

$$\begin{bmatrix} X_k & \star \\ \Psi_s(A_l X_k + B_l Y_k) & \beta_k \overline{\psi}_s^2 \end{bmatrix} \geq 0, \; l \in \{1, 2, \dots, L\}, \; s \in \{1, 2, \dots, q\}, \tag{10.10}$$

where $X = P^{-1}$, $\beta = \gamma^{-1}$, and $F = YX^{-1}$. $\begin{bmatrix} \gamma_k & \star \\ x_{k|k} & X_k \end{bmatrix} \geq 0$ has been changed into (10.7). It seems that (10.7)–(10.10) are more intuitive than (10.1)–(10.4). Why we need (10.1)–(10.4)? The reason is that in the literature, (10.3) and (10.4) are usually changed into

$$\begin{bmatrix} Q_k & Y_k^T \\ Y_k & Z_k \end{bmatrix} \geq 0, \; Z_{jj,k} \leqslant \bar{u}_j^2, \; j \in \{1, \dots, m\}, \tag{10.11}$$

$$\begin{bmatrix} Q_k & \star \\ \Psi(A_l Q_k + B_l Y_k) & \Gamma_k \end{bmatrix} \geq 0, \; \Gamma_{ss,k} \leqslant \overline{\psi}_s^2, \; l \in \{1, 2, \dots, L\}, \; s \in \{1, 2, \dots, q\}, \tag{10.12}$$

where Z_k and Γ_k are the additional unknown variables; Z_{jj} and Γ_{ss} represent, respectively, the jth and sth diagonal elements of Z and Γ. In general, (10.11) and (10.12) involve lighter computational burden than (10.3) and (10.4) (e.g., with relatively large m and q), but (10.3) and (10.4) are less conservative (at least theoretically) than (10.11) and (10.12). With the inclusion of Z and Γ, one needs to change (10.9) and (10.10) as

$$\begin{bmatrix} X_k & Y_k^T \\ Y_k & \beta_k Z_k \end{bmatrix} \geq 0, \; Z_{jj,k} \leqslant \bar{u}_j^2, \; j \in \{1, \dots, m\}, \tag{10.13}$$

$$\begin{bmatrix} X_k & \star \\ \Psi(A_l X_k + B_l Y_k) & \beta_k \Gamma_k \end{bmatrix} \geq 0, \; \Gamma_{ss,k} \leqslant \overline{\psi}_s^2, \; l \in \{1, 2, \dots, L\}, \; s \in \{1, 2, \dots, q\}. \tag{10.14}$$

As such, $\{\beta, Z, \Gamma\}_k$ is not linearly appear in (10.13) and (10.14).

(7) β_k appears in (10.7)–(10.10) for several times, while γ_k appears in (10.1)–(10.4) only once. Hence, the small trouble brought by the turnover and counter-intuitiveness is worthy.

10.1.2 KBM Controller

Let us give KBM optimization as

$$\min_{\{\gamma, Y, Q\}_k} \gamma_k, \; \text{s.t. } (10.1) - (10.4),$$

and KBM controller as

$$u_k = Y_k Q_k^{-1} x_k.$$

In Kothare et al. [1996], it is proved the closed-loop stability by implementing KBM controller in a receding-horizon manner. The proof in Kothare et al. [1996] has been widely cited in the literature, for which one should take care of one proof detail, as shown below.

(1) For the robust MPC of LPV model, if the controller optimization is solved online, then it usually takes the min–max optimization. The basic reason for taking this kind of optimization is that, since the model is uncertain, so are the prediction of $x_{k+i|k}$ and the calculation of $J_{0,k}^{\infty} = \sum_{i=0}^{\infty} \left[\|x_{k+i|k}\|_{\mathscr{Q}}^2 + \|u_{k+i|k}\|_{\mathscr{R}}^2\right]$. Moreover, in order to avoid encountering "blind alley," the control actions and constraints are all "infinite horizon," so the minimization of $J_{0,k}^{\infty}$ is an infinite-dimensional optimization problem. The so-called min–max optimization is to minimize the maximum of $J_{0,k}^{\infty}$. γ_k gives the upper-bound of $J_{0,k}^{\infty}$ for all uncertainty realizations, so the minimization of γ_k approximates the min–max optimization. Since γ_k is minimized, it is suggested to adopt the optimal value of γ_k, γ_k^*, as the Lyapunov function.
(2) What if, instead of applying γ_k^*, it applies $x_k^T P_k^* x_k$ as the Lyapunov function? $P_k^* = \gamma_k^* Q_k^{*-1}$. It does achieve $x_k^T P_k^* x_k \leqslant \gamma_k^*$, but does it achieve $x_k^T P_k^* x_k = \gamma_k^*$? If it does not achieve $\gamma_k^* = x_k^T P_k^* x_k$, then is it true that $x_k^T P_k^* x_k$ is minimized?
(3) Denote Lyapunov function of robust MPC stability as $\mathscr{L}_k^*$, and the usual approach is to first prove $\mathscr{L}_{k+1} \leqslant \mathscr{L}_k^*$ (where the equality only holds if the steady state is reached). According to the principle of optimality, there always exists $\mathscr{L}_{k+1}^* \leqslant \mathscr{L}_{k+1}$ (the minimized value must not be greater than the one that has not been minimized), and $\mathscr{L}_{k+1}^* \leqslant \mathscr{L}_k^*$ is obtained. It is easy to ignore the latter point, i.e., ignore investigating $\mathscr{L}_{k+1}^* \leqslant \mathscr{L}_{k+1}$. If $\mathscr{L}_{k+1}$ is not minimized at all, why the latter point is true?
(4) In order to avoid encountering "blind alley," the control actions and constraints are all "infinite horizon"; otherwise it is unclear that when k is large, whether or not the constraints are no longer satisfied.
(5) How is $\mathscr{L}_{k+1}$ obtained? This is not the result of optimization, but the result by appropriately constructing the decision variables that satisfy all the constraints of the optimization problem. In the context of KBM controller, it requires to construct the solution that satisfies four KBM formulas.
(6) The property of being able to find decision variables satisfying all constraints of the optimization problem after initial time (controller startup) is called recursive feasibility.

Considering that the above analysis is deeply involved with various issues of MPC stability research (not limited to this chapter and this book), we give the following suggestions.

(1) The item just being minimized (maybe some items, which are not affected by optimization, are also added) should be chosen as the Lyapunov function. For example, in the robust control, γ_k^* is often used as the Lyapunov function;
(2) In the proof of recursive feasibility, it is necessary to construct decision variables that "satisfy all constraints of the optimization problem." For example, for LMI-based robust control, the constructed decision variables must satisfy the corresponding LMI.

10.1.3 Example: Generalizing to Networked Control

Consider the system described by LTI state-space model in Figure 10.1, where it transmits data via sensor-controller link and controller-actuator link. The data are transmitted in a single packet at each time step. In Figure 10.1, $\breve{u} \in \mathbb{R}^m$ and $\breve{x} \in \mathbb{R}^n$ are the output and the input to the controller, respectively. The sensor is clock driven, while the controller and actuator are event driven. The sensor and controller only send data at each sampling time $k \geq 0$. If the actuator does not receive new data at sampling time $k \geq 0$, then the control move at time $k-1$ is implemented (zero-order hold effect).

Let $\mathscr{J} := \{j_1, j_2, \dots\}$, a subsequence of $\{0, 1, 2, 3, \dots\}$, denotes the sequence of time points of successful data transmissions from the sensor to controller. Let $\mathscr{I} := \{i_1, i_2, \dots\}$, a subsequence of $\{0, 1, 2, 3, \dots\}$, denotes the sequence of time points of successful data transmissions from the controller to actuator. Let $d_1 := \max_{j_l \in \mathscr{J}} (j_{l+1} - j_l)$ and $d_2 := \max_{i_l \in \mathscr{I}} (i_{l+1} - i_l)$ be the maximum packet loss upper bounds.

Definition 10.1 The packet-loss process is defined as $\{\eta_{j_l} = j_{l+1} - j_l | j_l \in \mathscr{J}\}$, $\{\rho_{i_l} = i_{l+1} - i_l | i_l \in \mathscr{I}\}$, where η_{j_l} and ρ_{i_l} take values in the finite sets $\mathscr{D}_1 := \{1, 2, \dots, d_1\}$ and $\mathscr{D}_2 := \{1, 2, \dots, d_2\}$, respectively.

Definition 10.2 The packet-loss process in Definite 10.1 is arbitrary, if η_{j_l} and ρ_{i_l} take values in $\mathscr{D}_1$ and $\mathscr{D}_2$, respectively, arbitrarily.

10.1.3.1 Closed-Loop Model for Double-Sided, Finite-Bounded, Arbitrary Packet Loss

Only a part of the successful data transmissions can affect the closed-loop system.

Definition 10.3 The operator ord$\{\cdot\}$ is the ordering operator. For any integer sequence $\{m_1, m_2, \dots, m_r, \dots\}$, ord$\{m_1, m_2, \dots, m_r, \dots\}$=$\{\overline{m}_1, \overline{m}_2, \dots, \overline{m}_r, \dots\}$ rearranges $\{m_1, m_2, \dots, m_r, \dots\}$, without removing or adding any element. $\{\overline{m}_1, \overline{m}_2, \dots, \overline{m}_r, \dots\}$ is called ordered sequence.

Let $\mathscr{S} = \text{ord}\{\mathscr{J} \bigcup \mathscr{I}\}$ where, if $i_{l_2} = j_{l_1}$ for two integers l_1 and l_2, then j_{l_1} goes before i_{l_2} in $\mathscr{S}$. There is one, and only one, ordered sequence $\mathbb{S} = \{\mathbb{j}_1, \mathbb{i}_1, \mathbb{j}_2, \mathbb{i}_2, \dots\} \subseteq \mathscr{S}$, such that

(a) $\{\mathbb{j}_1, \mathbb{j}_2, \dots\} \subseteq \mathscr{J}$;
(b) $\{\mathbb{i}_1, \mathbb{i}_2, \dots\} \subseteq \mathscr{I}$;
(c) for any $j_{l_1} \leqslant i_{l_2}$, if j_{l_1} and i_{l_2} are adjacent in $\mathscr{S}$, then $\{j_{l_1}, i_{l_2}\} \subset \mathbb{S}$.

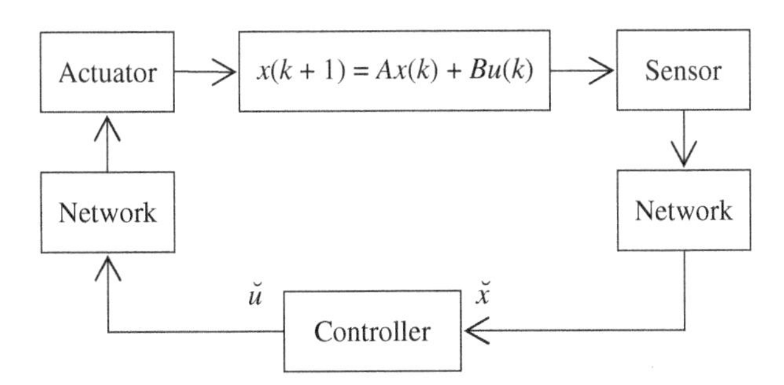

Figure 10.1 The networked control system.

Theorem 10.1 For the packet loss process in Definitions 10.1 and 10.2, it holds that $\mathbb{i}_r - \mathbb{j}_r + 1 \in \mathcal{D}_2$ and $\mathbb{i}_{r+1} - \mathbb{j}_r \in \mathcal{D}$, where $\mathcal{D} := \{1, 2, \dots, d_1 + d_2 - 1\}$.

Suppose the controller is $\check{u} = K\check{x}$. At each j_l, $\check{u}_{j_l} = K\check{x}_{j_l} = Kx_{j_l}$ is calculated. For all k satisfying $j_l \leqslant k < j_{l+1}$, $\check{u}_k = \check{u}_{j_l}$ is sent to the actuator. $\check{u}_k$ may or may not be able to arrive at the actuator, which the controller does not know a priori. By considering the definition of $\mathbb{S}$, the closed-loop system for all $k \geq \mathbb{i}_1$ is

$$x_{k+1} = A^{k-\mathbb{i}_r+1}x_{\mathbb{i}_r} + B_{k-\mathbb{i}_r+1}Kx_{\mathbb{j}_r}, \quad \mathbb{i}_r \leqslant k < \mathbb{i}_{r+1}, \ r \geq 1, \tag{10.15}$$

where $B_j = \sum_{s=0}^{j-1} A^s B$ for any integer $j > 0$. In order to analyze stability of (10.15), it is advisable to define $z_k = \left[x_k^T, x_{k-1}^T, \dots, x_{k-d_2+1}^T\right]^T$, and the following system with augmented state $z_{\mathbb{i}_r}$:

$$z_{\mathbb{i}_{r+1}} = \Phi_{\mathbb{i}_r} z_{\mathbb{i}_r}, \ \Phi_{\mathbb{i}_r} = \sum_{t\in\mathcal{D}_2} \sum_{l\in\mathcal{D}(t)} \varpi_{t,\mathbb{i}_r} \omega_{l,\mathbb{i}_r} \Phi_{lt},$$

$$\varpi_{t,\mathbb{i}_r} = \begin{cases} 1, & \mathbb{i}_r - \mathbb{j}_r + 1 = t \\ 0, & \text{otherwise} \end{cases}, \ \omega_{l,\mathbb{i}_r} = \begin{cases} 1, & \mathbb{i}_{r+1} - \mathbb{i}_r = l \\ 0, & \text{otherwise} \end{cases}, \tag{10.16}$$

where $\mathcal{D}(t) := \{1, 2, \dots, d_1 + d_2 - t\}$, $\Phi_{lt} \in \mathbb{R}^{(d_2 n)\times(d_2 n)}$, $\Phi_{lt} = \begin{bmatrix} \Psi_{lt} \\ [I \ \ 0] \end{bmatrix}$,

$\Psi_{lt} \in \mathbb{R}^{(\min\{l,d_2\}n)\times(d_2 n)}$, $\Psi_{lt} = \begin{bmatrix} \Psi_{lt}^1 & \Psi_{lt}^2 & \cdots & \Psi_{lt}^{d_2} \end{bmatrix}$, $\Psi_{lt}^s \in \mathbb{R}^{(\min\{l,d_2\}n)\times n}$, $s \in \mathcal{D}_2$;

$\Psi_{l1}^s = 0 \ (s \neq 1)$,

$$\Psi_{l1}^1 = \begin{bmatrix} A^l + B_l K \\ A^{l-1} + B_{l-1} K \\ \vdots \\ A^{\max\{1,l-d_2+1\}} + B_{\max\{1,l-d_2+1\}} K \end{bmatrix};$$

if $t \neq 1$, $\Psi_{lt}^s = 0 \ (s \notin \{1, t\})$,

$$\begin{bmatrix} \Psi_{lt}^1 & \Psi_{lt}^t \end{bmatrix} = \begin{bmatrix} A^l & B_l K \\ A^{l-1} & B_{l-1} K \\ \vdots & \vdots \\ A^{\max\{1,l-d_2+1\}} & B_{\max\{1,l-d_2+1\}} K \end{bmatrix}.$$

According to Theorem 10.1, stability of (10.16) is equivalent to stability of (10.15).

10.1.3.2 MPC for Double-Sided, Arbitrary Packet Loss

Consider the following input and state constraints:

$$\begin{aligned} -\underline{u} &\leqslant u_{\mathbb{i}_1+p} \leqslant \bar{u}, \\ -\underline{\psi} &\leqslant \Psi x_{\mathbb{i}_1+p+1} \leqslant \overline{\psi}, \quad p \geq 0, \end{aligned} \tag{10.17}$$

where $\Psi \in \mathbb{R}^{q\times n}$. Define $\mathbb{m} := \{1, 2, \dots, m\}$ and $\mathbb{q} := \{1, 2, \dots, q\}$. In MPC, at each j_l, an optimization problem is solved. The controller does not know a priori whether or not the current j_l will be equal to $\mathbb{j}_r$ $(r \geq 1)$. However, the controller predicts (assumes) that $j_l = \mathbb{j}_{s_l}$

where $s_l \geq 1$ (the exact value of s_l is not required) and optimizes $\{\check{u}_{j_l}, \check{u}_{\mathbb{j}_{s_l+1}|j_l}, \check{u}_{\mathbb{j}_{s_l+2}|j_l}, \dots\}$. At time k satisfying $j_l \leqslant k < j_{l+1}$, $\check{u}_k = \check{u}_{j_l}$ is sent to the actuator. This $\check{u}_{j_l}$ may or may not be able to arrive at the actuator. At time j_{l+1}, the optimization is performed again to obtain $\{\check{u}_{j_{l+1}}, \check{u}_{\mathbb{j}_{s_{l+1}+1}|j_{l+1}}, \check{u}_{\mathbb{j}_{s_{l+1}+2}|j_{l+1}}, \dots\}$. As KBM controller, a state feedback control law $\check{u} = F\check{x}$ will be utilized, where F is to be optimized.

Assumption 10.1 The controller knows, at each j_l, whether or not the previously sent values have been received by the actuator.

The controller does not know a *priori* whether or not the sent value will be received by the actuator. The sensor sends $S_{\text{act}} = [s_{\text{act},k-1}, s_{\text{act},k-2}, \dots, s_{\text{act},k-d_1}]$ (along with x_k) to the controller, at each time k. At time $k-i$ ($i \in \mathscr{D}_1$), take $s_{\text{act},k-i} = 1$ or $s_{\text{act},k-i} = 0$, depending on whether or not the actuator received new data from the controller.

In the networked MPC, (10.15) is utilized for the prediction model. For KBM controller, the output of the controller is $\check{u}_{k|j_l} = \check{u}_{j_{l+\tau}|j_l} = F_{j_l}\check{x}_{j_{l+\tau}|j_l} = F_{j_l}x_{j_{l+\tau}|j_l}$ for all $j_{l+\tau} \leqslant k < j_{l+\tau+1}$ and all $\tau \geq 0$; on the plant side, the predicted control input is

$$u_{k|j_l} = u_{\mathbb{i}_{s_l+\tau}|j_l} = \check{u}_{\mathbb{j}_{s_l+\tau}|j_l} = F_{j_l}x_{\mathbb{j}_{s_l+\tau}|j_l}, \quad \mathbb{i}_{s_l+\tau} \leqslant k < \mathbb{i}_{s_l+\tau+1}, \tag{10.18}$$

for all $\tau \geq 0$, where $j_l = \mathbb{j}_{s_l}$ is assumed (predicted). Hence, the closed-loop state prediction becomes

$$\begin{aligned} x_{k+1|j_l} =& A^{k-\mathbb{i}_{s_l+\tau}+1}x_{\mathbb{i}_{s_l+\tau}|j_l} + B_{k-\mathbb{i}_{s_l+\tau}+1}F_{j_l}x_{\mathbb{j}_{s_l+\tau}|j_l}, \\ & \mathbb{i}_{s_l+\tau} \leqslant k < \mathbb{i}_{s_l+\tau+1}, \end{aligned} \tag{10.19}$$

for all $\tau \geq 0$,

$$\begin{aligned} x_{\mathbb{j}_{s_l}|j_l} &= x_{j_l|j_l} = x_{j_l}, \\ x_{\mathbb{i}_{s_l}|j_l} &= A^{\mathbb{i}_{s_l}-j_l}x_{j_l} + B_{\mathbb{i}_{s_l}-j_l}u_{j_l|j_l}. \end{aligned} \tag{10.20}$$

At each j_l solve

$$\min_{F_{j_l}} \max_{\bigstar} J_{j_l} = \sum_{\tau=0}^{\infty} \left[\|x_{\mathbb{i}_{s_l+\tau}|j_l}\|_{\mathscr{Q}}^2 + \|u_{\mathbb{i}_{s_l+\tau}|j_l}\|_{\mathscr{R}}^2 \right], \tag{10.21}$$

$$\text{s.t.} \ -\underline{u} \leqslant u_{k|j_l} \leqslant \bar{u}, -\underline{\psi} \leqslant \Psi x_{k+1|j_l} \leqslant \overline{\psi}, \quad k \geq \mathbb{i}_{s_l}, \tag{10.22}$$

$$(10.18), \ (10.19), (10.20), \quad \tau \geq 0, \tag{10.23}$$

where $\bigstar = \{\mathbb{i}_{s_l+\tau} - \mathbb{j}_{s_l+\tau} + 1 \in \mathscr{D}_2, \ \mathbb{i}_{s_l+\tau+1} - \mathbb{j}_{s_l+\tau} \in \mathscr{D}, \ \tau \geq 0\}$; $\mathscr{Q}$ and $\mathscr{R}$ are symmetric positive-definite weight matrices.

10.1.3.3 Solution of MPC for Double-Sided Packet Loss

In order to derive an upper bound on $\max_{\bigstar} J_{j_l}$, let us impose the stability constraint

$$V(z_{\mathbb{i}_{s_l+\tau+1}|j_l}) - V(z_{\mathbb{i}_{s_l+\tau}|j_l}) \leqslant -\|x_{\mathbb{i}_{s_l+\tau}|j_l}\|_{\mathscr{Q}}^2 - \|u_{\mathbb{i}_{s_l+\tau}|j_l}\|_{\mathscr{R}}^2, \quad \tau \geq 0, \tag{10.24}$$

where

$$V(z_{\mathbb{i}_{s_l+\tau}|j_l}) = \|z_{\mathbb{i}_{s_l+\tau}|j_l}\|_{P_{\mathbb{i}_{s_l+\tau}|j_l}}^2, \quad P_{\mathbb{i}_{s_l+\tau}|j_l} = \sum_{t \in \mathscr{D}_2} \sum_{r \in \mathscr{D}(t)} \varpi_{t,\mathbb{i}_{s_l+\tau}} \omega_{r,\mathbb{i}_{s_l+\tau}} P_{rt},$$

$$\varpi_{t,\mathbb{i}_{s_l+\tau}} = \begin{cases} 1, & \mathbb{i}_{s_l+\tau} - \mathbb{j}_{s_l+\tau} + 1 = t \\ 0, & \text{otherwise} \end{cases}, \quad \omega_{r,\mathbb{i}_{s_l+\tau}} = \begin{cases} 1, & \mathbb{i}_{s_l+\tau+1} - \mathbb{i}_{s_l+\tau} = r \\ 0, & \text{otherwise} \end{cases}.$$

Define $F := YG^{-1}$ and $Q_{rt} := \gamma P_{rt}^{-1}$, where γ is a scalar. Denote

$$Q_{rt} = \begin{bmatrix} Q_{rt}^{11} & (Q_{rt}^{12})^T & \cdots & (Q_{rt}^{1,d_2})^T \\ Q_{rt}^{12} & Q_{rt}^{22} & \ddots & \vdots \\ \vdots & \ddots & \ddots & (Q_{rt}^{d_2-1,d_2})^T \\ Q_{rt}^{1,d_2} & \cdots & Q_{rt}^{d_2-1,d_2} & Q_{rt}^{d_2,d_2} \end{bmatrix},$$

$$G_{rt} = \begin{bmatrix} G_{rt}^{11} & G_{rt}^{12} & \cdots & G_{rt}^{1,d_2} \\ G_{rt}^{21} & G_{rt}^{22} & \ddots & \vdots \\ \vdots & \ddots & \ddots & G_{rt}^{d_2-1,d_2} \\ G_{rt}^{d_2,1} & \cdots & G_{rt}^{d_2,d_2-1} & G_{rt}^{d_2,d_2} \end{bmatrix}, \quad G_{rt}^{ts} = G, \quad s \in \mathscr{D}_2.$$

All the blocks in Q_{rt} and G_{rt} have the same dimension. It is shown that (10.24) is guaranteed by

$$\begin{bmatrix} G_{rt}^T + G_{rt} - Q_{rt} & \star & \star & \star \\ \Phi_{rt} G_{rt} & Q_{hp} & \star & \star \\ \mathscr{Q}^{1/2} G_{rt}^1 & 0 & \gamma I & \star \\ \mathscr{R}^{1/2} \mathcal{Y} & 0 & 0 & \gamma I \end{bmatrix} \geq 0, \; t, p \in \mathscr{D}_2, \; r \in \mathscr{D}(t), \; h \in \mathscr{D}(p), \tag{10.25}$$

where $G_{rt}^1 = \begin{bmatrix} G_{rt}^{11} & G_{rt}^{12} & \cdots & G_{rt}^{1,d_2} \end{bmatrix}$ and $\mathcal{Y} = \begin{bmatrix} Y & Y & \cdots & Y \end{bmatrix}$. Each Φ_{rt} is shown in (10.16), with K replaced by F. Note that in $\Phi_{rt} G_{rt}$, each F is multiplied with G, which is changed as Y.

For a stable closed-loop system, $\lim_{\tau\to\infty} V(z_{\mathbb{i}_{s_l+\tau}|j_l}) = 0$ holds. By summing (10.24) from $\tau = 0$ to $\tau = \infty$, it follows that $J_{j_l} \leqslant V(z_{\mathbb{i}_{s_l}|j_l})$, where

$$z_{\mathbb{i}_{s_l}|j_l} = \begin{bmatrix} x_{\mathbb{i}_{s_l}|j_l}^T & \cdots & x_{j_l+1|j_l}^T & x_{j_l|j_l}^T & x_{j_l-1|j_l}^T & \cdots & x_{\mathbb{i}_{s_l}-d_2+1|j_l}^T \end{bmatrix}^T.$$

By minimizing γ satisfying $V(z_{\mathbb{i}_{s_l}|j_l}) \leqslant \gamma$, the upper bound on the performance cost is minimized. If $\mathbb{i}_{s_l} - j_l + 1 = t$, then denote $z_{t,\mathbb{i}_{s_l}|j_l} := z_{j_l+t-1|j_l}$. By observing $z_{\mathbb{i}_{s_l}|j_l} = \sum_{t\in\mathscr{D}_2} \varpi_t(\mathbb{i}_{s_l}) z_{t,\mathbb{i}_{s_l}|j_l}$, it is easy to show that $V(z_{\mathbb{i}_{s_l}|j_l}) \leqslant \gamma$ is guaranteed by

$$\begin{bmatrix} 1 & \star \\ z_{t,\mathbb{i}_{s_l}|j_l} & Q_{rt} \end{bmatrix} \geq 0, \quad t \in \mathscr{D}_2, \; r \in \mathscr{D}(t). \tag{10.26}$$

Due to Assumption 10.1, the controller can obtain the values of $\{u_{j_l-d_2+1|j_l}, \dots, u_{j_l-2|j_l}, u_{j_l-1|j_l}\}$ from known information of $\{u_{j_{l-1}-d_2+1|j_{l-1}}, \dots, u_{j_{l-1}-2|j_{l-1}}, u_{j_{l-1}-1|j_{l-1}}\}$ and known information of $\{\breve{u}_{j_{l-1}-d_2+1}, \dots, \breve{u}_{j_{l-1}-2}, \breve{u}_{j_{l-1}-1}, \breve{u}_{j_{l-1}}\}$, i.e.,

$$u_{j_l-i|j_l} = \begin{cases} \breve{u}_{j_l-i}, & s_{\mathrm{act,j_l-i}} = 1 \\ u_{j_l-i-1|j_{l-1}}, & s_{\mathrm{act,j_l-i}} = 0 \end{cases}, \quad i \in \{1, 2, \dots, d_2 - 1\}.$$

With $u_{\mathbb{i}_{s_l}-1|j_l} = u_{\mathbb{i}_{s_l}-2|j_l} = \cdots = u_{j_l+1|j_l} = u_{j_l|j_l}$ (for $t > 1$, $u_{j_l|j_l} = u_{j_l-1|j_l}$), it is easy to exactly calculate $z_{t,\mathbb{i}_{s_l}|j_l}$ in (10.26).

Lemma 10.1 Suppose there exists a scalar γ, the symmetric matrices $\{Z, \Gamma, Q_{rt}\}$ and any matrices $\{G_{rt}, Y\}$ such that (10.25), (10.26), and the following are satisfied:

$$\begin{bmatrix} G^T + G - Q_{rt}^{tt} & \star \\ Y & Z \end{bmatrix} \geq 0, \quad t \in \mathscr{D}_2, \ r \in \mathscr{D}(t), \tag{10.27}$$

$$\begin{bmatrix} G^T + G - Q_{r1}^{11} & \star \\ \Psi(A^r G + B_r Y) & \Gamma \end{bmatrix} \geq 0, \quad r \in \mathscr{D}, \tag{10.28}$$

$$\begin{bmatrix} (G_{rt}^{11})^T + G_{rt}^{11} - Q_{rt}^{11} & \star & \star \\ (G_{rt}^{1t})^T + G - Q_{rt}^{1t} & G^T + G - Q_{rt}^{tt} & \star \\ \Psi(A^r G_{rt}^{11} + B_r Y) & \Psi(A^r G_{rt}^{1t} + B_r Y) & \Gamma \end{bmatrix} \geq 0, \ t \in \{2, \ldots, d_2\}, \ r \in \mathscr{D}(t), \tag{10.29}$$

$$Z_{ss} \leqslant u_{s,\inf}^2, \quad s \in \mathfrak{m}, \quad \Gamma_{ss} \leqslant \psi_{s,\inf}^2, \quad s \in \mathfrak{q}, \tag{10.30}$$

where $u_{s,\inf} = \min\{\underline{u}_s, \bar{u}_s\}$, $\psi_{s,\inf} = \min\{\underline{\psi}_s, \overline{\psi}_s\}$, Z_{ss} (Γ_{ss}) is the sth diagonal element of Z (Γ), then (10.22) is satisfied.

In summary, (10.21)–(10.23) are approximated by the following LMI optimization:

$$\min_{\gamma, Q_{rt}, G_{rt}, Y, Z, \Gamma} \gamma \ \text{s.t.,} \ (10.26), \ (10.25), \ (10.27) - (10.30). \tag{10.31}$$

Theorem 10.2 Assume that (10.31) is feasible at $j_{\overline{n}}$ ($\overline{n} \geq 1$) ((10.31) may be infeasible at $j_1, j_2, \ldots, j_{\overline{n}-1}$) and $\breve{u}_{j_{\overline{n}}}$ is the first control move received by the actuator. Then, (10.31) will be feasible at $j_{\overline{n}+\tau}$ for all $\tau > 0$, and the receding horizon sending of the control move $\breve{u}_k = Y_{j_l} G_{j_l}^{-1} x_{j_l}$ ($j_l \leqslant k < j_{l+1}$, $l \geq 1$) guarantees satisfaction of (10.17), and the closed-loop system is exponentially stable.

Proof: Denote $\ell := l + 1$. Assume that (10.31) is feasible at time $\mathbb{j}_{s_\ell-1}$ ($\mathbb{j}_{s_\ell-1} = j_l$ if and only if $\breve{u}_{j_l}$ has been received by the actuator). At time j_ℓ, due to Assumption 10.1, there is no uncertainty in predicting $z_{p,\mathbb{i}_{s_\ell}|j_\ell}$ for all $p = \mathbb{i}_{s_\ell} - j_\ell + 1 \in \mathscr{D}_2$. However, at time $\mathbb{j}_{s_\ell-1}$, in the prediction of $z^*_{j_\ell+p-1|\mathbb{j}_{s_\ell-1}}$ for all $p = \mathbb{i}_{s_\ell} - j_\ell + 1 \in \mathscr{D}_2$, there is uncertainty since the controller can only exactly predict $z^*_{t,\mathbb{i}_{s_\ell-1}|\mathbb{j}_{s_\ell-1}}$ for all $t = \mathbb{i}_{s_\ell-1} - \mathbb{j}_{s_\ell-1} + 1 \in \mathscr{D}_2$. Hence, any possible $z_{p,\mathbb{i}_{s_\ell}|j_\ell}$ is included in the realizations of $z^*_{j_\ell+p-1|\mathbb{j}_{s_\ell-1}}$ for all $p = \mathbb{i}_{s_\ell} - j_\ell + 1 \in \mathscr{D}_2$.

By applying (10.24),

$$\begin{aligned} &\|z^*_{j_\ell+p-1|\mathbb{j}_{s_\ell-1}}\|^2_{P^*_{hp,\mathbb{j}_{s_\ell-1}}} - \|z_{\mathbb{j}_{s_\ell-1}+t-1|\mathbb{j}_{s_\ell-1}}\|^2_{P^*_{rt,\mathbb{j}_{s_\ell-1}}} \\ &\leqslant -\|x_{\mathbb{j}_{s_\ell-1}+t-1|\mathbb{j}_{s_\ell-1}}\|^2_{\mathscr{Q}} - \|\breve{u}^*_{\mathbb{j}_{s_\ell-1}}\|^2_{\mathscr{R}}, \ t, p \in \mathscr{D}_2, \ r \in \mathscr{D}(t), \ h \in \mathscr{D}(p). \end{aligned} \tag{10.32}$$

Since any possible $z_{j_\ell+p-1|j_\ell}$ is included in the realizations of $z^*_{j_\ell+p-1|\mathbb{j}_{s_\ell-1}}$, (10.32) leads to

$$\begin{aligned}&\|z_{j_\ell+p-1|j_\ell}\|^2_{P^*_{hp,j_{s_\ell-1}}} - \|z_{j_{s_\ell-1}+t-1|j_{s_\ell-1}}\|^2_{P^*_{rt,j_{s_\ell-1}}}\\ &\leqslant -\|x_{j_{s_\ell-1}+t-1|j_{s_\ell-1}}\|^2_{\mathscr{Q}} - \|\check{u}^*_{j_{s_\ell-1}}\|^2_{\mathscr{R}},\ t,p\in\mathscr{D}_2,\ r\in\mathscr{D}(t),\ h\in\mathscr{D}(p).\end{aligned} \tag{10.33}$$

Note that

$$\begin{aligned}&\|z_{j_{s_\ell-1}+t-1|j_{s_\ell-1}}\|^2_{P^*_{rt,j_{s_\ell-1}}} \leqslant \gamma^*_{j_{s_\ell-1}},\\ &\|z_{j_\ell+p-1|j_\ell}\|^2_{P_{hp,j_\ell}} \leqslant \gamma_{j_\ell}.\end{aligned} \tag{10.34}$$

According to (10.33)–(10.34), it is admissible to choose

$$\begin{aligned}\gamma_{j_\ell} &= \gamma^*_{j_{s_\ell-1}} - \max_{t\in\mathscr{D}_2}\left\{\|x_{j_{s_\ell-1}+t-1|j_{s_\ell-1}}\|^2_{\mathscr{Q}}\right\} - \|\check{u}^*_{j_{s_\ell-1}}\|^2_{\mathscr{R}},\\ \{Q_{rt},G_{rt},Y,Z,\Gamma\}_{j_\ell} &= \frac{\gamma_{j_\ell}}{\gamma^*_{j_{s_\ell-1}}}\{Q_{rt},G_{rt},Y,Z,\Gamma\}^*_{j_{s_\ell-1}},\end{aligned}$$

as the solution to (10.31). It means feasibility of (10.31) at $j_{s_\ell-1}$ leads to its feasibility at j_ℓ. By induction, feasibility of (10.31) at time $j_{\overline{n}}$ means its feasibility at any $j_{\overline{n}+\tau}$, $\tau>0$.

By re-optimization at time j_ℓ, it must result in $\gamma^*_{j_\ell} \leqslant \gamma_{j_\ell}$. Therefore,

$$\begin{aligned}\gamma^*_{j_\ell} - \gamma^*_{j_{s_\ell-1}} &\leqslant -\max_{t\in\mathscr{D}_2}\left\{\|x_{j_{s_\ell-1}+t-1|j_{s_\ell-1}}\|^2_{\mathscr{Q}}\right\} - \|\check{u}^*_{j_{s_\ell-1}}\|^2_{\mathscr{R}}\\ &\leqslant -\lambda_{\min}(\mathscr{Q})\max_{t\in\mathscr{D}_2}\left\{\|x_{j_{s_\ell-1}+t-1|j_{s_\ell-1}}\|^2\right\},\end{aligned} \tag{10.35}$$

where $\lambda_{\min}(\mathscr{Q})$ is the minimum eigenvalue of $\mathscr{Q}$. Note that $\max_{t\in\mathscr{D}_2}\left\{\|x_{j_{s_\ell-1}+t-1|j_{s_\ell-1}}\|^2\right\}\neq 0$ for any $z_{j_{s_\ell-1}|j_{s_\ell-1}}\neq 0$. Therefore, (10.35) means that $\gamma^*_{j_l}$, $l\geq 1$ is the Lyapunov function for proving the exponential stability of the closed-loop system. Note that $j_{\overline{n}} = j_1$ and refer to Lemma 10.1 for proving satisfaction of input and state constraints.

10.2 Invariant Set Trap

In Kothare et al. [1996], it defines the quadratic function $V(x)=x^TP_kx$, $P_k>0$, and uses appropriate techniques to make $\mathscr{E}_{Q_k^{-1}}$ $(Q_k=\gamma_kP_k^{-1})$ an invariant set. If the optimization problem is feasible at time k, then at time $k+1$, one of KBM formulas,

$$\begin{bmatrix}1 & \star\\ x_{k+1|k+1} & Q_{k+1}\end{bmatrix}\geq 0,$$

is feasible taking $Q_{k+1}=Q_k$.

There have been many improvements in the literatures on the quadratic functions (Lyapunov functions in most cases) recently. Like $V(x)=x^TP_kx$ in Kothare et al. [1996], the way of taking P_k as a single positive-definite matrix (for MPC, means P_k is a single positive-definite matrix at each time k and can be different at different time) is thought to be simple. Actually, beyond MPC, Lyapunov function can be taken as

$$V(x)=\sum_{l_1=1}^{L}\omega_{l_1}\left(\sum_{l_2=1}^{L}\omega_{l_2}\left(\cdots\left(\sum_{l_p=1}^{L}\omega_{l_p}P_{l_1l_2\cdots l_p}\right)\right)\right), \tag{10.36}$$

or as

$$V(x) = \left(\sum_{l_1=1}^{L} \omega_{l_1} \left(\sum_{l_2=1}^{L} \omega_{l_2} \left(\cdots \left(\sum_{l_p=1}^{L} \omega_{l_p} Q_{l_1 l_2 \cdots l_p} \right) \right) \right) \right)^{-1}. \tag{10.37}$$

There are many other ways. Taking (10.36) and (10.37) (see [Lee et al., 2010]), if ω is the function of x (changing with x), it is better not to call it a quadratic function of x, but to call it the parameter-dependent positive-definite function. Note that if it is applied properly, all matrices $P_{l_1 l_2 \cdots l_p}/Q_{l_1 l_2 \cdots l_p}$ are not necessarily positive definite. For ease of calculation and application, (10.36) and (10.37) are written as (see [Oliveira and Peres, 2007, Ding, 2010c])

$$V(x) = \sum_{l_1=1}^{L} \omega_{l_1} \left(\sum_{l_2=l_1}^{L} \omega_{l_2} \left(\cdots \left(\sum_{l_p=l_{p-1}}^{L} \omega_{l_p} \overline{P}_{l_1 l_2 \cdots l_p} \right) \right) \right), \tag{10.38}$$

$$V(x) = \left(\sum_{l_1=1}^{L} \omega_{l_1} \left(\sum_{l_2=l_1}^{L} \omega_{l_2} \left(\cdots \left(\sum_{l_p=l_{p-1}}^{L} \omega_{l_p} \overline{Q}_{l_1 l_2 \cdots l_p} \right) \right) \right) \right)^{-1}, \tag{10.39}$$

which is called the homogeneous polynomial parameter-dependent positive-definite function. If we apply these functions appropriately, then (10.36) is equivalent to (10.38), and (10.37) is equivalent to (10.39). Compared with (10.36) and (10.37), (10.38) and (10.39) have some obvious advantages.

Take (10.38) as an example, $\overline{P}_{l_1 l_2 \cdots l_p}$ is the sum of all $P_{l'_1 l'_2 \cdots l'_p}$, $\{l'_1, l'_2, \ldots, l'_p\} \in \mathcal{P}\{l_1, l_2, \ldots, l_p\}$, and $\mathcal{P}$ is the permutation operation. For each $\overline{P}_{l_1 l_2 \cdots l_p}$, set the corresponding parameters as $\omega_1^{r_1} \omega_2^{r_2} \cdots \omega_L^{r_L}$, $r_1 + r_2 + \cdots + r_L = p$, then the number of $\{l'_1, l'_2, \ldots, l'_p\}$ is $C(l_1, l_2, \ldots, l_p) = \frac{p!}{r_1! r_2! \ldots r_L!}$. Hence, we usually rewrite (10.38) and (10.39) as

$$V(x) = \sum_{l_1=1}^{L} \omega_{l_1} \left(\sum_{l_2=l_1}^{L} \omega_{l_2} \left(\cdots \left(\sum_{l_p=l_{p-1}}^{L} \omega_{l_p} C(l_1, l_2, \ldots, l_p) P_{l_1 l_2 \cdots l_p} \right) \right) \right) \tag{10.40}$$

$$V(x) = \left(\sum_{l_1=1}^{L} \omega_{l_1} \left(\sum_{l_2=l_1}^{L} \omega_{l_2} \left(\cdots \left(\sum_{l_p=l_{p-1}}^{L} \omega_{l_p} C(l_1, l_2, \ldots, l_p) Q_{l_1 l_2 \cdots l_p} \right) \right) \right) \right)^{-1}. \tag{10.41}$$

According to Oliveira and Peres [2007] and Ding [2010c], with the increase in p, Lyapunov matrix in (10.36)–(10.41) can play the role of any positive-definite matrix continuous on ω. ω is a short for $\{\omega_1, \omega_2, \ldots, \omega_L\}$.

Pay attention to the differences below.

- Positive-definite functions like $x^T P x$ and (10.36)–(10.41) are used to guarantee invariance.
- Minimized value like γ is used to prove the closed-loop stability.

Cuzzola et al. [2002] aim at the l-th vertex model of the polytope and choose $V(x) = x^T P_{l,k} x$, $F_k = Y_k G_k^{-1}$. Accordingly, KBM formula is written as

$$\begin{bmatrix} 1 & x_{k|k}^T \\ x_{k|k} & Q_{l,k} \end{bmatrix} \geq 0, \; l \in \{1, \ldots, L\} \tag{10.42}$$

$$\begin{bmatrix} G_k^T + G_k - Q_{l,k} & \star & \star & \star \\ A_l G_k + B_l Y_k & Q_{l,k} & \star & \star \\ \mathcal{Q}^{1/2} G_k & 0 & \gamma_k I & \star \\ \mathcal{R}^{1/2} Y_k & 0 & 0 & \gamma_k I \end{bmatrix} \geq 0, \ l \in \{1, \dots, L\}, \tag{10.43}$$

$$\begin{bmatrix} G_k^T + G_k - Q_{l,k} & Y_{j,k}^T \\ Y_{j,k} & \bar{u}_j^2 \end{bmatrix} \geq 0, \ l \in \{1, \dots, L\}, \ j \in \{1, \dots, m\}, \tag{10.44}$$

$$\begin{bmatrix} G_k^T + G_k - Q_{l,k} & \star \\ \Psi_s (A_l G_k + B_l Y_k) & \overline{\psi}_s^2 \end{bmatrix} \geq 0, \ l \in \{1, 2, \dots, L\}, \ s \in \{1, 2, \dots, q\}. \tag{10.45}$$

This approach falls into the invariant set trap, which cannot prove the recursive feasibility. Considering the 2×2 block in the left upper corner of (10.43) yields

$$\begin{aligned} & \begin{bmatrix} G_k^T + G_k - Q_{l,k} & \star \\ A_l G_k + B_l Y_k & Q_{l,k} \end{bmatrix} \geq 0 \\ \Rightarrow & \begin{bmatrix} G_k^T Q_{l,k}^{-1} G_k & \star \\ A_l G_k + B_l Y_k & Q_{l,k} \end{bmatrix} \geq 0 \\ \Rightarrow & \begin{bmatrix} Q_{l,k}^{-1} & \star \\ A_l + B_l F_k & Q_{l,k} \end{bmatrix} \geq 0 \Rightarrow \begin{bmatrix} P_{l,k} & \star \\ A_l + B_l F_k & P_{l,k}^{-1} \end{bmatrix} \geq 0 \\ \Rightarrow & \begin{bmatrix} P_{l,k} & \star \\ P_{l,k}[A_l + B_l F_k] & P_{l,k} \end{bmatrix} \geq 0. \end{aligned}$$

Except that $\omega_{l,k} = \omega_{l,k+1} \in \{0, 1\}$ (LTI switch system), it holds

$$\begin{aligned} & \begin{bmatrix} P_{l,k} & \star \\ P_{l,k}[A_l + B_l F_k] & P_{l,k} \end{bmatrix} \geq 0 \\ \nRightarrow & \sum_{l=1}^{L} \omega_{l,k} \sum_{j=1}^{L} \omega_{j,k+1} \begin{bmatrix} P_{l,k} & \star \\ P_{j,k}[A_l + B_l F_k] & P_{j,k} \end{bmatrix} \geq 0 \\ \Rightarrow & x_{k+1}^T \sum_{l=1}^{L} \omega_{l,k+1} Q_{l,k}^{-1} x_{k+1} \leqslant x_k^T \sum_{l=1}^{L} \omega_{l,k} Q_{l,k}^{-1} x_k \leqslant 1. \end{aligned}$$

When $\omega_{l,k} = \omega_{l,k+1} \in \{0, 1\}$, the above $\nRightarrow$ becomes $\Rightarrow$, and (10.42) is feasible with $k \to k+1$; for other $\omega_{l,k}$ and $\omega_{l,k+1}$, we cannot show that (10.42) is feasible with $k \to k+1$. This implies that, even for $\omega_{l,k}$ being invariant with time k, the result in Cuzzola et al. [2002] still falls into the invariant set trap.

Mao [2003] corrects the deficiency of [Cuzzola et al., 2002], where (10.43) is modified as

$$\begin{bmatrix} G_k^T + G_k - Q_{l,k} & \star & \star & \star \\ A_l G_k + B_l Y_k & Q_{j,k} & \star & \star \\ \mathscr{Q}^{1/2} G_k & 0 & \gamma_k I & \star \\ \mathscr{R}^{1/2} Y_k & 0 & 0 & \gamma_k I \end{bmatrix} \geq 0,\ j, l \in \{1, \ldots, L\}, \tag{10.46}$$

i.e., the subscript of (2, 2)-block in the 4×4 matrix is changed. In order to compare with Type II xxx below, {(10.42), (10.46), (10.44), (10.45)} are combined and called Type I improved KBM formula. Although (10.46) is much more conservative than (10.43), it avoids the invariant set trap. Considering the 2×2 block in the left upper corner of (10.43), similarly, yields

$$\begin{bmatrix} G_k^T + G_k - Q_{l,k} & \star \\ A_l G_k + B_l Y_k & Q_{j,k} \end{bmatrix} \geq 0$$

$$\Rightarrow \begin{bmatrix} P_{l,k} & \star \\ P_{j,k}[A_l + B_l F_k] & P_{j,k} \end{bmatrix} \geq 0$$

$$\Rightarrow \sum_{l=1}^{L} \omega_{l,k} \sum_{j=1}^{L} \omega_{j,k+1} \begin{bmatrix} P_{l,k} & \star \\ P_{j,k}[A_l + B_l F_k] & P_{j,k} \end{bmatrix} \geq 0$$

$$\Rightarrow x_{k+1}^T \sum_{l=1}^{L} \omega_{l,k+1} Q_{l,k}^{-1} x_{k+1} \leqslant x_k^T \sum_{l=1}^{L} \omega_{l,k} Q_{l,k}^{-1} x_k \leqslant 1.$$

Hence, it is suitable for the general time-varying uncertain systems and is also suitable for the time-invariant uncertain systems.

The above invariant set trap mainly reflects a fact, i.e., the invariance requirement on the recursive feasibility of MPC may be higher than that on other control strategies without receding horizon optimization. For example, if the receding horizon optimization is not used, (10.43) can be the stability and invariance condition of the general time-invariant uncertain systems; but this invariance cannot guarantee the recursive feasibility of MPC.

If we change [Mao, 2003] a little and take the positive-definite function

$$V(x) = x^T G^{-T} \left(\sum_{j=1}^{L} \omega_l P_l \right) G^{-1} x,$$

then KBM formula can be written as

$$\begin{bmatrix} 1 & x_{k|k}^T \\ x_{k|k} & G_k^T + G_k - Q_{l,k} \end{bmatrix} \geq 0,\ l \in \{1, \ldots, L\}, \tag{10.47}$$

$$\begin{bmatrix} Q_{l,k} & \star & \star & \star \\ A_l G_k + B_l Y_k & G_k^T + G_k - Q_{j,k} & \star & \star \\ \mathscr{Q}^{1/2} G_k & 0 & \gamma_k I & \star \\ \mathscr{R}^{1/2} Y_k & 0 & 0 & \gamma_k I \end{bmatrix} \geq 0,\ j, l \in \{1, \ldots, L\}, \tag{10.48}$$

$$\begin{bmatrix} Q_{l,k} & Y_{j,k}^T \\ Y_{j,k} & \bar{u}_j^2 \end{bmatrix} \geq 0, \ l \in \{1,\ldots,L\}, \ j \in \{1,\ldots,m\}, \tag{10.49}$$

$$\begin{bmatrix} Q_{l,k} & \star \\ \Psi_s(A_l G_k + B_l Y_k) & \overline{\psi}_s^2 \end{bmatrix} \geq 0, \ l \in \{1,2,\ldots,L\}, \ s \in \{1,2,\ldots,q\}, \tag{10.50}$$

where the lower right corner of the first KBM formula should be similar as the (2, 2) block of the second KBM formula; the upper left corner of the second, third, and fourth KBM formulas should be similar. The above four formulas are combined and called Type II improved KBM formula. Garone and Casavola [2012] adopt Type II, but based on the stronger assumption: $\omega_{l,k}$ is exactly known, and thus

$$u = \left(\sum_{j=1}^{L} \omega_l Y_l\right) \left(\sum_{j=1}^{L} \omega_l G_l\right)^{-1} x$$

can be calculated in real-time. Accordingly,

$$V(x) = x^T \left(\sum_{j=1}^{L} \omega_l G_l\right)^{-T} \left(\sum_{j=1}^{L} \omega_l P_l\right) \left(\sum_{j=1}^{L} \omega_l G_l\right)^{-1} x.$$

For $\omega_{l,k}$ exactly known, Ding [2011d] adopts Type I and defines the same control law as Garone and Casavola [2012] and

$$V(x) = x^T \left(\sum_{j=1}^{L} \sum_{l=1}^{L} \omega_j \omega_l Q_{jl}\right)^{-1} x.$$

Assume that $\omega_{l,k}$ is exactly known (like Takagi-Sugeno fuzzy model), which can bring many benefits. For example, for Garone and Casavola [2012] and Ding [2011d], there is a good improvement. Equation (19) in Garone and Casavola [2012] is revised (with notations being consistent with [Garone and Casavola, 2012]), i.e.,

$$\begin{bmatrix} 1 & \star \\ \tilde{x} & \overline{G}_i^T + \overline{G}_i - \overline{P}_i \end{bmatrix} \geq 0, \ i = 1,2,\ldots,l \ \rightarrow \begin{bmatrix} 1 & \star \\ \tilde{x} & \sum_{i=1}^{l} p_i(\overline{G}_i^T + \overline{G}_i - \overline{P}_i) \end{bmatrix} \geq 0,$$

which by using the symbols in this book, is

$$\begin{bmatrix} 1 & \star \\ x_k & G_{l,k}^T + G_{l,k} - Q_{l,k} \end{bmatrix} \geq 0, \ l \in \{1,2,\ldots,L\} \ \rightarrow \begin{bmatrix} 1 & \star \\ x_k & \sum_{l=1}^{L} \omega_{l,k}(G_{l,k}^T + G_{l,k} - Q_{l,k}) \end{bmatrix} \geq 0.$$

Equation (2a) in Ding [2011d] is revised (with notations being consistent with Ding [2011d]), i.e.,

$$\begin{bmatrix} 1 & \star \\ x_{k|k} & \hat{S}_{lj} \end{bmatrix} \geq 0, \ l,j \in S \ \rightarrow \begin{bmatrix} 1 & \star \\ x_{k|k} & \sum_{l\in S} h_l(\theta_k) \sum_{j\in S} h_j(\theta_k)\hat{S}_{lj} \end{bmatrix} \geq 0,$$

which by using the symbols in this book, is

$$\begin{bmatrix} 1 & \star \\ x_{k|k} & Q_{lj,k} \end{bmatrix} \geq 0,\ l,j \in \{1,2,\ldots,L\} \rightarrow \begin{bmatrix} 1 & \star \\ x_{k|k} & \sum_{l=1}^{L} \omega_{l,k} \sum_{j=1}^{L} \omega_{j,k} Q_{lj,k} \end{bmatrix} \geq 0.$$

With these revisions, the computational burden is lower, and the result is less conservative.

Compared with (10.36)–(10.41), Mao [2003] and Garone and Casavola [2012] correspond to $p = 1$ (roughly), while Ding [2011d] corresponds to $p = 2$. For an arbitrary p, it is admissible to give a unified result directly; one can refer to Ding [2010c]. Ding [2010c] gives any feedback control gain matrix continuous on ω, and any Lyapunov matrix continuous on ω (notations must be consistent with Ding [2010c]), i.e.,

$$\begin{aligned} u(t) &= -Y_{p,z} S_{p,z}^{-1} x(t), \\ V(x(t)) &= x(t)^T S_{p,z}^{-1} x(t), \\ X_{p,z} &= \sum_{\mathbf{p} \in \mathcal{K}(p)} \mathbf{h}^{\mathbf{p}} X_{\mathbf{p}},\ X \in \{Y, S\}, \end{aligned}$$

which by using the notations in this book, is,

$$\begin{aligned} u_k &= -Y_{p,\omega,k} S_{p,\omega,k}^{-1} x_k, \\ V(x_k) &= x_k^T S_{p,\omega,k}^{-1} x_k, \\ X_{p,\omega} &= \sum_{l_1=1}^{L} \omega_{l_1} \left(\sum_{l_2=l_1}^{L} \omega_{l_2} \left(\cdots \left(\sum_{l_p=l_{p-1}}^{L} \omega_{l_p} X_{l_1 l_2 \cdots l_p} \right) \right) \right),\ X \in \{Y, S\}. \end{aligned}$$

For the robust MPC, Ding [2010c] only provides the invariance condition. Although Ding [2010c] does not study MPC, by using the principle of KBM formula, it is easy to generalize the invariance condition in Ding [2010c] to the optimality condition and to obtain the input/state LMI constraints based on the invariance condition. As for the current augmented state condition, with $\omega_{l,k}$ known, it should be

$$\begin{bmatrix} 1 & \star \\ x_{k|k} & S_{p,\omega,k} \end{bmatrix} \geq 0,$$

and when $\omega_{l,k}$ is unknown, be

$$\begin{bmatrix} 1 & \star \\ x_{k|k} & \frac{S_{l_1 l_2 \cdots l_p}}{C(l_1, l_2, \ldots, l_p)} \end{bmatrix} \geq 0,\ l_1 \leqslant l_2 \leqslant \cdots \leqslant l_p,\ l_1, l_2, \ldots, l_p \in \{1, 2, \ldots, L\}.$$

Note that $\frac{S_{l_1 l_2 \cdots l_p}}{C(l_1, l_2, \ldots, l_p)}$ in the lower right corner should not be replaced by $S_{l_1 l_2 \cdots l_p}$, because

$$\sum_{l_1=1}^{L} \omega_{l_1} \sum_{l_2=l_1}^{L} \omega_{l_2} \cdots \sum_{l_p=l_{p-1}}^{L} \omega_{l_p} = 1$$

does not hold, while

$$\sum_{l_1=1}^{L} \omega_{l_1} \sum_{l_2=1}^{L} \omega_{l_2} \cdots \sum_{l_p=1}^{L} \omega_{l_p} = 1$$

holds.

Example 10.1 When using Lyapunov matrix such as (10.39), Polya theorem can be further used to introduce parameters $\{d, d_+\}$ to give a less conservative condition (see [Ding, 2010c]). Let us consider the following nonlinear system:

$$\begin{aligned} x_{1,k+1} =& x_{1,k} - x_{1,k}x_{2,k} + (5 + x_{1,k})u_k, \\ x_{2,k+1} =& - x_{1,k} - 0.5x_{2,k} + 2x_{1,k}u_k. \end{aligned}$$

Define $F_1^1(x_{1,k}) = (\beta + x_{1,k})/(2\beta)$, $F_1^2(x_{1,k}) = (\beta - x_{1,k})/(2\beta)$, $\beta > 0$. If $x_{1,k} \in [-\beta, \beta]$, then $x_{1,k} = \beta F_1^1(x_{1,k}) - \beta F_1^2(x_{1,k})$. Then, for $x_{1,k} \in [-\beta, \beta]$, the nonlinear system can be exactly represented by the following two rules of discrete-time T–S fuzzy model ($\omega_{1,k} = F_1^1(x_{1,k})$ and $\omega_{2,k} = F_1^2(x_{1,k})$):

Rule 1: IF $x_{1,k}$ is β, THEN

$$x_{k+1} = \begin{bmatrix} 1 & -\beta \\ -1 & -0.5 \end{bmatrix} x_k + \begin{bmatrix} 5+\beta \\ 2\beta \end{bmatrix} u_k;$$

Rule 1: IF $x_{1,k}$ is $-\beta$, THEN

$$x_{k+1} = \begin{bmatrix} 1 & \beta \\ -1 & -0.5 \end{bmatrix} x_k + \begin{bmatrix} 5-\beta \\ -2\beta \end{bmatrix} u_k.$$

Each rule corresponds to a vertex model of the LPV model. Applying the stability result in Ding [2010c] yields Tables 10.1 and 10.2.

According to Table 10.1, we have the following conclusions.

(1) when $p = 1$ and $d_+ = 0$, taking $d = 1$ yields less-conservative conditions than $d = 0$;
(2) when $p = 1$ and $d_+ = 0$, taking $d = 2$ yields less-conservative conditions than $d = 1$;
(3) when $p = 2$ and $d_+ = 0$, raking $d = 1$ yields less-conservative conditions than $d = 0$;
(4) when $p = 2$ and $d = 1$, taking $d_+ = 1$ yields less-conservative conditions than $d_+ = 0$;
(5) when $d = 1$ and $d_+ = 0$, taking $p = 2$ yields less-conservative conditions than $p = 1$;
(6) when $d = 1$ and $d_+ = 1$, taking $p = 2$ yields less-conservative conditions than $p = 1$;
(7) when $p = 1$ and $d = 1$, taking $d_+ = 1$ yields the same-level-conservative conditions as $d_+ = 0$;
(8) when $p = 1$ and $d = 2$, taking $d_+ = 2$ yields the same-level-conservative conditions as $d_+ = 0$.

Table 10.1 For several simple cases of $\{p, d, d_+\}$, find the largest values of β_0

p	1	1	1	1	1	2	2	2
d	0	1	1	2	2	0	1	1
d_+	0	0	1	0	2	0	0	1
β_0	1.48	1.62	1.62	1.64	1.64	1.66	1.67	1.68

a) The corresponding stability condition is feasible for all $\beta \leqslant \beta_0$.

Table 10.2 For several simple cases of p ($d = d_+ = p$), find the largest values of β_0

$p = d = d_+$	2	3	4	5	6	7	8	9
β_0	1.71	1.74	1.78	1.80	1.81	1.81	1.82	1.82

a) The corresponding stability condition is feasible for all $\beta \leqslant \beta_0$.

10.3 Prediction Horizon: Zero or One

10.3.1 One Over Zero

For the robust MPC based on (9.1), another important job is Lu and Arkun [2000]. Kothare et al. [1996], for all $i \geq 0$, define $u_{k+i|k} = F_k x_{k+i|k}$, and optimize F_k, which is called the constraint robust LQR. Lu and Arkun [2000], for all $i \geq 1$, define $u_{k+i|k} = F_k x_{k+i|k}$ and optimize u_k and F_k. This section discusses the following formulas (designed by [Lu and Arkun, 2000], but slightly modified):

$$\begin{bmatrix} 1 & \star & \star & \star \\ A_l x_{k|k} + B_l u_{k|k} & Q_k & \star & \star \\ \mathcal{Q}^{1/2} x_{k|k} & 0 & \gamma_k I & \star \\ \mathcal{R}^{1/2} u_{k|k} & 0 & 0 & \gamma_k I \end{bmatrix} \geq 0, \ l \in \{1, \dots, L\}, \tag{10.51}$$

$$\begin{bmatrix} Q_k & \star & \star & \star \\ A_l Q_k + B_l Y_k & Q_k & \star & \star \\ \mathcal{Q}^{1/2} Q_k & 0 & \gamma_k I & \star \\ \mathcal{R}^{1/2} Y_k & 0 & 0 & \gamma_k I \end{bmatrix} \geq 0, \ l \in \{1, \dots, L\}, \tag{10.52}$$

$$-\underline{u} \leqslant u_{k|k} \leqslant \bar{u}, \tag{10.53}$$

$$\begin{bmatrix} Q_k & Y_{j,k}^T \\ Y_{j,k} & u_{j,\inf}^2 \end{bmatrix} \geq 0, \ j \in \{1, \dots, m\}, \tag{10.54}$$

$$-\underline{\psi} \leqslant \Psi[A_l x_{k|k} + B_l u_{k|k}] \leqslant \overline{\psi}, \ l \in \{1, 2, \dots, L\}, \tag{10.55}$$

$$\begin{bmatrix} Q_k & \star \\ \Psi_s(A_l Q_k + B_l Y_k) & \psi_{s,\inf}^2 \end{bmatrix} \geq 0, \ l \in \{1, 2, \dots, L\}, \ s \in \{1, 2, \dots, q\}. \tag{10.56}$$

At each sampling time, γ_k is minimized satisfying (10.51)–(10.56), and the corresponding controller is called Type I LA (Lu-Arkun) controller. Equation (10.51) is rewritten as

$$\begin{bmatrix} 1 & \star \\ A_l x_{k|k} + B_l u_{k|k} & Q_k \end{bmatrix} \geq 0, \ l \in \{1, \dots, L\}, \tag{10.57}$$

$$\begin{bmatrix} 1 & \star & \star \\ \mathcal{Q}^{1/2} x_{k|k} & \gamma_{1,k} I & \star \\ \mathcal{R}^{1/2} u_{k|k} & 0 & \gamma_{1,k} I \end{bmatrix} \geq 0. \tag{10.58}$$

At each sampling time, $\gamma_k + \gamma_{1,k}$ is minimized satisfying {(10.57)–(10.58), (10.52)–(10.56)}, and the corresponding controller is called Type II LA controller. The necessary and sufficient condition for recursive feasibility of (10.51) is that (10.57)–(10.58) are feasible. There are numerical differences between Type I and Type II LA controllers. It is easily shown that Type II LA controller follows from KBM controller, i.e., {(10.57), (10.52), (10.54), (10.56)} is the application of KBM formula for the synthesis MPC.

This section clarifies the following problem. Compared with the control horizon N mentioned in Section 9.1, how to identify the N for Kothare et al. [1996] and Lu and Arkun [2000]? There are three ideas, i.e., $N = \infty$, $N = 1$, and $N = 0$.

The routine strategy for synthesis MPC (MPC with guaranteed stability) is decomposing the infinite horizon performance index into two parts, i.e.,

$$J_{0,k}^{\infty} = J_{0,k}^{N-1} + J_{N,k}^{\infty}$$

where

$$J_{0,k}^{\infty} = \sum_{i=0}^{\infty}[\|x_{k+i|k}\|_{\mathscr{Q}}^2 + \|u_{k+i|k}\|_{\mathscr{R}}^2],$$

$$J_{0,k}^{N-1} = \sum_{i=0}^{N-1}[\|x_{k+i|k}\|_{\mathscr{Q}}^2 + \|u_{k+i|k}\|_{\mathscr{R}}^2],$$

$$J_{N,k}^{\infty} = \sum_{i=N}^{\infty}[\|x_{k+i|k}\|_{\mathscr{Q}}^2 + \|u_{k+i|k}\|_{\mathscr{R}}^2].$$

For $J_{0,k}^{N-1}$, the decision variables are the control move sequence

$$\tilde{u}_{k|k} = \{u_{k|k}, u_{k+1|k}, \dots, u_{k+N-1|k}\}.$$

For $J_{N,k}^{\infty}$, $i \geq N$, the synthesis MPC usually define

$$u_{k+i|k} = F_k x_{k+i|k},$$

so that similar to LQR, it obtains

$$J_{N,k}^{\infty} \leqslant x_{k+N|k}^T P_k x_{k+N|k}.$$

The term $x_{k+N|k}^T P_k x_{k+N|k}$ is defined as the terminal performance index which can be Lyapunov function for proving the prediction invariance of the closed-loop state after N steps. Denote

$$J_{0,k}^{\infty} \leqslant \bar{J}_{0,k}^{N} \triangleq \sum_{i=0}^{N-1}[\|x_{k+i|k}\|_{\mathscr{Q}}^2 + \|u_{k+i|k}\|_{\mathscr{R}}^2] + \|x_{k+N|k}\|_{P_k}^2.$$

In summary,

I. The synthesis MPC hopes to minimize $J_{0,k}^{\infty}$, but for many reasons, except for nominal linear model, $\bar{J}_{0,k}^{N}$ is minimized instead. The terminal cost $\|x_{k+N|k}\|_{P_k}^2$ of $\bar{J}_{0,k}^{N}$ is well-designed and usually $\bar{J}_{0,k}^{N} \neq J_{0,k}^{\infty}$.
Then, whether or not the synthesis MPC can always minimize $\bar{J}_{0,k}^{N}$ directly? Here are the conclusions.
II. $\bar{J}_{0,k}^{N}$ can be directly minimized for nominal (including linear and nonlinear) closed-loop models, but cannot be directly minimized for uncertain (including linear and nonlinear) closed-loop models. An alternative choice is adding constraint $\bar{J}_{0,k}^{N} \leqslant \gamma_k$ and minimizing γ_k; $\bar{J}_{0,k}^{N} = \gamma_k$ holds for only some special cases.

Moreover, we find that

III. For $N = 1$, as well as the nominal (including linear and nonlinear) closed-loop models, one can directly take

$$\tilde{u}_{k|k} = \{u_{k|k}, u_{k+1|k}, \dots, u_{k+N-1|k}\}$$

as the decision variables. For the uncertain (including linear and nonlinear) closed-loop models, if $\tilde{u}_{k|k}$ are the decision variables, it is hard to guarantee the recursive feasibility for $N > 1$.

Due to III, the robust MPC has developed in diversity. This section aims at the robust MPC of LPV, which is divided into four categories in the light of how to parameterize $\tilde{u}_{k|k}$.

A1. The feedback MPC (closed-loop MPC): Define

$$u_{k+j|k} = F_{k+j|k}x_{k+j|k},\ j \in \{0, 1, \dots, N-1\},$$

and take N control gains $\{F_{k|k}, F_{k+1|k}, \dots, F_{k+N-1|k}\}$ as the decision variables. For $N > 2$, the convex optimization is still unsolved.

A2. The variant feedback MPC (variant closed-loop MPC): Add the constraint $x_{k+i|k} \in \mathscr{E}_{Q_i^{-1}}$, $i \in \{1, 2, \dots, N-1\}$ implicitly, which can be solved by convex optimization. See Li et al. [2009] and Cychowski and O'Mahony [2010].

A3. The parameter-dependent open-loop MPC: Use the control moves in Section 9.1, i.e., take $\{u_{k|k}, u^{l_0}_{k+1|k}, \dots, u^{l_{N-2}\cdots l_1 l_0}_{k+N-1|k}\}$ as the decision variables, which can be solved by convex optimization. See Pluymers et al. [2005].

A4. The partial feedback MPC: Define

$$u_{k+j|k} = F_{k+j|k}x_{k+j|k} + c_{k+j|k},\ j \in \{0, 1, \dots, N-1\},$$

and take the perturbation terms $\{c_{k|k}, c_{k+1|k}, \dots, c_{k+N-1|k}\}$ as the decision variables, which can be solved by the convex optimization. $\{F_{k|k}, F_{k+1|k}, \dots, F_{k+N-1|k}\}$ are not taken as the decision variables at time k; they may be solved at previous time or may be determined off-line. In the past, the partial feedback MPC played a mainstream role in robust MPC of LPV.

From the above categories, LA controller of [Lu and Arkun, 2000] is exactly regarded as $N = 1$; it barely belongs to category A3 (non-parameter-dependence); however, it also belongs to A1, A2, and A4. This reflects the importance of [Lu and Arkun, 2000]. Since $N = 1$ in Lu and Arkun [2000], it is only suitable to consider [Kothare et al., 1996] as $N = 0$. If [Kothare et al., 1996] took $N = 1$, then it would conflict with Lu and Arkun [2000]; more importantly, $N = 1$ for Kothare et al. [1996] does not conform to the definition of $J_{0,k}^{N-1}$. Notice that A1 and A2 can only be distinguished for $N \geq 2$, and they are equivalent for $N \in \{0, 1\}$. Regarding Kothare et al. [1996] as A1 can reduce misunderstanding.

One may argue that $N = \infty$ for Kothare et al. [1996], which is not wrong. However, under this argument, by decomposing $J_{0,k}^{\infty}$ into two parts, it should take $N = \infty$, which is hard to distinguish in terms of N. In fact, $N = \infty$ is artificial for proving stability based on the theory of optimal control. When we identify N, it is better to begin with the "truly solved optimization."

For the synthesis MPC, i.e., MPC with guaranteed stability, $N = 0$ implies taking one F_k such that, for all $i \geq 0$, $u_{k+i|k} = F_k x_{k+i|k}$. The synthesis MPC decomposes the infinite horizon control move sequence into two parts, i.e., $\tilde{u}_{k|k}$ and $\{u_{k+i|k} | i \geq N\}$. A1–A4 are just based on classification of $\tilde{u}_{k|k}$, which cannot cover all the syntheses of MPC. A more desirable classification also depends on how to deal with $\{u_{k+i|k} | i \geq N\}$, which involves three elements and four conditions (sometimes, even being called four axioms) of synthesis MPC (see [Mayne et al., 2000, Mayne, 2014]). For the robust MPC of LPV, the three elements are $\{Q_k, P_k, F_k\}$ which can be constructed based on KBM controller (usually $P = \gamma Q^{-1}$, $F = YQ^{-1}$). Four KBM formulas, serve as "axioms," change in different cases. Hence, the three elements and the four conditions of the general MPC are not detailed here. The synthesis of MPC with the quadratic performance index can be divided into three categories.

B1. The standard method: $\{P_k, F_k\}$ are calculated off-line, and $\bar{J}_{0,k}^{N}$ or γ_k is optimized on-line. The standard method is commonly used in the synthesis MPC, especially for nonlinear systems. A typical example is referred to Chen and Allgöwer [1998]. At present, MPC applied in industry is usually called the standard method.
B2. The on-line method: $\{P_k, F_k\}$ are calculated on-line, and $\bar{J}_{0,k}^{N}$ or γ_k is optimized on-line. See Bloemen et al. [2002] where the linear nominal system is considered. The on-line method is very common in robust MPC of LPV; for the nonlinear model, however, because of computational burden, the on-line method is hard to work.
B3. The off-line method: $\{P_k, F_k\}$ are calculated off-line, and $\bar{J}_{0,k}^{N}$ or γ_k is optimized off-line. The off-line method does not involve with the on-line optimization problem, thus the computation is low.

For example, KBM controller in Kothare et al. [1996] belongs to A1B2, the controller in Lu and Arkun [2000] belongs to A*B2, and many (some) controllers for nominal nonlinear models belong to A*B1 (to A*B3). Which category does the current industrial MPC belong to? Apparently, it does not belong to A*B*, because it cannot guarantee the stability. However, the current mainstream industrial MPC can also be regarded as A4B1, because the bottom controllers (e.g., PID) usually pre-stabilizes the process, and MPC then controls the generalized process (i.e., process including PID). It appears that the controller like PID is similar as the feedback, and MPC is similar as optimizing the perturbation terms. Hence, if PID is regarded as a part of MPC, it is reasonable for the industrial MPC belonging to A4B1.

10.3.2 One: Generalizing to Networked Control

Refer to Ding [2011a]. Continue with the previous example in Section 10.1.3. Add a free control move and apply LA controller. For all $j_l \leqslant k < j_{l+1}$, let the output of the controller be $\breve{u}_{k|j_l} = \breve{u}_{j_l}$, where $\breve{u}_{j_l}$ is the "free control input" (i.e., degree-of-freedom for optimization); on the controller side, then

$$u_{k|j_l} = u_{\mathbb{i}_{s_l}|j_l} = \breve{u}_{j_l}, \quad \mathbb{i}_{s_l} \leqslant k < \mathbb{i}_{s_l+1}, \tag{10.59}$$

where it is assumed that $j_l = \mathbb{j}_{s_l}$. Hence, the closed-loop state prediction satisfies

$$x_{k+1|j_l} = A^{k-\mathbb{i}_{s_l}+1} x_{\mathbb{i}_{s_l}|j_l} + B_{k-\mathbb{i}_{s_l}+1} \breve{u}_{j_l}, \quad \mathbb{i}_{s_l} \leqslant k < \mathbb{i}_{s_l+1}. \tag{10.60}$$

At each j_l, solve the optimization problem

$$\begin{aligned} &\min_{\breve{u}_{j_l}, F_{j_l}} \max_{\star} J_{j_l} \\ &\text{s.t. } (10.22),\ (10.59),\ (10.60), (10.18) - (10.20), \quad \tau \geqslant 1, \end{aligned} \tag{10.61}$$

where $\star$ appears in (10.21).

10.3.2.1 Algorithm

In order to derive an upper bound on $\max_{\star} J_{j_l}$, let us impose the stability constraint (10.24) for all $\tau \geqslant 1$ ($\tau \neq 0$). For a stable closed-loop system, by summing (10.24) from $i = 1$ to $i = \infty$, it follows that $J_{j_l} \leqslant \|x_{\mathbb{i}_{s_l}|j_l}\|^2_{\mathscr{Q}} + \|u_{\mathbb{i}_{s_l}|j_l}\|^2_{\mathscr{R}} + V(z_{\mathbb{i}_{s_l+1}|j_l})$. Let

$$V(z_{\mathbb{i}_{s_l+1}|j_l}) \leqslant \gamma, \tag{10.62}$$

$$\|u_{\mathbb{i}_{s_l}|j_l}\|^2_{\mathscr{R}} \leqslant \gamma_1, \tag{10.63}$$

where $\gamma_1 > 0$ is a scalar. By defining $Q_{\mathbb{i}_{s_l+1}|j_l} := \gamma P^{-1}_{\mathbb{i}_{s_l+1}|j_l}$, applying (10.59) and the Schur complement, it is shown that (10.62) and (10.63) are equivalent to

$$\begin{bmatrix} 1 & \star \\ z_{\mathbb{i}_{s_l+1}|j_l} & Q_{\mathbb{i}_{s_l+1}|j_l} \end{bmatrix} \geqslant 0, \tag{10.64}$$

$$\begin{bmatrix} \gamma_1 & \star \\ \mathscr{R}^{1/2}\breve{u}_{j_l} & I \end{bmatrix} \geqslant 0. \tag{10.65}$$

By applying (10.60), it is easy to show that

$$z_{\mathbb{i}_{s_l+1}|j_l} = \sum_{t\in\mathscr{D}_2}\sum_{r\in\mathscr{D}(t)} \varpi_{t,\mathbb{i}_{s_l}}\omega_{r,\mathbb{i}_{s_l}}\left[\hat{\Phi}_r z^1_{rt,\mathbb{i}_{s_l}|j_l} + \hat{\Gamma}_r\breve{u}_{j_l}\right],$$

where the superscript "1" shows the switching horizon, $t = \mathbb{i}_{s_l} - j_l + 1$, $r = \mathbb{i}_{s_l+1} - \mathbb{i}_{s_l}$,

$$\hat{\Phi}_r = \begin{bmatrix} \begin{bmatrix} \begin{bmatrix} A^r \\ A^{r-1} \\ \vdots \\ A^{\max\{1,r-d_2+1\}} \end{bmatrix} & 0 \end{bmatrix} \\ I \end{bmatrix}, \quad \hat{\Gamma}_r = \begin{bmatrix} B_r \\ B_{r-1} \\ \vdots \\ B_{\max\{1,r-d_2+1\}} \\ 0 \end{bmatrix},$$

$$z^1_{rt,\mathbb{i}_{s_l}|j_l} = \begin{bmatrix} x_{j_l+t-1|j_l} \\ x_{j_l+t-2|j_l} \\ \vdots \\ x_{j_l+t-\max\{1,d_2-r\}|j_l} \end{bmatrix}.$$

Note that $Q_{\mathbb{i}_{s_l+1}|j_l} = \sum_{p\in\mathscr{D}_2}\sum_{h\in\mathscr{D}(p)} \varpi_{p,\mathbb{i}_{s_l+1}}\omega_{h,\mathbb{i}_{s_l+1}}Q_{hp}$, it is easy to show that (10.64) is guaranteed by

$$\begin{bmatrix} 1 & \star \\ \hat{\Phi}_r z^1_{rt,\mathbb{i}_{s_l}|j_l} + \hat{\Gamma}_r\breve{u}_{j_l} & Q_{hp} \end{bmatrix} \geqslant 0, \quad t,p \in \mathscr{D}_2, \ r \in \mathscr{D}(t), \ h \in \mathscr{D}(p). \tag{10.66}$$

According to Assumption 10.1, it is easy to exactly calculate $z^1_{rt,\mathbb{i}_{s_l}|j_l}$.

Lemma 10.2 Suppose there exist a scalar γ, a vector $\breve{u}_{j_l}$, the symmetric matrices $\{Z, \Gamma, Q_{rt}\}$, and any matrices $\{G_{rt}, Y\}$ such that $\{(10.66), (10.25), (10.27)\text{–}(10.30)\}$ and the following are satisfied:

$$-\underline{u} \leqslant \breve{u}_{j_l} \leqslant \bar{u}, \tag{10.67}$$

$$-\underline{\psi} \leqslant \Psi x_{\mathbb{i}_{s_l+1}|j_l} \leqslant \overline{\psi}, \quad \mathbb{i}_{s_l+1} - j_{s_l} \in \mathscr{D}. \tag{10.68}$$

Then (10.22) is satisfied.

Applying (10.60) and (10.20) yields

$$x_{\mathbb{i}_{s_l+1}|j_l} = \sum_{t\in\mathscr{D}_2} \sum_{r\in\mathscr{D}(t)} \varpi_{t,\mathbb{i}_{s_l}} \omega_{r,\mathbb{i}_{s_l}} \left[A^{r+t-1}x_{j_l} + A^r B_{t-1} u_{j_l|j_l} + B_r \check{u}_{j_l}\right].$$

Since $u_{j_l|j_l}$ is exactly known to the controller for $t > 1$, (10.68) is guaranteed by

$$-\underline{\psi} \leqslant \Psi \left[A^r x_{j_l} + B_r \check{u}_{j_l}\right] \leqslant \overline{\psi}, \quad r \in \mathscr{D}, \tag{10.69}$$

$$-\underline{\psi} \leqslant \Psi \left[A^{r+t-1}x_{j_l} + A^r B_{t-1} u_{j_l|j_l} + B_r \check{u}_{j_l}\right] \leqslant \overline{\psi},\ t \in \{2, 3, \dots, d_2\},\ r \in \mathscr{D}(t). \tag{10.70}$$

In summary, (10.61) is approximated by the following LMI optimization problem:

$$\begin{aligned} &\min_{\gamma_1,\gamma,\check{u}(j_l),Q_{rt},G_{rt},Y,Z,\Gamma} (\gamma_1 + \gamma) \\ &\text{s.t. } (10.65),\ (10.67),\ (10.69),\ (10.70),\ (10.66),\ (10.25), (10.27) - (10.30). \end{aligned} \tag{10.71}$$

Theorem 10.3 Assume that (10.71) is feasible at $j_{\overline{n}}$, $\overline{n} \geqslant 1$ ((10.71) may be infeasible at $j_1, j_2, \dots, j_{\overline{n}-1}$) and $\check{u}(j_{\overline{n}})$ is the first control move received by the actuator. Then, (10.71) will be feasible at $j_{\overline{n}+\tau}$ for all $\tau > 0$, and the receding horizon sending of the control move $\check{u}_k = \check{u}_{j_l}, j_l \leqslant k < j_{l+1}, l \geqslant 1$ guarantees satisfaction of (10.17), and the closed-loop system is exponentially stable.

Proof: Denote $\ell := l + 1$ and

$$L_{j_l} := \max_{t\in\mathscr{D}_2} \left\{ \|x_{j_l+t-1|j_l}\|^2_{\mathcal{Q}} \right\} + \|\check{u}_{j_l}\|^2_{\mathscr{R}} + \gamma_{j_l}.$$

At each j_l, the minimization of $\gamma_{1j_l} + \gamma_{j_l}$ is equivalent to the minimization of L_{j_l}. Assume that (10.71) is feasible at time $\mathbb{j}_{s_\ell-1}$. By applying (10.24),

$$\begin{aligned} &\|z^*_{j_\ell+p+h-1|\mathbb{j}_{s_\ell-1}}\|^2_{P^*_{hp,\mathbb{j}_{s_\ell-1}}} - \|z^*_{\mathbb{j}_{s_\ell-1}+t+r-1|\mathbb{j}_{s_\ell-1}}\|^2_{P^*_{rt,\mathbb{j}_{s_\ell-1}}} \\ &\leqslant -\|x^*_{\mathbb{j}_{s_\ell-1}+t+r-1|\mathbb{j}_{s_\ell-1}}\|^2_{\mathcal{Q}} - \|\check{u}^*_{j_\ell|\mathbb{j}_{s_\ell-1}}\|^2_{\mathscr{R}},\ t, p \in \mathscr{D}_2,\ r \in \mathscr{D}(t),\ h \in \mathscr{D}(p). \end{aligned} \tag{10.72}$$

Since any possible $z_{j_\ell+p+h-1|j_\ell}$ is included in the realizations of $z^*_{j_\ell+p+h-1|\mathbb{j}_{s_\ell-1}}$ (similar to Theorem 10.2), (10.72) leads to

$$\begin{aligned} &\|z_{j_\ell+p+h-1|j_\ell}\|^2_{P^*_{hp,\mathbb{j}_{s_\ell-1}}} - \|z^*_{\mathbb{j}_{s_\ell-1}+t+r-1|\mathbb{j}_{s_\ell-1}}\|^2_{P^*_{rt,\mathbb{j}_{s_\ell-1}}} \\ &\leqslant -\|x^*_{\mathbb{j}_{s_\ell-1}+t+r-1|\mathbb{j}_{s_\ell-1}}\|^2_{\mathcal{Q}} - \|\check{u}^*_{j_\ell|\mathbb{j}_{s_\ell-1}}\|^2_{\mathscr{R}},\ t, p \in \mathscr{D}_2,\ r \in \mathscr{D}(t),\ h \in \mathscr{D}(p). \end{aligned} \tag{10.73}$$

Note that

$$\gamma^*_{\mathbb{j}_{s_\ell-1}} \geq \|z^*_{\mathbb{j}_{s_\ell-1}+t+r-1|\mathbb{j}_{s_\ell-1}}\|^2_{P^*_{rt,\mathbb{j}_{s_\ell-1}}}, \tag{10.74}$$

$$\gamma_{j_\ell} \geq \|z_{j_\ell+p+h-1|j_\ell}\|^2_{P_{hp,j_\ell}}, \tag{10.75}$$

$$J_{\mathbb{j}_{s_\ell-1}} \leqslant L^*_{\mathbb{j}_{s_\ell-1}} = \max_{t\in\mathscr{D}_2} \left\{ \|x_{\mathbb{j}_{s_\ell-1}+t-1|\mathbb{j}_{s_\ell-1}}\|^2_{\mathcal{Q}} \right\} + \|\check{u}^*_{\mathbb{j}_{s_\ell-1}}\|^2_{\mathscr{R}} + \gamma^*_{\mathbb{j}_{s_\ell-1}}, \tag{10.76}$$

$$J_{j_\ell} \leqslant L_{j_\ell} = \max_{p\in\mathscr{D}_2} \left\{ \|x_{j_\ell+p-1|j_\ell}\|^2_{\mathcal{Q}} \right\} + \|\breve{u}_{j_\ell}\|^2_{\mathcal{R}} + \gamma_{j_\ell}. \tag{10.77}$$

Since any possible $x_{\mathbb{j}_{s_\ell}|j_\ell}$ is included in the realizations of $x^*_{\mathbb{j}_{s_\ell}|\mathbb{j}_{s_\ell-1}}$ (similar to Theorem 10.2), it holds that

$$\max_{p\in\mathscr{D}_2} \left\{ \|x_{j_\ell+p-1|j_\ell}\|^2_{\mathcal{Q}} \right\} \leqslant \max_{r\in\mathscr{D}(t)} \max_{t\in\mathscr{D}_2} \left\{ \|x^*_{\mathbb{j}_{s_\ell-1}+t+r-1|\mathbb{j}_{s_\ell-1}}\|^2_{\mathcal{Q}} \right\}. \tag{10.78}$$

According to (10.73)–(10.78), we can show that it is admissible to choose

$$\breve{u}_{j_\ell} = \breve{u}^*_{j_\ell|\mathbb{j}_{s_\ell-1}} = F^*_{\mathbb{j}_{s_\ell-1}} x_{j_\ell},$$

$$L_{j_\ell} = \gamma^*_{\mathbb{j}_{s_\ell-1}} = L^*_{\mathbb{j}_{s_\ell-1}} - \max_{t\in\mathscr{D}_2} \left\{ \|x_{\mathbb{j}_{s_\ell-1}+t-1|\mathbb{j}_{s_\ell-1}}\|^2_{\mathcal{Q}} \right\} - \|\breve{u}^*_{\mathbb{j}_{s_\ell-1}}\|^2_{\mathcal{R}},$$

$$\{Q_{rt}, G_{rt}, Y, Z, \Gamma\}_{j_\ell} = \frac{\gamma(j_\ell)}{\gamma^*_{\mathbb{j}_{s_\ell-1}}} \{Q_{rt}, G_{rt}, Y, Z, \Gamma\}^*_{\mathbb{j}_{s_\ell-1}},$$

as the solution to (10.71). Note that the existence of L_{j_ℓ} is equivalent to the existence of $\gamma_{1j_\ell} + \gamma_{j_\ell}$. Clearly, feasibility of (10.71) at $\mathbb{j}_{s_\ell-1}$ leads to its feasibility at j_ℓ. By induction, feasibility of (10.71) at time $j_{\overline{n}}$ means its feasibility at any $j_{\overline{n}+\tau}$, $\tau > 0$.

By re-optimization at time j_ℓ, it must result in $L^*_{j_\ell} \leqslant L_{j_\ell}$. Therefore,

$$L^*_{j_\ell} - L^*_{\mathbb{j}_{s_\ell-1}} \leqslant -\max_{t\in\mathscr{D}_2} \left\{ \|x_{\mathbb{j}_{s_\ell-1}+t-1|\mathbb{j}_{s_\ell-1}}\|^2_{\mathcal{Q}} \right\} - \|\breve{u}^*_{\mathbb{j}_{s_\ell-1}}\|^2_{\mathcal{R}},$$

which means that $L^*_{\mathbb{j}_l}, l \geqslant 1$ is Lyapunov function for proving the exponential stability of the closed-loop system (similar to Theorem 10.2). For proving satisfaction of input and state constraints, note that $j_{\overline{n}} = \mathbb{j}_1$ and refer to Lemma 10.2.

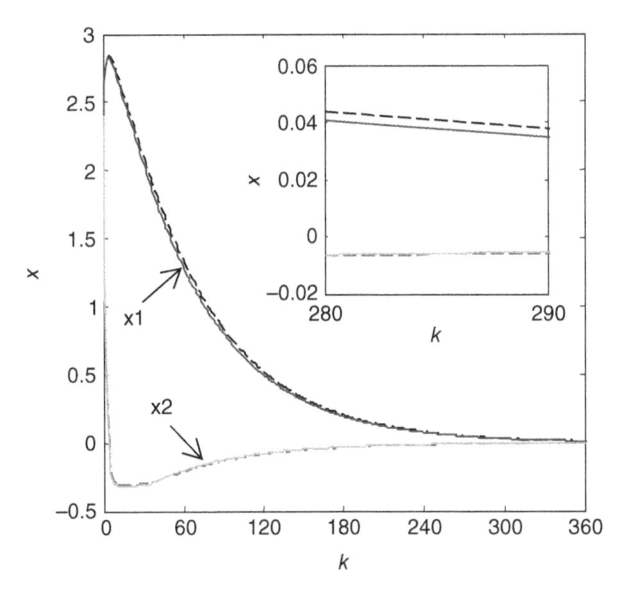

Figure 10.2 The state responses of the closed-loop systems.

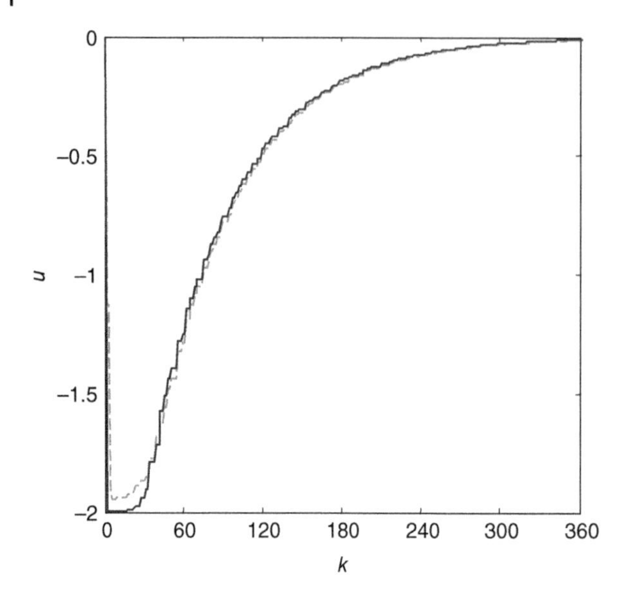

Figure 10.3 The control moves.

10.3.2.2 A Numerical Example

We consider the following model for the angular positioning system (see [Kothare et al., 1996]):

$$x_{k+1} = \begin{bmatrix} \theta_{k+1} \\ \dot{\theta}_{k+1} \end{bmatrix} = \begin{bmatrix} 1 & 0.1 \\ 0 & 1-0.1\epsilon \end{bmatrix} x_k + \begin{bmatrix} 0 \\ 0.0787 \end{bmatrix} u_k,$$

$$x(0) = \varepsilon \begin{bmatrix} \frac{\pi}{4} \\ 0 \end{bmatrix},$$

where θ (rad) is the angular position of the antenna, $\dot{\theta}$ (rad s^{-1}) is the angular velocity of the antenna, u (V) is the input voltage to the motor, ϵ (s^{-1}) is propositional to the coefficient of viscous friction in the rotating parts of the antenna, and $\epsilon \geq 0.1\ s^{-1}$ and $\varepsilon > 0$ are scalars for testing feasibility of the networked MPC. The constraints are $|u| \leqslant 2$ V and $|\theta| \leqslant \pi$ rad. More details for the control problem can be found in Kothare et al. [1996]. The actual input of the system is zero before any control move is received by the actuator. Moreover, $d_1 = d_2 = 3$, $j_1 = 1$, $j_2 > 2$ and $i_1 = 2$. For $\varepsilon = 1$, by applying (10.31) (applying (10.71)), the optimization problem at $j_1 = \mathrm{j}_1$ is feasible for any $\epsilon \leqslant 7.5\ s^{-1}$ (any $\epsilon \leqslant 7.6\ s^{-1}$); for $\epsilon = 5\ s^{-1}$, by applying (10.31) (applying (10.71)), the optimization problem at $j_1 = \mathrm{j}_1$ is feasible for any $\varepsilon \leqslant 2.4$ (any $\varepsilon \leqslant 2.6$). Hence, (10.71) is easier to be feasible than (10.31).

Take $\epsilon = 5\ s^{-1}$ and $\varepsilon = 2.4$, and randomly generate $\mathcal{J} = \{1, 4, 6, 8, 11, 12, 13, 16, 19, 20, 22, 24, 25, 26, \dots\}$, $\mathcal{I} = \{2, 4, 5, 6, 9, 11, 13, 15, 18, 20, 21, 23, 25, 28, \dots\}$. By applying (10.31) ((10.71)), the resultant state response and the control move are shown in Figures 10.2 and 10.3, respectively, in dashed lines (solid lines). Compared with applying (10.31), more aggressive control moves and slightly faster state responses are observed by applying (10.71).

10.4 Variant Feedback MPC

The constraints of open-loop MPC have the following forms:

$$\begin{bmatrix} 1 & \star \\ x_{k+N|k}^{l_{N-1}\cdots l_1 l_0} & Q_k \end{bmatrix} \geqslant 0,\ l_i \in \{1,\ldots,L\},\ i\in\{0,\ldots,N-1\}, \tag{10.79}$$

$$\begin{bmatrix} 1 & \star & \star \\ \mathscr{Q}^{1/2} x_{k+i|k}^{l_{i-1}\cdots l_1 l_0} & \gamma_{i,k} I & \star \\ \mathscr{R}^{1/2} u_{k+i|k} & 0 & \gamma_{i,k} I \end{bmatrix} \geqslant 0,\ l_i \in \{1,\ldots,L\},\ i\in\{0,\ldots,N-1\}, \tag{10.80}$$

$$\begin{bmatrix} Q_k & \star & \star & \star \\ A_l Q_k + B_l Y_k & Q_k & \star & \star \\ \mathscr{Q}^{1/2} Q_k & 0 & \gamma_k I & \star \\ \mathscr{R}^{1/2} Y_k & 0 & 0 & \gamma_k I \end{bmatrix} \geqslant 0,\ l \in \{1,\ldots,L\}, \tag{10.81}$$

$$-\underline{u} \leqslant u_{k+i|k} \leqslant \bar{u},\ i\in\{0,\ldots,N-1\}, \tag{10.82}$$

$$\begin{bmatrix} Q_k & Y_{j,k}^T \\ Y_{j,k} & u_{j,\inf}^2 \end{bmatrix} \geqslant 0,\ j \in \{1,\ldots,m\}, \tag{10.83}$$

$$-\underline{\psi}^s \leqslant \Psi^{\mathrm{d}} \begin{bmatrix} x_{k+1|k}^{l_0} \\ x_{k+2|k}^{l_1 l_0} \\ \vdots \\ x_{k+N|k}^{l_{N-1}\cdots l_1 l_0} \end{bmatrix} \leqslant \overline{\psi}^s,\ l_i \in \{1,\ldots,L\},\ i\in\{0,\ldots,N-1\}, \tag{10.84}$$

$$\begin{bmatrix} Q_k & \star \\ \Psi_s(A_l Q_k + B_l Y_k) & \psi_{s,\inf}^2 \end{bmatrix} \geq 0,\ l \in \{1,2,\ldots,L\},\ s\in\{1,2,\ldots,q\}, \tag{10.85}$$

where the definition of vertex state prediction $[x_{k+1|k}^{l_0}; x_{k+2|k}^{l_1 l_0}; \vdots; x_{k+N|k}^{l_{N-1}\cdots l_1 l_0}]$ is similar with Section 9.1, which is substituted by

$$\begin{bmatrix} x_{k+1|k}^{l_0} \\ x_{k+2|k}^{l_1 l_0} \\ \vdots \\ x_{k+N|k}^{l_{N-1}\cdots l_1 l_0} \end{bmatrix} = \begin{bmatrix} A_{l_0} \\ A_{l_1} A_{l_0} \\ \vdots \\ \prod_{i=0}^{N-1} A_{l_{N-1-i}} \end{bmatrix} x_k + \begin{bmatrix} B_{l_0} & 0 & \cdots & 0 \\ A_{l_1} B_{l_0} & B_{l_1} & \ddots & \vdots \\ \vdots & \vdots & \ddots & 0 \\ \prod_{i=0}^{N-2} A_{l_{N-1-i}} B_{l_0} & \prod_{i=0}^{N-3} A_{l_{N-1-i}} B_{l_1} & \cdots & B_{l_{N-1}} \end{bmatrix} \begin{bmatrix} u_{k|k} \\ u_{k+1|k} \\ \vdots \\ u_{k+N-1|k} \end{bmatrix}.$$

The control moves are optimized via solving $\min \left(\sum_{i=0}^{N-1} \gamma_{i,k} + \gamma_k \right)$ satisfying constraints (10.79)–(10.85), which is called the open-loop MPC.

It is easily shown that (10.79)–(10.85) are extensions of constraints of Type II LA controller (see Section 10.3), i.e., $N = 1$ is generalized to $N \geqslant 1$. The set {(10.79), (10.81), (10.83), (10.85)} implies the application of KBM formula in MPC. In proving the recursive feasibility of the constraints (10.79)–(10.85), the open-loop MPC cannot obtain the exact value of $u_{k+N|k+1}$ at time $k+1$; $u_{k+N|k+1} = F_k^* x_{k+N|k}^*$ is not an exact value, because $x_{k+N|k}^*$ is the parameter-dependent value with $x_{k+N|k}^{*l_{N-1}\cdots l_1 l_0}$ as the vertex.

For the general feedback MPC, it is hard to apply the convex optimization with $N > 2$. Take Li et al. [2009] as an example, we give the following constraints of the variant feedback MPC:

$$\begin{bmatrix} 1 & \star \\ A_l x_{k|k} + B_l u_{k|k} & Q_{1,k} \end{bmatrix} \geqslant 0, \; l \in \{1, \ldots, L\}, \tag{10.86}$$

$$\begin{bmatrix} Q_{i,k} & \star & \star & \star \\ A_l Q_{i,k} + B_l Y_{i,k} & Q_{i+1,k} & \star & \star \\ \mathscr{Q}^{1/2} Q_{i,k} & 0 & \gamma_k I & \star \\ \mathscr{R}^{1/2} Y_{i,k} & 0 & 0 & \gamma_k I \end{bmatrix} \geqslant 0,$$
$$i \in \{1, 2, \ldots, N-1\}, \; l \in \{1, \ldots, L\}, \; Q_N = Q, \tag{10.87}$$

$$\begin{bmatrix} Q_k & \star & \star & \star \\ A_l Q_k + B_l Y_k & Q_k & \star & \star \\ \mathscr{Q}^{1/2} Q_k & 0 & \gamma_k I & \star \\ \mathscr{R}^{1/2} Y_k & 0 & 0 & \gamma_k I \end{bmatrix} \geqslant 0, \; l \in \{1, \ldots, L\}, \tag{10.88}$$

$$-\underline{u} \leqslant u_{k|k} \leqslant \bar{u}, \tag{10.89}$$

$$\begin{bmatrix} Q_{i,k} & Y_{ij,k}^T \\ Y_{ij,k} & u_{j,\inf}^2 \end{bmatrix} \geqslant 0, \; i \in \{1, 2, \ldots, N-1\}, \; j \in \{1, \ldots, m\}, \tag{10.90}$$

$$\begin{bmatrix} Q_k & Y_{j,k}^T \\ Y_{j,k} & u_{j,\inf}^2 \end{bmatrix} \geqslant 0, \; j \in \{1, \ldots, m\}, \tag{10.91}$$

$$-\underline{\psi} \leqslant \Psi[A_l x_{k|k} + B_l u_{k|k}] \leqslant \overline{\psi}, \; l \in \{1, 2, \ldots, L\}, \tag{10.92}$$

$$\begin{bmatrix} Q_{i,k} & \star \\ \Psi_s(A_l Q_{i,k} + B_l Y_{i,k}) & \psi_{s,\inf}^2 \end{bmatrix} \geqslant 0,$$
$$i \in \{1, 2, \ldots, N-1\}, \; l \in \{1, 2, \ldots, L\}, \; s \in \{1, 2, \ldots, q\}, \tag{10.93}$$

$$\begin{bmatrix} Q_k & \star \\ \Psi_s(A_l Q_k + B_l Y_k) & \psi_{s,\inf}^2 \end{bmatrix} \geqslant 0, \; l \in \{1, 2, \ldots, L\}, \; s \in \{1, 2, \ldots, q\}. \tag{10.94}$$

The control moves are optimized via solving $\min\left(\|u_{k|k}\|_{\mathscr{R}}^2 + \gamma_k\right)$, which is called the variant feedback MPC. Taking $Q_1 = Q_2 = \cdots = Q_N = Q$ retrieves the type II LA controller.

Equation (10.87) guarantees $x_{k+i|k} \in \mathscr{E}_{Q_{i,k}}$, $i \in \{1,2,\dots,N\}$, so the variant feedback MPC implicitly imposes constraint $x_{k+i|k} \in \mathscr{E}_{Q_{i,k}}$, $i \in \{1,2,\dots,N\}$ which is its difference with the general feedback MPC. Applying the periodic invariance tool provided by Lee and Kouvaritakis [2006] yields the following constraints:

$$\begin{bmatrix} 1 & \star \\ A_l x_{k|k} + B_l u_{k|k} & Q_{1,k} \end{bmatrix} \geqslant 0,\ l \in \{1,\dots,L\}, \tag{10.95}$$

$$\begin{bmatrix} Q_{i,k} & \star & \star & \star \\ A_l Q_{i,k} + B_l Y_{i,k} & Q_{i+1,k} & \star & \star \\ \mathscr{Q}^{1/2} Q_{i,k} & 0 & \gamma_k I & \star \\ \mathscr{R}^{1/2} Y_{i,k} & 0 & 0 & \gamma_k I \end{bmatrix} \geqslant 0,$$

$$i \in \{1,2,\dots,N\},\ l \in \{1,\dots,L\},\ Q_{N+1} = Q_1 \tag{10.96}$$

$$-\underline{u} \leqslant u_{k|k} \leqslant \bar{u}, \tag{10.97}$$

$$\begin{bmatrix} Q_{i,k} & Y_{ij,k}^T \\ Y_{ij,k} & u_{j,\inf}^2 \end{bmatrix} \geqslant 0,\ i \in \{1,2,\dots,N\},\ j \in \{1,\dots,m\}, \tag{10.98}$$

$$-\underline{\psi} \leqslant \Psi[A_l x_{k|k} + B_l u_{k|k}] \leqslant \overline{\psi},\ l \in \{1,2,\dots,L\}, \tag{10.99}$$

$$\begin{bmatrix} Q_{i,k} & \star \\ \Psi_s(A_l Q_{i,k} + B_l Y_{i,k}) & \psi_{s,\inf}^2 \end{bmatrix} \geqslant 0,$$

$$i \in \{1,2,\dots,N\},\ l \in \{1,2,\dots,L\},\ s \in \{1,2,\dots,q\}. \tag{10.100}$$

The control moves are optimized via solving min $\left(\|u_{k|k}\|_{\mathscr{R}}^2 + \gamma_k\right)$ satisfying constraints (10.95)–(10.100). Note that the difference between (10.95)–(10.100) and (10.86)–(10.94) is that, the former is $Q_{N+1} = Q_1$ and the latter is $Q_{N+1} = Q_N = Q$, i.e., the former does not take $\mathscr{E}_{Q_N} = \mathscr{E}_Q$, but $\{\mathscr{E}_{Q_1}, \mathscr{E}_{Q_2}, \dots, \mathscr{E}_{Q_N}, \mathscr{E}_{Q_1}, \mathscr{E}_{Q_2}, \dots, \mathscr{E}_{Q_N}, \dots\}$, as the terminal constraint set. In terms of expression, the difference between (10.95)–(10.100) and (10.86)–(10.94) is just a subscript. This chapter, however, calls the approach based on (10.95)–(10.100) LA controller based on periodic invariant set, i.e., the annular LA controller.

Example 10.2 Rewrite Type II improved KBM formula in Section 10.2 as

$$\begin{bmatrix} 1 & x_{k|k}^T \\ x_{k|k} & G_k^T + G_k - Q_{l,k} \end{bmatrix} \geqslant 0,\ l \in \{1,\dots,L\}, \tag{10.101}$$

$$\begin{bmatrix} Q_{l,k} & \star & \star & \star \\ A_l G_k + B_l Y_k & G_k^T + G_k - Q_{j,k} & \star & \star \\ \mathscr{Q}^{1/2} G_k & 0 & \gamma_k I & \star \\ \mathscr{R}^{1/2} Y_k & 0 & 0 & \gamma_k I \end{bmatrix} \geqslant 0,\ j,l \in \{1,\dots,L\}, \tag{10.102}$$

$$\begin{bmatrix} Q_{l,k} & Y_{j,k}^T \\ Y_{j,k} & u_{j,\inf}^2 \end{bmatrix} \geqslant 0,\ l \in \{1,\dots,L\},\ j \in \{1,\dots,m\}, \tag{10.103}$$

$$\begin{bmatrix} Q_{l,k} & \star \\ \Psi_s(A_l G_k + B_l Y_k) & \psi_{s,\inf}^2 \end{bmatrix} \geqslant 0,\ l \in \{1,2,\dots,L\},\ s \in \{1,2,\dots,q\}. \tag{10.104}$$

In (10.95)–(10.100), remove the free control move. Letting $N = L$ and substituting $F_i = Y_i Q_i^{-1}$ by $F_i = YG^{-1}$ yields

$$\begin{bmatrix} 1 & x_{k|k}^T \\ x_{k|k} & G_k^T + G_k - Q_{1,k} \end{bmatrix} \geqslant 0, \tag{10.105}$$

$$\begin{bmatrix} Q_{i,k} & \star & \star & \star \\ A_l G_k + B_l Y_k & G_k^T + G_k - Q_{i+1,k} & \star & \star \\ \mathscr{Q}^{1/2} G_k & 0 & \gamma_k I & \star \\ \mathscr{R}^{1/2} Y_k & 0 & 0 & \gamma_k I \end{bmatrix} \geqslant 0,\ i,l \in \{1,\dots,L\},\ Q_{L+1} = Q_1, \tag{10.106}$$

$$\begin{bmatrix} Q_{i,k} & Y_{j,k}^T \\ Y_{j,k} & u_{j,\inf}^2 \end{bmatrix} \geqslant 0,\ i \in \{1,\dots,L\},\ j \in \{1,\dots,m\}, \tag{10.107}$$

$$\begin{bmatrix} Q_{i,k} & \star \\ \Psi_s(A_l G_k + B_l Y_k) & \psi_{s,\inf}^2 \end{bmatrix} \geqslant 0,\ i,l \in \{1,2,\dots,L\},\ s \in \{1,2,\dots,q\}, \tag{10.108}$$

which can be called the Type III improved KBM formula. We compare (10.101)–(10.104) with (10.105)–(10.108) as follows:

(1) (10.105) is just the case of (10.101) where $l = 1$, thus (10.105) is better;
(2) (10.106) and (10.102) both include L^2 LMIs, in which L LMIs are the same. Thus, it is hard to compare them;
(3) (10.107) and (10.103) are equivalent;
(4) (10.108) includes L^2 LMI, where L LMI is in (10.104), thus (10.104) is better;
(5) (10.105)–(10.108) use $N = L$, and N is the additional degree of freedom.

Pluymers et al. [2005] provide the parameter-dependent open-loop MPC; according to notations of this book, constraints of this strategy are

$$\begin{bmatrix} 1 & \star \\ x_{k+N|k}^{l_{N-1}\cdots l_1 l_0} & Q_k \end{bmatrix} \geqslant 0,\ l_i \in \{1 \dots L\},\ i \in \{0 \dots N-1\}, \tag{10.109}$$

$$\begin{bmatrix} 1 & \star & \star \\ \mathscr{Q}^{1/2} x_{k+i|k}^{l_{i-1}\cdots l_1 l_0} & \gamma_{i,k} I & \star \\ \mathscr{R}^{1/2} u_{k+i|k}^{l_{i-1}\cdots l_1 l_0} & 0 & \gamma_{i,k} I \end{bmatrix} \geqslant 0,\ l_i \in \{1,\dots,L\},\ i \in \{0,\dots,N-1\}, \tag{10.110}$$

$$\begin{bmatrix} Q_k & \star & \star & \star \\ A_l Q_k + B_l Y_k & Q_k & \star & \star \\ \mathcal{Q}^{1/2} Q_k & 0 & \gamma_k I & \star \\ \mathcal{R}^{1/2} Y_k & 0 & 0 & \gamma_k I \end{bmatrix} \geqslant 0,\ l \in \{1, \dots, L\}, \tag{10.111}$$

$$-\underline{u} \leqslant u_{k|k} \leqslant \bar{u}, -\underline{u} \leqslant u_{k+i|k}^{l_{i-1}\cdots l_1 l_0} \leqslant \bar{u},\ i \in \{1, \dots, N-1\},\ l_{i-1} \in \{1, \dots, L\}, \tag{10.112}$$

$$\begin{bmatrix} Q_k & Y_{j,k}^T \\ Y_{j,k} & u_{j,\inf}^2 \end{bmatrix} \geqslant 0,\ j \in \{1, \dots, m\}, \tag{10.113}$$

$$-\underline{\psi}^s \leqslant \Psi^d \begin{bmatrix} x_{k+1|k}^{l_0} \\ x_{k+2|k}^{l_1 l_0} \\ \vdots \\ x_{k+N|k}^{l_{N-1}\cdots l_1 l_0} \end{bmatrix} \leqslant \overline{\psi}^s,\ l_i \in \{1, \dots, L\},\ i \in \{0, \dots, N-1\}, \tag{10.114}$$

$$\begin{bmatrix} Q_k & \star \\ \Psi_s(A_l Q_k + B_l Y_k) & \psi_{s,\inf}^2 \end{bmatrix} \geqslant 0,\ l \in \{1, 2, \dots, L\},\ s \in \{1, 2, \dots, q\}, \tag{10.115}$$

where the vertex state prediction $[x_{k+1|k}^{l_0}; x_{k+2|k}^{l_1 l_0}; \vdots; x_{k+N|k}^{l_{N-1}\cdots l_1 l_0}]$ is the same as Section 9.1. At each time the control moves are optimized via solving

$$\min \left(\sum_{i=0}^{N-1} \gamma_{i,k} + \gamma_k \right)$$

satisfying constraints (10.109)–(10.115). Comparing (10.79)–(10.85) with (10.109)–(10.115), the difference is small, i.e.,

$$\begin{bmatrix} u_{k|k} \\ u_{k+1|k} \\ \vdots \\ u_{k+N-1|k} \end{bmatrix} \rightarrow \begin{bmatrix} u_{k|k} \\ u_{k+1|k}^{l_0} \\ \vdots \\ u_{k+N-1|k}^{l_{N-2}\cdots l_1 l_0} \end{bmatrix}.$$

Although the modification is "not big", the parameter-dependent open-loop MPC has many advantages.

(1) The initial admissible set of the open-loop MPC may not increase with N, but that of the parameter-dependent open-loop MPC will increase;
(2) The performance of the open-loop MPC may not be improved as N increases, but that of the parameter-dependent open-loop MPC will be improved;
(3) Compared with the open-loop MPC, the parameter-dependent open-loop MPC can guarantee the recursive feasibility, and so the stability can be proved;

(4) The feedback MPC seems to be the best approach in terms of feasibility and optimality; however, it is the parameter-dependent open-loop MPC that plays this role;
(5) When $N > 2$, the feedback MPC cannot be solved by convex optimization, but the parameter-dependent open-loop MPC can be.

Example 10.3 To what extent is the parameter-dependent open-loop MPC equivalent to, or different from, the feedback MPC? The relation between the feedback MPC and the parameter-dependent open-loop MPC is shown as (see [Ding, 2010b])

$$\bar{u}_{k+i|k}^{l_{i-1}\cdots l_0} = F_{k+i|k} \times [A_{l_{i-1}} + B_{l_{i-1}} F_{k+i-1|k}] \times \cdots \times [A_{l_1} + B_{l_1} F_{k+1|k}] \times [A_{l_0} x_k + B_{l_0} u_{k|k}]$$
$$l_0, \ldots, l_{i-1} \in \{1, \ldots, L\}. \tag{10.116}$$

Compared with the parameter-dependent open-loop MPC, the feedback MPC has more constraints. In the constraints of feedback MPC, only (10.116) is related to $F_{k+i|k}$, while the other constraints do not have $F_{k+i|k}$, i.e., the existence of $F_{k+i|k}$ is only related to (10.116). $u_{k+i|k}^{l_{i-1}\cdots l_0}$ can be obtained via solving the optimization problem for the parameter-dependent open-loop MPC. If, by taking $u_{k+i|k}^{l_{i-1}\cdots l_0}$ as the solution to the parameter-dependent open-loop MPC, one can calculate $F_{k+i|k}$ from (10.116), then (10.116) does not affect the recursive feasibility and optimality of the feedback MPC; otherwise (e.g., when $A_{l_0} x_k + B_{l_0} u_{k|k} = 0$, then for the same l_0, it holds $\bar{u}_{k+1|k}^{l_0} = 0$), then (10.116) affects the recursive feasibility and optimality of the feedback MPC. The dimension of $F_{k+i|k}$ is $n \times m$. For the common case $n \times m > L$, the feedback MPC is often equivalent to the parameter-dependent open-loop MPC.

The partial feedback MPC is the mainstream technique of the robust MPC of LPV. The following are the constraints being presented in Schuurmans and Rossiter [2000] (notations being rewritten following this book):

$$\begin{bmatrix} 1 & \star \\ x_{k+N|k}^{l_{N-1}\cdots l_1 l_0} & Q_k \end{bmatrix} \geqslant 0,\ l_i \in \{1 \ldots L\},\ i \in \{0, \ldots, N-1\}, \tag{10.117}$$

$$\begin{bmatrix} 1 & \star & \star \\ \mathcal{Q}^{1/2} x_{k+i|k}^{l_{i-1}\cdots l_1 l_0} & \gamma_{i,k} I & \star \\ \mathcal{R}^{1/2}[F_{k+i|k} x_{k+i|k}^{l_{i-1}\cdots l_1 l_0} + c_{k+i|k}] & 0 & \gamma_{i,k} I \end{bmatrix} \geqslant 0,\ l_i \in \{1, \ldots, L\},\ i \in \{0, \ldots, N-1\}, \tag{10.118}$$

$$\begin{bmatrix} Q_k & \star & \star & \star \\ A_l Q_k + B_l Y_k & Q_k & \star & \star \\ \mathcal{Q}^{1/2} Q_k & 0 & \gamma_k I & \star \\ \mathcal{R}^{1/2} Y_k & 0 & 0 & \gamma_k I \end{bmatrix} \geqslant 0,\ l \in \{1, \ldots, L\}, \tag{10.119}$$

$$-\underline{u} \leqslant F_{k|k} x_{k|k} + c_{k|k} \leqslant \bar{u},\ -\underline{u} \leqslant F_{k+i|k} x_{k+i|k}^{l_{i-1}\cdots l_1 l_0} + c_{k+i|k} \leqslant \bar{u},$$
$$i \in \{1, \ldots, N-1\},\ l_{i-1} \in \{1, \ldots, L\}, \tag{10.120}$$

$$\begin{bmatrix} Q_k & Y_{j,k}^T \\ Y_{j,k} & u_{j,\inf}^2 \end{bmatrix} \geqslant 0,\ j \in \{1, \ldots, m\}, \tag{10.121}$$

$$-\underline{\psi}^s \leqslant \Psi^d \begin{bmatrix} x_{k+1|k}^{l_0} \\ x_{k+2|k}^{l_1 l_0} \\ \vdots \\ x_{k+N|k}^{l_{N-1}\cdots l_1 l_0} \end{bmatrix} \leqslant \overline{\psi}^s,\ l_i \in \{1, \ldots, L\},\ i \in \{0, \ldots, N-1\}, \tag{10.122}$$

$$\begin{bmatrix} Q_k & \star \\ \Psi_s(A_l Q_k + B_l Y_k) & \psi_{s,\inf}^2 \end{bmatrix} \geqslant 0,\ l \in \{1, 2, \ldots, L\},\ s \in \{1, 2, \ldots, q\}, \tag{10.123}$$

where the definition of vertex state prediction is similar with Section 9.1, which is substituted by

$$\begin{bmatrix} x_{k+1|k}^{l_0} \\ x_{k+2|k}^{l_1 l_0} \\ \vdots \\ x_{k+N|k}^{l_{N-1}\cdots l_1 l_0} \end{bmatrix} = \begin{bmatrix} \mathcal{A}_{l_0,k|k} \\ \mathcal{A}_{l_1,k+1|k}\mathcal{A}_{l_0,k|k} \\ \vdots \\ \mathcal{A}_{l_{N-1},k+N-1|k}\cdots\mathcal{A}_{l_1,k+1|k}\mathcal{A}_{l_0,k|k} \end{bmatrix} x_k$$

$$+ \begin{bmatrix} B_{l_0} & 0 & \cdots & 0 \\ \mathcal{A}_{l_1,k+1|k}B_{l_0} & B_{l_1} & \ddots & \vdots \\ \vdots & \vdots & \ddots & 0 \\ \mathcal{A}_{l_{N-2},k+N-2|k}\cdots\mathcal{A}_{l_1,k+1|k}B_{l_0} & \heartsuit & \cdots & B_{l_{N-1}} \end{bmatrix} \begin{bmatrix} u_{k|k} \\ u_{k+1|k} \\ \vdots \\ u_{k+N-1|k} \end{bmatrix},$$

$$\begin{aligned} \heartsuit &= \mathcal{A}_{l_{N-2},k+N-2|k}\cdots\mathcal{A}_{l_2,k+2|k}B_{l_1}, \\ \mathcal{A}_{l_i,k+i|k} &= A_{l_i} + B_{l_i}F_{k+i|k},\ i \in \{0, 1, \ldots, N-1\}, \\ F_{k+i|k} &= F_{k+i+1|k-1},\ k > 0,\ i \in \{0, 1, \ldots, N-2\}, \\ F_{k+N-1|k} &= F_{k-1}^* = Y_{k-1}^* Q_{k-1}^{*-1},\ k > 0, \\ F_{i|0} &= 0,\ i \in \{0, 1, \ldots, N-1\}. \end{aligned}$$

At each time, the control moves are optimized via solving min $\left(\sum_{i=0}^{N-1} \gamma_{i,k} + \gamma_k\right)$ satisfying constraints (10.117)–(10.123). Comparing (10.109)–(10.115) with (10.117)–(10.123), the difference is small, i.e.,

$$\begin{bmatrix} u_{k|k} \\ u_{k+1|k}^{l_0} \\ \vdots \\ u_{k+N-1|k}^{l_{N-2}\cdots l_1 l_0} \end{bmatrix} \rightarrow \begin{bmatrix} F_{k|k}x_{k|k} + c_{k|k} \\ F_{k+1|k}x_{k+1|k}^{l_0} + c_{k+1|k} \\ \vdots \\ F_{k+N-1|k}x_{k+N-1|k}^{l_{N-2}\cdots l_1 l_0} + c_{k+N-1|k} \end{bmatrix}.$$

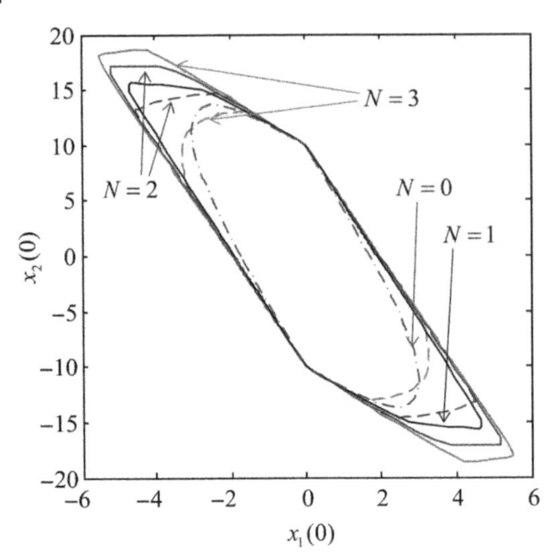

Figure 10.4 Comparisons of RoAs between parameter-dependent open-loop MPC and partial feedback MPC.

At time $k = 0$, the above partial MPC is the open-loop MPC. However, since the partial MPC updates $F_{k+i|k}$, $i \in \{0, 1, \dots, N-1\}$ appropriately, the recursive feasibility and stability are guaranteed.

Example 10.4 See Ding [2010b]. Consider the system

$$\begin{bmatrix} x_{1,k+1} \\ x_{2,k+1} \end{bmatrix} = \begin{bmatrix} 1 & 0.1 \\ H_k & 1 \end{bmatrix} \begin{bmatrix} x_{1,k} \\ x_{2,k} \end{bmatrix} + \begin{bmatrix} 1 \\ 0 \end{bmatrix} u_k, \; H_k \in [0.5, \; 2.5].$$

The input constraint is $|u_k| \le 1$, and the weight matrices are $\mathcal{Q} = I$ and $\mathcal{R} = 1$. Take $N = 1, 2, 3$. In Figure 10.4, RoA of the partial feedback MPC (dashed line) and the parameter-dependent open-loop MPC are depicted. If $N = 1$, the dashed area is exactly the same as the solid area; if $N = 2$ or $N = 3$, the solid area is larger than the dashed area, and it demonstrates the advantage of parameter-dependent open-loop MPC. For the partial MPC, RoA with $N = 3$ ($N = 2$) does not contain RoA with $N = 2$ ($N = 1$). For comparison, Figure 10.4 also gives RoA of KBM controller which is depicted by dash-dotted line.

10.5 About Optimality

In Sections 10.1–10.4, γ_k is the upper bound of the upper bound of the upper bound of the performance index. Why we say the “upper bound” three times? The performance index is

$$J_{0,k}^{\infty} = \sum_{i=0}^{\infty} [\|x_{k+i|k}\|_{\mathcal{Q}}^2 + \|u_{k+i|k}\|_{\mathcal{R}}^2].$$

Taking $\|x_{k+N|k}\|_{P_k}^2$ as the upper bound of $J_{N,k}^{\infty} = \sum_{i=N}^{\infty} [\|x_{k+i|k}\|_{\mathcal{Q}}^2 + \|u_{k+i|k}\|_{\mathcal{R}}^2]$, we obtain

$$J_{0,k}^{\infty} \leqslant \bar{J}_{0,k}^{N} \triangleq \sum_{i=0}^{N-1} [\|x_{k+i|k}\|_{\mathcal{Q}}^2 + \|u_{k+i|k}\|_{\mathcal{R}}^2] + \|x_{k+N|k}\|_{P_k}^2.$$

This is the first "upper bound." Usually, we use KBM formula or its improved version to determine the upper bound of $\|x_{k+N|k}\|^2_{P_k}$ and the upper bound of

$$J_{0,k}^{N-1} = \sum_{i=0}^{N-1}[\|x_{k+i|k}\|^2_{\mathcal{Q}} + \|u_{k+i|k}\|^2_{\mathcal{R}}].$$

This is the second "upper bound" which is put in the constraints. Since more constraints (e.g., LMI constraints of input and state, current state constraints) or more conservative constraints (most constraints, especially those being converted into LMI, are handled conservatively) need to be satisfied, we can only take a larger value which deviates the second upper bound. This is the third "upper bound." The third upper bound is often the main factor, which cannot be ignored. The three upper bounds, especially the third, make the optimality of robust MPC unclear, sometimes γ_k quite large.

10.5.1 Constrained Linear Time-Varying Quadratic Regulation with Near-Optimal Solution

Relatively speaking, the upper bound on the nominal system is closed to the performance index itself. Suppose $[A_k, B_k]$ is known, is the bounded function of k, and is uniformly stabilizable, and

$$[A_i|B_i] \in \Omega, \ i \geqslant N_0.$$

The objective is to solve CLTVQR (the Constrained Linear Time-Varying Quadratic Regulation), i.e., solve the following problem:

$$\min_{u_0^\infty} \Phi\left(x_{0|0}\right) = \sum_{i=0}^{\infty}\left[\left\|x_{i|0}\right\|^2_{\mathcal{Q}} + \left\|u_{i|0}\right\|^2_{\mathcal{R}}\right], \tag{10.124}$$

$$\text{s.t. } x_{i+1|0} = A_i x_{i|0} + B_i u_{i|0}, x_{0|0} = x_0, i \geqslant 0, \tag{10.125}$$

$$-\underline{u} \leqslant u_{i|0} \leqslant \bar{u}, i \geqslant 0, \tag{10.126}$$

$$-\underline{\psi} \leqslant \Psi x_{i+1|0} \leqslant \overline{\psi}, i \geqslant 0, \tag{10.127}$$

where $\{\mathcal{Q}, \mathcal{R}\}$ are the nonnegative weight matrices, and $u_0^\infty = \{u_{0|0}, u_{1|0}, \dots, u_{\infty|0}\}$ are the decision variables.

Let $\Phi^* = \min_{u_0^\infty} \Phi\left(x_{0|0}\right)$. The idea is to find the suboptimal solution of CLTVQR such that

$$\left(\Phi^f - \Phi^*\right)/\Phi^* \leqslant \delta, \tag{10.128}$$

where $\delta > 0$ is a given scalar. Since δ can be chosen arbitrarily small, (10.128) means that the suboptimal solution can be arbitrarily close to the optimal solution. For satisfying (10.128), $\Phi\left(x_{0|0}\right)$ can be divided into two parts, i.e.,

$$\Phi\left(x_{0|0}\right) = \sum_{i=0}^{N-1}\left[\left\|x_{i|0}\right\|^2_{\mathcal{Q}} + \left\|u_{i|0}\right\|^2_{\mathcal{R}}\right] + \Phi_{\text{tail}}\left(x_{N|0}\right), \tag{10.129}$$

where $N \geqslant N_0$ and

$$\Phi_{\text{tail}}\left(x_{N|0}\right) = \sum_{i=N}^{\infty}\left[\left\|x_{i|0}\right\|^2_{\mathcal{Q}} + \left\|u_{i|0}\right\|^2_{\mathcal{R}}\right]. \tag{10.130}$$

For the minimization of (10.130), applying KBM controller yields

$$u_{i|0} = Fx_{i|0}, \quad i \geqslant N, \tag{10.131}$$

and the bound of (10.130) can be derived as

$$\Phi_{\text{tail}}\left(x_{N|0}\right) \leqslant x_{N|0}^T Q_N x_{N|0} \leqslant \gamma,$$

where $\gamma > 0$ is a scalar; $Q_N > 0$ is the symmetric weight matrix. Hence,

$$\Phi\left(x_{0|0}\right) \leqslant \sum_{i=0}^{N-1} \left[\left\|x_{i|0}\right\|_{\mathcal{Q}}^2 + \left\|u_{i|0}\right\|_{\mathcal{R}}^2\right] + x_{N|0}^T Q_N x_{N|0} = \overline{\Phi}_{Q_N}\left(x_{0|0}\right).$$

If $Q_N = 0$, then $\overline{\Phi}_{Q_N}\left(x_{0|0}\right) = \overline{\Phi}_0\left(x_{0|0}\right)$. $\mathcal{X}_{N|0} \subset \mathbb{R}^n$ denotes the set of $x_{N|0}$ such that (10.131) is feasible. Let us discuss the following useful problems.

Problem 10.1 CLTVQR, without the terminal cost, solves

$$\overline{\Phi}_0^* = \min_{u_0^{N-1}} \overline{\Phi}_0\left(x_{0|0}\right), \quad \text{s.t. } (10.125), (10.126), (10.127), 0 \leqslant i \leqslant N-1.$$

Denote the terminal state as $x_{N|0}^0$ by the optimal solution to Problem 10.1.

Problem 10.2 CLTVQR, with the terminal cost, solves

$$\overline{\Phi}_{Q_N}^* = \min_{u_0^{N-1}} \overline{\Phi}_{Q_N}\left(x_{0|0}\right), \quad \text{s.t. } (10.125), (10.126), (10.127), 0 \leqslant i \leqslant N-1.$$

Denote the terminal state as $x_{N|0}^*$ by the optimal solution to Problem 10.2.

10.5.1.1 Solving KBM Controller

Define the quadratic function

$$V_i = x_{i|0}^T S^{-1} x_{i|0}, \quad i \geqslant N.$$

In order to give the upper bound of the performance index (10.130), impose

$$V_{i+1} - V_i \leqslant -1/\gamma \left[\left\|x_{i|0}\right\|_{\mathcal{Q}}^2 + \left\|u_{i|0}\right\|_{\mathcal{R}}^2\right], \quad \left[A_i | B_i\right] \in \Omega, \quad i \geqslant N,$$

which is guaranteed by

$$\begin{bmatrix} S & \star & \star & \star \\ A_l S + B_l Y & S & \star & \star \\ \mathcal{Q}^{\frac{1}{2}} S & 0 & \gamma I & \star \\ \mathcal{R}^{\frac{1}{2}} Y & 0 & 0 & \gamma I \end{bmatrix} \geqslant 0, \quad l = 1, 2, \dots, L, \tag{10.132}$$

where $F = YS^{-1}$. The feasible state feedback law is required to satisfy (10.126)–(10.127). Equations (10.126) and (10.127) are guaranteed by the following LMI:

$$\begin{bmatrix} Z & Y \\ Y^T & S \end{bmatrix} \geqslant 0, \quad Z_{jj} \leqslant z_{j,\inf}^2, j = 1, 2, \dots, m, \tag{10.133}$$

$$\begin{bmatrix} S & \star \\ \Psi\left(A_l S + B_l Y\right) & \Gamma \end{bmatrix} \geqslant 0,\ \Gamma_{ss} \leqslant \psi^2_{s,\inf},\ l = 1, 2, \ldots, L; s = 1, 2, \ldots, q, \tag{10.134}$$

where $z_{j,\inf} = \min\left\{\underline{u}_j, \bar{u}_j\right\}, \psi_{s,\inf} = \min\left\{\underline{\psi}_s, \overline{\psi}_s\right\}$. Hence, by restricting $x_{N|0} \in \mathcal{X}_{N|0}$, KBM controller is calculated by

$$\min_{\gamma,S,Y,Z,\Gamma} \gamma,\ \text{s.t. } (10.132), (10.133), (10.134), \begin{bmatrix} 1 & \star \\ x_{N|0} & S \end{bmatrix} \geqslant 0. \tag{10.135}$$

Lemma 10.3 Consider the minimization problem (10.135). Any feasible solution defines the set $\mathcal{X}_{N|0} = \left\{x | x^T S^{-1} x \leqslant 1\right\}$ in which the local controller $Fx = YS^{-1}x$ is adopted, and starting from N, the bound of the closed-loop cost of the infinite horizon is

$$\Phi_{\text{tail}}\left(x_{N|0}\right) \leqslant x^T_{N|0} \gamma S^{-1} x_{N|0}.$$

10.5.1.2 Solving Problem Without Terminal Cost

Define

$$\tilde{x} = \left[x^T_{0|0}, x^T_{1|0}, \ldots, x^T_{N-1|0}\right]^T,$$

$$\tilde{u} = \left[u^T_{0|0}, u^T_{1|0}, \ldots, u^T_{N-1|0}\right]^T,$$

then

$$\tilde{x} = \tilde{A}\tilde{x} + \tilde{B}\tilde{u} + \tilde{x}_0, \tag{10.136}$$

where

$$\tilde{A} = \begin{bmatrix} 0 & 0 \\ \text{diag}\{A_0, A_1, \ldots, A_{N-2}\} & 0 \end{bmatrix},$$

$$\tilde{B} = \begin{bmatrix} 0 & 0 \\ \text{diag}\{B_0, B_1, \ldots, B_{N-2}\} & 0 \end{bmatrix},$$

$$\tilde{x}_0 = \left[x^T_{0|0}, 0, \ldots, 0\right]^T.$$

Equation (10.136) can be written as

$$\tilde{x} = \tilde{W}\tilde{u} + \tilde{V}_0,$$

where

$$\tilde{W} = (I - \tilde{A})^{-1}\tilde{B},\ \tilde{V}_0 = (I - \tilde{A})^{-1}\tilde{x}_0.$$

Hence, the cost function of Problem 10.1 can be written as

$$\overline{\Phi}_0\left(x_{0|0}\right) = \|\tilde{x}\|^2_{\tilde{\mathcal{Q}}} + \|\tilde{u}\|^2_{\tilde{\mathcal{R}}} = \tilde{u}^T W \tilde{u} + W_v \tilde{u} + V_0 \leqslant \eta^0, \tag{10.137}$$

where η^0 is a scalar,

$$\tilde{\mathcal{Q}} = \text{diag}\{\mathcal{Q}, \mathcal{Q}, \ldots, \mathcal{Q}\},\ \tilde{\mathcal{R}} = \text{diag}\{\mathcal{R}, \mathcal{R}, \ldots, \mathcal{R}\},$$

$$W = \tilde{W}^T \tilde{\mathcal{Q}} \tilde{W} + \tilde{\mathcal{R}},\ W_v = 2\tilde{V}_0^T \tilde{\mathcal{Q}} \tilde{W},\ V_0 = \tilde{V}_0{}^T \tilde{\mathcal{Q}} \tilde{V}_0.$$

Equation (10.137) is represented by the following LMI:

$$\begin{bmatrix} \eta^0 - W_v \tilde{u} - V_0 & \star \\ W^{\frac{1}{2}} \tilde{u} & I \end{bmatrix} \geqslant 0. \tag{10.138}$$

Moveover, define $\tilde{x}^+ = \left[x_{1|0}^T, \dots, x_{N-1|0}^T, x_{N|0}^T\right]^T$, then $\tilde{x}^+ = \tilde{A}^+\tilde{x} + \tilde{B}^+\tilde{u}$, where

$$\tilde{A}^+ = \text{diag}\{A_0, A_1, \dots, A_{N-1}\},$$

$$\tilde{B}^+ = \text{diag}\{B_0, B_1, \dots, B_{N-1}\}.$$

The constraints of Problem 10.1 are converted into

$$-\underline{u}^s \leqslant \tilde{u} \leqslant \bar{u}^s, \; -\underline{\psi}^s \leqslant \Psi^d(\tilde{A}^+\tilde{W}\tilde{u} + \tilde{B}^+\tilde{u} + \tilde{A}^+\tilde{V}_0) \leqslant \overline{\psi}^s. \tag{10.139}$$

Problem 10.1 is converted into

$$\min_{\eta^0, \tilde{u}} \eta^0, \quad \text{s.t.} \;\; (10.138), (10.139). \tag{10.140}$$

The optimal solution $\tilde{u}$ is calculated via solving (10.140).

10.5.1.3 Solving Problem with Terminal Cost

The performance index of Problem 10.2 is denoted as

$$\begin{aligned}\overline{\Phi}_{Q_N}\left(x_{0|0}\right) &= \|\tilde{x}\|_{\tilde{Q}}^2 + \|\tilde{u}\|_{\tilde{\mathcal{R}}}^2 + \left\|\mathcal{A}_{N,0}x_{0|0} + \overline{B}\tilde{u}\right\|_{Q_N}^2 \\ &= \left(\tilde{u}^T W\tilde{u} + W_v\tilde{u} + V_0\right) + \left\|\mathcal{A}_{N,0}x_{0|0} + \overline{B}\tilde{u}\right\|_{Q_N}^2 \leqslant \eta,\end{aligned} \tag{10.141}$$

where η is a scalar, $\mathcal{A}_{j,i} = \prod_{l=i}^{j-1} A_{j-1+i-l}$,

$$\overline{B} = \left[\mathcal{A}_{N,1}B_0, \dots, \mathcal{A}_{N,N-1}B_{N-2}, B_{N-1}\right].$$

Equation (10.141) can be represented by the following LMI:

$$\begin{bmatrix} \eta - W_v\tilde{u} - V_0 & \star & \star \\ W^{\frac{1}{2}}\tilde{u} & I & \star \\ \mathcal{A}_{N,0}x_{0|0} + \overline{B}\tilde{u} & 0 & Q_N^{-1} \end{bmatrix} \geqslant 0. \tag{10.142}$$

Problem 10.2 is converted into

$$\min_{\eta, \tilde{u}} \eta, \quad \text{s.t.} \;\; (10.142), (10.139). \tag{10.143}$$

The optimal solution of $\tilde{u}$, denoted as $\tilde{u}^*$, is calculated via solving (10.143).

10.5.1.4 Overall Algorithm and Analysis

First all, give the following conclusion.

Lemma 10.4 If $\overline{\Phi}_{Q_N}^*$ acts as Φ^f and

$$\left(\overline{\Phi}_{Q_N}^* - \overline{\Phi}_0^*\right) / \overline{\Phi}_0^* \leqslant \delta,$$

then (10.128) is satisfied.

Denote $\mathcal{X}_{0|0}$ as the set of state $x_{0|0}$ that makes CLTVQR feasible. Then the following theorem describes the recursive feasibility and stability of the suboptimal CLTVQR.

Theorem 10.4 Apply Algorithm 10.1. If (10.135) has the feasible solution for the appropriate $x_{N|0}$, then for all $x_{0|0} \in \mathcal{X}_{0|0}$, there exists a finite N and feasible $\tilde{u}^*$ such that (10.128) is satisfied, and the closed-loop system is asymptotically stable.

10.5.1.5 Numerical Example

The two-mass-spring model is adopted, i.e.,

$$\begin{bmatrix} x_{1,k+1} \\ x_{2,k+1} \\ x_{3,k+1} \\ x_{4,k+1} \end{bmatrix} = \begin{bmatrix} 1 & 0 & 0.1 & 0 \\ 0 & 1 & 0 & 0.1 \\ -0.1K_k/m_1 & 0.1K_k/m_1 & 1 & 0 \\ 0.1K_k/m_2 & -0.1K_k/m_2 & 0 & 1 \end{bmatrix} \begin{bmatrix} x_{1,k} \\ x_{2,k} \\ x_{3,k} \\ x_{4,k} \end{bmatrix} + \begin{bmatrix} 0 \\ 0 \\ 0.1/m_1 \\ 0 \end{bmatrix} u_k.$$

Assume that $m_1 = m_2 = 1$ and $K_k = 1.5 + 2e^{-0.1k}(1 + \sin k) + 0.973\sin(k\pi/11)$. The initial state is $x_0 = \alpha \times [5, 5, 0, 0]^T$ where α is a constant. The weight matrices are $\mathcal{Q} = I$ and $\mathcal{R} = 1$, and the input constraint is $|u_k| \leqslant 1$.

Let us first consider Algorithm 10.1. The control objective is to find a sequence of control input signals such that (10.128) is satisfied with $\delta \leqslant 10^{-4}$. As $k = 50$, it holds $2e^{-0.1k} \approx 0.0135$. Hence, approximately $0.527 \leqslant K_k \leqslant 2.5$ for $k \geq 50$. We choose $N_0 = 50$,

$$[A_1|B_1] = \left[\begin{array}{cccc|c} 1 & 0 & 0.1 & 0 & 0 \\ 0 & 1 & 0 & 0.1 & 0 \\ -0.0527 & 0.0527 & 1 & 0 & 0.1 \\ 0.0527 & -0.0527 & 0 & 1 & 0 \end{array}\right],$$

Algorithm 10.1 (CLTVQR)

Step 1. Choose the initial (larger) $x_{N|0} = \hat{x}_{N|0}$ satisfying $\left\|\hat{x}_{N|0}\right\| > \Delta$,where Δ is a given scalar. Note that N is unknown.

Step 2. Solve (10.135) and obtain $\{\gamma^*, S^*, F^*\}$.

Step 3. If (10.135) is infeasible, then decrease $x_{N|0}$ (replace it by $rx_{N|0}$, where $0 < r < 1$ is a given scalar, satisfying $\left\|rx_{N|0}\right\| > \Delta$), and goto Step 2. However, if (10.135) is infeasible and $\left\|x_{N|0}\right\| \leqslant \Delta$, then the whole algorithm is infeasible and stop.

Step 4. Let $Q_N = \gamma^* S^{*-1}$ and $\mathcal{X}_{N|0} = \left\{x | x^T S^{*-1} x \leqslant 1\right\}$.

Step 5. Choose the initial $N > N_0$.

Step 6. Solve (10.143) to obtain $\tilde{u}^*$ and $x^*_{N|0} = \mathcal{A}_{N,0}x_0 + \bar{B}\tilde{u}^*$.

Step 7. If $x^*_{N|0} \notin \mathcal{X}_{N|0}$, then increase N and goto Step 6.

Step 8. Choose $\tilde{u}^*$ as the initial solution of $\tilde{u}$ in Step 6. Solve (10.140) to obtain $\tilde{u}^0$ and $x^0_{N|0} = \mathcal{A}_{N,0}x_0 + \bar{B}\tilde{u}^0$.

Step 9. If $x^0_{N|0} \notin \mathcal{X}_{N|0}$, then increase N and goto Step 6.

Step 10. If $\left(\bar{\Phi}^*_{Q_N} - \bar{\Phi}^*_0\right) / \bar{\Phi}^*_0 > \delta$, then increase N and goto Step 6.

Step 11. Implement $\tilde{u}^*$ and F^*.

$$[A_2|B_2] = \left[\begin{array}{cccc|c} 1 & 0 & 0.1 & 0 & 0 \\ 0 & 1 & 0 & 0.1 & 0 \\ -0.25 & 0.25 & 1 & 0 & 0.1 \\ 0.25 & -0.25 & 0 & 1 & 0 \end{array}\right].$$

Choose $\hat{x}_{N|0} = 0.02 \times [1, 1, 1, 1]^T$. Then problem (10.135) has feasible solution, $F = \left[-8.7199\ 6.7664\ -4.7335\ -2.4241 \right]$. Algorithm 10.1 has a feasible solution whenever $\alpha \leqslant 22.5$. Choose $\alpha = 1$ and $N = 132$, then $\overline{\Phi}^*_{Q_{132}} = 1475.91$, $\overline{\Phi}^*_0 = 1475.85$ and the desired optimality requirement (10.128) is achieved.

10.5.2 Alternatives with Nominal Performance Cost

There is a way to make γ_k smaller, i.e., to use the nominal performance index, i.e., the state prediction in the performance index is based on the nominal model. Rewrite KBM formula as

$$\begin{bmatrix} 1 & x_{k|k}^T \\ x_{k|k} & Q_k \end{bmatrix} \geqslant 0, \tag{10.144}$$

$$\begin{bmatrix} Q_k & \star \\ A_l Q_k + B_l Y_k & Q_k \end{bmatrix} \geqslant 0,\ l \in \{1, \dots, L\},$$

$$\begin{bmatrix} Q_k & \star & \star & \star \\ A_0 Q_k + B_0 Y_k & Q_k & \star & \star \\ \mathcal{Q}^{1/2} Q_k & 0 & \gamma_k I & \star \\ \mathcal{R}^{1/2} Y_k & 0 & 0 & \gamma_k I \end{bmatrix} \geqslant 0, \tag{10.145}$$

$$\begin{bmatrix} Q_k & Y_{j,k}^T \\ Y_{j,k} & \bar{u}_j^2 \end{bmatrix} \geqslant 0,\ j \in \{1, \dots, m\}, \tag{10.146}$$

$$\begin{bmatrix} Q_k & \star \\ \Psi_s(A_l Q_k + B_l Y_k) & \overline{\psi}_s^2 \end{bmatrix} \geqslant 0,\ l \in \{1, 2, \dots, L\},\ s \in \{1, 2, \dots, q\}, \tag{10.147}$$

where $[A_0|B_0]$ is the nominal model. By applying (10.144)–(10.147), however, there is no evidence that the online robust MPC with guaranteed stability can be obtained.

10.5.2.1 Problem Formulation

At each time k, solve the following optimization problem of robust MPC:

$$\min_{\tilde{u}_k} \max_{[A_{k+i}|B_{k+i}], i \geqslant 0} J_{\infty,k} = \sum_{i=0}^{\infty} \left[\left\| \hat{x}_{k+i|k} \right\|_{\mathcal{Q}}^2 + \left\| u_{k+i|k} \right\|_{\mathcal{R}}^2 \right], \tag{10.148}$$

$$\text{s.t. } \hat{x}_{k+i+1|k} = \hat{A} \hat{x}_{k+i|k} + \hat{B} u_{k+i|k},\ \forall i \geqslant 0,\ \hat{x}_{k|k} = x_k, \tag{10.149}$$

$$x_{k+i+1|k} = A_{k+i} x_{k+i|k} + B_{k+i} u_{k+i|k},\ \forall i \geqslant 1, \tag{10.150}$$

$$-\underline{u} \leqslant u_{k+i|k} \leqslant \bar{u}, -\underline{\psi} \leqslant \Psi x_{k+i+1|k} \leqslant \overline{\psi}, \ \forall i \geqslant 0, \tag{10.151}$$

where $\vec{u}_k = \left[u_{k|k}^T, u_{k+1|k}^T, u_{k+2|k}^T, \dots\right]^T$ are the decision variables. Let us discuss two kinds of $[\hat{A}|\hat{B}]$:

(a) $[\hat{A}|\hat{B}] = [A_0|B_0] \in \Omega$, where $[A_0|B_0]$ denotes the nominal model that is closer to the real system;
(b) $[\hat{A}|\hat{B}] = [A_k|B_k] \in \Omega$ which denotes the current model that is exactly known.

After the switching horizon, the control input is parameterized as

$$u_{k+i|k} = F_k x_{k+i|k}, \ i \geq N, \tag{10.152}$$

where F_k is the feedback gain. Define

$$J_{N,k} = \sum_{i=N}^{\infty} \left[\|\hat{x}_{k+i|k}\|_{\mathcal{Q}}^2 + \left\|u_{k+i|k}\right\|_{\mathcal{R}}^2 \right].$$

In order to obtain the bound of $J_{N,k}$, impose the following constraints:

$$\left[\hat{A} + \hat{B}F_k\right]^T P_k \left[\hat{A} + \hat{B}F_k\right] - P_k + \mathcal{Q} + F_k^T \mathcal{R} F_k \leqslant 0, \tag{10.153}$$

where $P_k > 0$ is the symmetric matrix. By (10.153), it yields

$$\max_{[A_{k+i}|B_{k+i}], i \geqslant N} J_{N,k} \leqslant \hat{x}_{k+N|k}^T P_k \hat{x}_{k+N|k}. \tag{10.154}$$

By applying the above method, the optimization problem (10.148)–(10.151) is easily substituted by

$$\begin{aligned} &\min_{\tilde{u}_k, F_k, P_k} \max_{[A_{k+i}|B_{k+i}], 0 \leqslant i \leqslant N-1} \bar{J}_k = \sum_{i=0}^{N-1} \left[\left\|\hat{x}_{k+i|k}\right\|_{\mathcal{Q}}^2 + \left\|u_{k+i|k}\right\|_{\mathcal{R}}^2 \right] + \left\|\hat{x}_{k+N|k}\right\|_{P_k}^2 \\ &\text{s.t. } (10.149), (10.150), (10.151), (10.152), (10.153). \end{aligned} \tag{10.155}$$

where $\tilde{u}_k = \left[u_{k|k}^T, u_{k+1|k}^T, \dots, u_{k+N-1|k}^T\right]^T$.

For solving (10.155), consider the two types of u_{k+i} with $1 \leqslant i \leqslant N-1$, i.e.,

(A) partial feedback MPC;
(B) parameter-dependent open-loop MPC.

For solving (10.155), we need the state prediction $x_{k+i|k}$ and $\hat{x}_{k+i|k}$ for $i > 0$. Obviously, consider (a)-(b) and (A)-(B), $x_{k+1|k}$ and $\hat{x}_{k+1|k}$ are deterministic. It is impossible to obtain the deterministic state prediction for $i > 1$ and $L > 1$.

10.5.2.2 Robust MPC Based on Partial Feedback Control

In the following, we will consider (b)-(A) and convert (10.155) into LMI optimization problem. Introduce γ_i, $0 \leqslant i \leqslant N$, satisfying

$$\gamma_i \geqslant \left\|\hat{x}_{k+i|k}\right\|_{\mathcal{Q}}^2 + \left\|u_{k+i|k}\right\|_{\mathcal{R}}^2, \ i = 0, \dots, N-1, \tag{10.156}$$

$$\gamma_N \geqslant \left\|\hat{x}_{k+N|k}\right\|_{P_k}^2. \tag{10.157}$$

We have $\bar{J}_k \leqslant \sum_{i=0}^{N} \gamma_i$. Define $Q := \gamma_N P_k^{-1}$. Equations (10.156) and (10.157) can be converted into the following LMI:

$$\begin{bmatrix} \gamma_i & \star & \star \\ \mathcal{Q}^{1/2}\hat{x}_{k+i|k} & I & \star \\ \mathcal{R}^{1/2}[F_{k+i|k}x_{k+i|k} + c_{k+i|k}] & 0 & I \end{bmatrix} \geqslant 0,\ i = 0, 1, \tag{10.158}$$

$$\begin{bmatrix} \gamma_i & \star & \star \\ \mathcal{Q}^{1/2}\hat{x}_{k+i|k}^{l_{i-1}\cdots l_1} & I & \star \\ \mathcal{R}^{1/2}[F_{k+i|k}x_{k+i|k}^{l_{i-1}\cdots l_1} + c_{k+i|k}] & 0 & I \end{bmatrix} \geqslant 0,\ i = 2\cdots N-1,\ \{l_1, \ldots, l_{i-1}\} = 1, \ldots, L, \tag{10.159}$$

$$\begin{bmatrix} 1 & \star \\ \hat{x}_{k+N|k}^{l_{N-1}\cdots l_1} & Q \end{bmatrix} \geqslant 0,\ \{l_1, \ldots, l_{N-1}\} = 1, \ldots, L. \tag{10.160}$$

Moreover, define $F_k := YQ^{-1}$ where Y is the matrix with a suitable dimension. Applying Schur complement, (10.153) is converted into the following LMI:

$$\begin{bmatrix} Q & \star & \star & \star \\ \hat{A}Q + \hat{B}Y & Q & \star & \star \\ \mathcal{Q}^{\frac{1}{2}}Q & 0 & \gamma_N I & \star \\ \mathcal{R}^{\frac{1}{2}}Y & 0 & 0 & \gamma_N I \end{bmatrix} \geqslant 0. \tag{10.161}$$

Within the switching horizon N, the input constraint is guaranteed by

$$-\underline{u} \leqslant u_{k|k} \leqslant \bar{u},\ -\underline{u} \leqslant u_{k+1|k} \leqslant \bar{u}, -\underline{u} \leqslant F_{k+i|k}x_{k+i|k}^{l_{i-1}\cdots l_1} + c_{k+i|k} \leqslant \bar{u},$$
$$i = 2, \ldots, N-1,\ \{l_1, \ldots, l_{i-1}\} = 1, \ldots, L. \tag{10.162}$$

Within the switching horizon N, the state constraint is guaranteed by

$$-\underline{\psi} \leqslant \Psi x_{k+1|k} \leqslant \overline{\psi}, -\underline{\psi} \leqslant \Psi x_{k+i|k}^{l_{i-1}\cdots l_1} \leqslant \overline{\psi},\ i = 2, \ldots, N,\ \{l_1, \ldots, l_{i-1}\} = 1, \ldots, L. \tag{10.163}$$

Since the input is parameterized as the state feedback law (10.152) after the switching horizon N, the following conclusion is used to handle the constraints after the switching horizon.

Lemma 10.5 Assume that there exist the symmetric matrices $\{Q, Z, \Gamma\}$ and matrix Y satisfying

$$\begin{bmatrix} Q & \star \\ A_l Q + B_l Y & Q \end{bmatrix} > 0,\ l = 1, \ldots, L, \tag{10.164}$$

$$\begin{bmatrix} Z & Y \\ Y^T & Q \end{bmatrix} \geqslant 0,\ Z_{jj} \leqslant z_{j,\inf}^2,\ j = 1, \ldots, m, \tag{10.165}$$

$$\begin{bmatrix} Q & \star \\ \Psi\left(A_l Q + B_l Y\right) & \Gamma \end{bmatrix} \geqslant 0, \Gamma_{ss} \leqslant \psi_{s,\inf}^2, l = 1, \ldots, L, \quad s = 1, \ldots, q, \tag{10.166}$$

then whenever $x_{k+N|k} \in \varepsilon_Q = \left\{\zeta | \zeta^T Q^{-1} \zeta \leqslant 1\right\}$, the closed-loop system is exponentially stable by the state feedback control law $u_{k+i+N|k} = YQ^{-1}x_{k+i+N|k}$, and for all $i \geqslant N$, (10.151) is satisfied and the state trajectory $x_{k+i+N|k}$, $i \geq 0$ always stay in the region ε_Q.

In order to guarantee the recursive feasibility, it requires to satisfy $x_{k+N|k} \in \varepsilon_Q$ which is converted to the following LMI:

$$\begin{bmatrix} 1 & \star \\ x_{k+N|k}^{l_{N-1}\cdots l_1} & Q \end{bmatrix} \geqslant 0, \quad \{l_1, \ldots, l_{N-1}\} = 1, \ldots, L. \tag{10.167}$$

Hence, the optimization problem is

$$\min_{\gamma_0,\ldots,\gamma_N,\tilde{c}_k,Y,Q,Z,\Gamma} \sum_{i=0}^{N} \gamma_i, \quad \text{s.t. } (10.158), (10.159), (10.161) - (10.166), (10.167), \tag{10.168}$$

where $\tilde{c}_k = \left[c_{k|k}^T, c_{k+1|k}^T, \ldots, c_{k+N-1|k}^T\right]^T$. Note that (10.168) ignores (10.160) since, according to the prediction model (10.149), (10.160) is contained in (10.167).

10.5.2.3 Introducing Vertex Control Moves

In order to keep linear, adopt the following vertex control input to substitute $\tilde{u}_k$:

$$\tilde{u}_k^{l_1\cdots l_{N-2}} := \{u_{k|k}, u_{k+1|k}, u_{k+2|k}^{l_1}, \ldots, u_{k+N-1|k}^{l_{N-2}\cdots l_1}\}, \{l_1, \ldots, l_{N-2}\} = 1, \ldots, L. \tag{10.169}$$

Apparently, for all $N \geqslant 2$, $\tilde{u}_k^{l_1\cdots l_{N-2}}$ is different from $\tilde{u}_k$. Then, the objective is to consider (b)-(B) and convert the problem (10.155) into LMI optimization problem. Introduce γ_i, $0 \leqslant i \leqslant N$ satisfying (10.156) and (10.157) and define $Q := \gamma_N P_k^{-1}$. Then inequalities in (10.156) are converted to the following LMIs:

$$\begin{bmatrix} \gamma_i & \star & \star \\ \mathcal{Q}^{1/2}\hat{x}_{k+i|k} & I & \star \\ \mathcal{R}^{1/2}u_{k+i|k} & 0 & I \end{bmatrix} \geqslant 0, \quad i = 0, 1, \tag{10.170}$$

$$\begin{bmatrix} \gamma_i & \star & \star \\ \mathcal{Q}^{1/2}\hat{x}_{k+i|k}^{l_{i-1}\cdots l_1} & I & \star \\ \mathcal{R}^{1/2}u_{k+i|k}^{l_{i-1}\cdots l_1} & 0 & I \end{bmatrix} \geqslant 0, \quad i = 2, \ldots, N-1, \quad \{l_1, \ldots, l_{i-1}\} = 1, \ldots, L. \tag{10.171}$$

Within the switching horizon N, the input constraint is guaranteed by

$$-\underline{u} \leqslant u_{k|k} \leqslant \bar{u}, \; -\underline{u} \leqslant u_{k+1|k} \leqslant \bar{u}, -\underline{u} \leqslant u_{k+i|k}^{l_{i-1}\cdots l_1} \leqslant \bar{u}. \tag{10.172}$$

Hence, similar to (10.168), the overall optimization problem is

$$\min_{\gamma_0,\ldots,\gamma_N,\tilde{u}_k^{l_1\cdots l_{N-2}},Y,Q,Z,\Gamma} \sum_{i=0}^{N} \gamma_i$$

$$\text{s.t. } (10.170), (10.171), (10.161), (10.172), (10.163) - (10.166), (10.167). \tag{10.173}$$

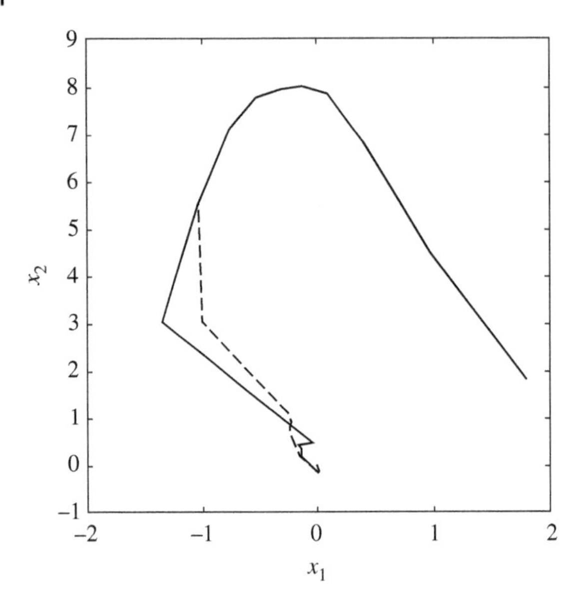

Figure 10.5 Closed-loop state trajectories.

10.5.2.4 Numerical Example

Consider

$$\begin{bmatrix} x_{1,k+1} \\ x_{2,k+1} \end{bmatrix} = \begin{bmatrix} 1 & 0.1 \\ K_k & 1 \end{bmatrix} \begin{bmatrix} x_{1,k} \\ x_{2,k} \end{bmatrix} + \begin{bmatrix} 1 \\ 0 \end{bmatrix} u_k.$$

where $K_k \in [0.5\ 2.5]$ is the uncertain parameter. Take the weight matrices as $\mathcal{Q} = I$ and $\mathcal{R} = 1$, and the input constraint is $|u_k| \leqslant 1$. Choose $N = 3$, $x(0) = [1.8, 1.8]^T$ and $K(k) = 1.5 + \sin(k)$, then the closed-loop state trajectories by solving (10.168) and (10.173) are shown in Figure 10.5 in dotted line and solid line, respectively.

In Figure 10.6, for $N = 1, 2, 3$, RoA of (10.168) is shown by the dashed line, and RoA of (10.173) is shown by the solid line. For $N = 1, 2$, algorithms (10.168) and (10.173) are equivalent, so their RoAs are same. For (10.168), RoA with $N = 3$ does not contain RoA with $N = 2$. For (10.173), however, RoA with a larger N contains RoA with smaller N.

Consider the system $A_k = \begin{bmatrix} 1 & 0 \\ K_k & 1 \end{bmatrix}, A_0 = \begin{bmatrix} 1 & 0 \\ 1.5 & 1 \end{bmatrix}$. Other details are the same as above. For $N = 1$, RoA of (10.168) and RoA of the partial feedback MPC are shown by the solid line and the dashed line in Figure 10.7, respectively. Obviously, applying the nominal performance index based on the current model enhances the recursive feasibility.

10.5.3 More Discussions

Look at (10.86)–(10.94), i.e., the constraints of the variant feedback MPC. A formula of unclear optimality is (10.87) which uses the same γ_k for every $F_{k+i|k}$, $i = 1, 2, \dots, N-1$ within the switching horizon. Moreover, this γ_k is applied to F_k in the terminal constraint set. The optimization of one-step stage cost (i.e, $\|x_{k+i|k}\|^2_{\mathcal{Q}} + \|u_{k+i|k}\|^2_{\mathcal{R}}$) may be ignored.

Figure 10.6 RoAs between the problems (10.168) and (10.173).

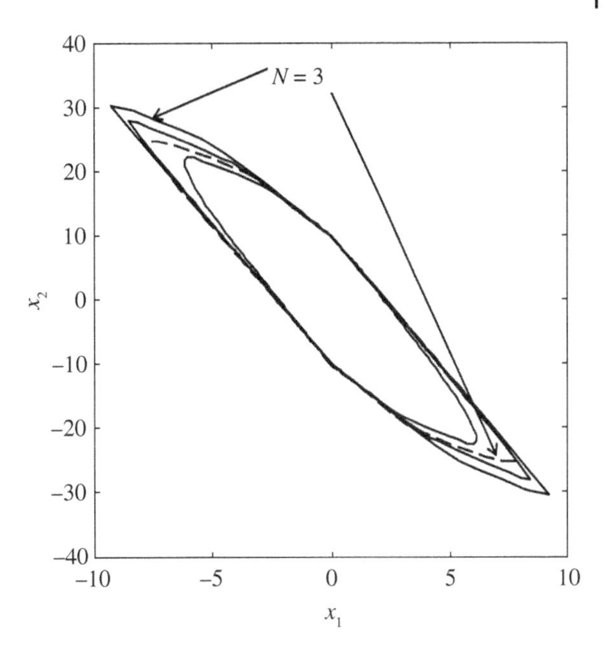

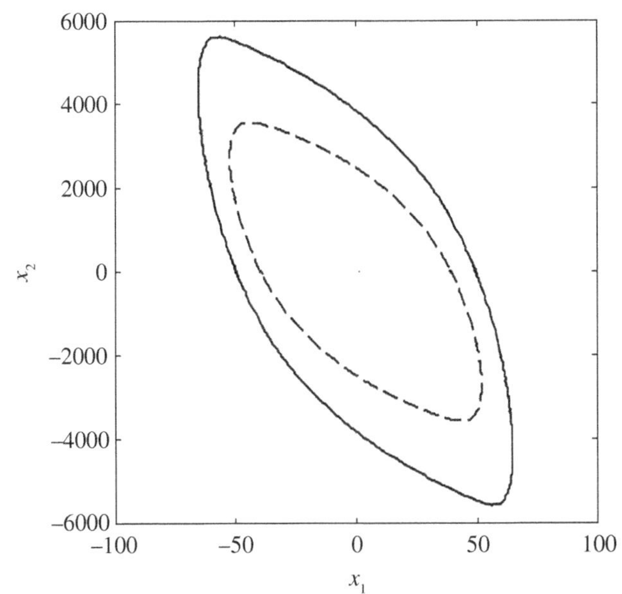

Figure 10.7 RoAs between the problems (10.168) and the partial feedback MPC ($N = 1$).

Equation (10.87) could be substituted by

$$\begin{bmatrix} Q_{i,k} & \star & \star & \star \\ A_l Q_{i,k} + B_l Y_{i,k} & \frac{\gamma_{i,k}}{\gamma_{i+1,k}} Q_{i+1,k} & \star & \star \\ \mathcal{Q}^{1/2} Q_{i,k} & 0 & \gamma_{i,k} I & \star \\ \mathcal{R}^{1/2} Y_{i,k} & 0 & 0 & \gamma_{i,k} I \end{bmatrix} \geqslant 0,$$

$$i \in \{1, 2, \ldots, N-1\},\ l \in \{1, \ldots, L\},\ Q_N = Q,\ \gamma_{N,k} = \gamma_k. \tag{10.174}$$

Solve

$$\min \left[\|u_{k|k}\|^2_{\mathcal{R}} + \sum_{i=1}^{N-1} \gamma_{i,k} + \gamma_k \right].$$

Note that optimizing $\frac{\gamma_{i,k}}{\gamma_{i+1,k}}$ directly may bring trouble. In the sequential off-line method, (10.174) is written as

$$\begin{bmatrix} Q_i & \star & \star & \star \\ A_l Q_i + B_l Y_i & \gamma_i P_{i+1}^{-1} & \star & \star \\ \mathcal{Q}^{1/2} Q_i & 0 & \gamma_i I & \star \\ \mathcal{R}^{1/2} Y_i & 0 & 0 & \gamma_i I \end{bmatrix} \geqslant 0,$$

$$i \in \{1, 2, \ldots, N-1\},\ l \in \{1, \ldots, L\},\ Q_N = Q, \tag{10.175}$$

where P_{i+1} is determined before solving Q_i.

If we use $\gamma_{i,k}$ to substitute γ_k directly in (10.87) and (10.96), then at each time we can solve $\min \left[\|u_{k|k}\|^2_{\mathcal{R}} + \sum_{i=1}^{N} \gamma_{i,k} \right]$ to obtain the control moves. This is the slight modification by a direct view of LMI.

11

Output Feedback Robust Model Predictive Control

Most notations are the same as in Chapter 10. For convenience, some notations in this chapter are different from Chapters 1–10. New notations are as follows:

$u \in \mathbb{R}^{n_u}$: the control input signal;
$w \in \mathbb{R}^{n_w}$: the disturbance;
$x \in \mathbb{R}^{n_x}$: the true state;
$x_c \in \mathbb{R}^{n_{x_c}}$: the estimator state or the controller state;
$y \in \mathbb{R}^{n_y}$: the output;
$|\xi|$: the component-wise absolute value of ξ;
$\xi_{i|k}$: the prediction of signal $\xi(k+i)$ at time k.

Consider LPV model (9.5) which is written as:

$$\begin{cases} x_{k+1} = A_k x_k + B_k u_k + D_k w_k \\ y_k = C_k x_k + E_k w_k \\ z_k = \mathcal{C}_k x_k + \mathcal{E}_k w_k \\ z'_k = \mathcal{F}_k x_k + \mathcal{G}_k w_k \end{cases}, \tag{11.1}$$

satisfying Assumptions 9.1 and 9.2 in Chapter 9 (with appropriate matrix substitutions). The hard physical constraint is

$$|u_k| \leqslant \bar{u}, \ |\Psi z_{k+1}| \leqslant \bar{\psi}, \quad k \geqslant 0, \tag{11.2}$$

where $\bar{u} = [\bar{u}_1, \bar{u}_2, \dots, \bar{u}_{n_u}]^T$; $\bar{\psi} = [\bar{\psi}_1, \bar{\psi}_2, \dots, \bar{\psi}_q]^T$; $\bar{u}_j > 0, j = 1, \dots, n_u$; $\bar{\psi}_j > 0, j = 1, \dots, q$; and $\Psi \in \mathbb{R}^{q \times n_z}$.

When x is fully measurable and $w_k \equiv 0$, KBM controller solves, at each time k, an LMI optimization problem with constraints (confinement of the current state, invariance/stability/optimality condition, input constraint, state/output constraint). In the following, we will generalize KBM controller to the cases when x is unmeasurable and $w_k \neq 0$.

Theorem 11.1 See Ding and Pan [2017]. Consider the system (11.1), with Assumptions 9.1 and 9.2 being satisfied. Adopt the dynamic output feedback controller, i.e.,

$$\begin{cases} x_{c,k+1} = A_{c,k} x_{c,k} + B_{c,k} y_k \\ u_k = C_{c,k} x_{c,k} + D_{c,k} y_k \end{cases}, \tag{11.3}$$

where the controller parameters are defined as parameter-dependent, i.e.,

$$\begin{cases} A_{c,k} = \sum_{l=1}^{L}\sum_{j=1}^{L} \omega_{l,k}\omega_{j,k}\bar{A}_{c,k}^{lj} \\ B_{c,k} = \sum_{l=1}^{L} \omega_{l,k}\bar{B}_{c,k}^{l} \\ C_{c,k} = \sum_{j=1}^{L} \omega_{j,k}\bar{C}_{c,k}^{j} \\ D_{c,k} = \bar{D}_{c,k} \end{cases}. \tag{11.4}$$

The controller parametric matrices $\{\bar{A}_c^{lj}, \bar{B}_c^l, \bar{C}_c^j, \bar{D}_c\}$ are taken as

$$\begin{cases} \bar{D}_c = \hat{D}_c \\ \bar{C}_c^j = \left(\hat{C}_c^j - \bar{D}_c C_j Q_1\right) Q_2^{-1} \\ \bar{B}_c^l = M_2^{-\mathrm{T}}\left(\hat{B}_c^l - M_1 B_l \bar{D}_c\right) \\ \bar{A}_c^{lj} = M_2^{-\mathrm{T}}\left(\hat{A}_c^{lj} - M_1 A_l Q_1 - M_1 B_l \bar{D}_c C_j Q_1 - M_2^T \bar{B}_c^l C_j Q_1 - M_1 B_l \bar{C}_c^j Q_2\right) Q_2^{-1} \end{cases}, \tag{11.5}$$

where "(k)" is omitted for brevity. In (11.5), the parameterized matrix $\{\hat{A}_c^{lj}, \hat{B}_c^l, \hat{C}_c^j, \hat{D}_c, M_1, Q_1\}_k$ is optimized by later Algorithm 11.1, where $U_k = -M_{1,k}^{-1}M_{2,k}^T$, and $\{M_1, M_2, Q_1, Q_2\}$ are from the following inverse matrices:

$$M = \begin{bmatrix} M_1 & M_2^T \\ M_2 & M_3 \end{bmatrix}, \quad Q = \begin{bmatrix} Q_1 & Q_2^T \\ Q_2 & Q_3 \end{bmatrix}.$$

Choose $x_{c,0}$ and assume $x_0 - U_0 x_{c,0} \in \varepsilon_{M_{e,0}}$. Apply the later Algorithm 11.2 and assume that (11.6)–(11.11) are feasible at time $k = 0$. Then

(1) (11.6)–(11.11) are feasible at every $k > 0$;
(2) $\{\gamma, z', u\}$ converge to the neighborhood of 0, and for all $k \geqslant 0$, the constraint (11.2) is satisfied.

The following problem is utilized in Algorithm 11.1:

$$\min_{\{\gamma, \alpha_{lj}, \varrho, Q_1, M_1, \hat{A}_c^{lj}, \hat{B}_c^l, \hat{C}_c^j, \hat{D}_c\}_k} \gamma_k, \tag{11.6}$$

$$\text{s.t. } M_{1,k} \leqslant \varrho_k M_{e,k}, \tag{11.7}$$

$$\begin{bmatrix} 1-\varrho_k & \star & \star \\ U_k x_{c,k} & Q_{1,k} & \star \\ 0 & I & M_{1,k} \end{bmatrix} \geqslant 0, \tag{11.8}$$

$$\sum_{l=1}^{L} C_l^{\ell}(d,2)\Upsilon_{ll,k}^{\mathrm{QB}} + \sum_{l=1}^{L-1}\sum_{j=l+1}^{L} C_{lj}^{\ell}(d,1,1)\left[\Upsilon_{lj,k}^{\mathrm{QB}} + \Upsilon_{jl,k}^{\mathrm{QB}}\right] \geqslant 0, \ell = 1, \ldots, |\mathcal{K}(d+2)|, \tag{11.9}$$

$$\begin{bmatrix} M_{1,k} & \star & \star & \star \\ I & Q_{1,k} & \star & \star \\ 0 & 0 & I & \star \\ \frac{1}{\sqrt{1-\eta_{1s}}}\xi_s \hat{D}_{c,k} C_j & \frac{1}{\sqrt{1-\eta_{1s}}}\xi_s \hat{C}_{c,k}^j & \frac{1}{\sqrt{\eta_{1s}}}\xi_s \hat{D}_{c,k} E_j & \bar{u}_s^2 \end{bmatrix} \geqslant 0, j = 1, \ldots, L, \ s = 1, \ldots, n_u, \tag{11.10}$$

Algorithm 11.1 Main optimization problem (part 1)

Solve the optimization problem (11.6)–(11.11).

Notations are shown as follows:

(1) $\{\mathcal{Q}_1, \mathcal{R}\}$ are weight matrices;

(2) ξ_s is the s-th row of n_u-order identity matrix, and Ψ_s is the s-th row of Ψ;

(3) $\{\eta_{1s}, \eta_{2s}, \eta_{3s}\} \in [0, 1)$ are the fixed scalars;

(4) d is the fixed nonnegative integer. $\mathcal{K}(d+2)$ is the set of L-tuples obtained from all possible combinations of $d_1 d_2 \cdots d_L$, $d_l \geqslant 0$, $l = 1, \ldots, L$ such that $d_1 + d_2 + \cdots + d_L = d + 2$. The number of elements of $\mathcal{K}(d+2)$ is given by $|\mathcal{K}(d+2)| = \frac{(L+d+1)!}{(d+2)!(L-1)!}$. The L-tuples of $\mathcal{K}(d+2)$ are lexically ordered $\ell = 1, \ldots, |\mathcal{K}(d+2)|$. Moreover,

$$C_l^{\ell}(d,2) = \begin{cases} \frac{d!}{d_1! \cdots d_{l-1}!(d_l-2)! d_{l+1}! \cdots d_L!}, & d_l \geqslant 2 \\ 0, & \text{otherwise} \end{cases},$$

$$C_{lj}^{\ell}(d,1,1) = \begin{cases} \frac{d!}{d_1! \cdots d_{l-1}!(d_l-1)! d_{l+1}! \cdots d_{j-1}!(d_j-1)! d_{j+1}! \cdots d_L!}, & d_l \geqslant 1, d_j \geqslant 1 \\ 0, & \text{otherwise} \end{cases}.$$

$$\sum_{l=1}^{L} C_l^{\ell}(d,2) \Upsilon_{hlls,k}^{z} + \sum_{l=1}^{L-1} \sum_{j=l+1}^{L} C_{lj}^{\ell}(d,1,1) \left[\Upsilon_{hljs,k}^{z} + \Upsilon_{hjls,k}^{z} \right] \geqslant 0,$$
$$\ell = 1, \ldots, |\mathcal{K}(d+2)|, \quad h = 1, \ldots, L, \quad s = 1, \ldots, q, \tag{11.11}$$

where

$$\Upsilon_{lj}^{\mathrm{QB}} = \begin{bmatrix} (1-\alpha_{lj})M_1 & \star & \star & \star & \star & \star & \star \\ (1-\alpha_{lj})I & (1-\alpha_{lj})Q_1 & \star & \star & \star & \star & \star \\ 0 & 0 & \alpha_{lj} I & \star & \star & \star & \star \\ A_l + B_l \hat{D}_c C_j & A_l Q_1 + B_l \hat{C}_c^j & B_l \hat{D}_c E_j + D_l & Q_1 & \star & \star & \star \\ M_1 A_l + \hat{B}_c^l C_j & \hat{A}_c^{lj} & \hat{B}_c^l E_j + M_1 D_l & I & M_1 & \star & \star \\ \mathcal{Q}_1^{1/2} \mathcal{F}_j & \mathcal{Q}_1^{1/2} \mathcal{F}_j Q_1 & \mathcal{Q}_1^{1/2} \mathcal{G}_j & 0 & 0 & \gamma I & \star \\ \mathcal{R}^{1/2} \hat{D}_c C_j & \mathcal{R}^{1/2} \hat{C}_c^j & \mathcal{R}^{1/2} \hat{D}_c E_j & 0 & 0 & 0 & \gamma I \end{bmatrix},$$

$$\Upsilon_{hljs}^{z} = \begin{bmatrix} M_1 & \star & \star & \star \\ I & Q_1 & \star & \star \\ 0 & 0 & I & \star \\ \spadesuit_1 & \spadesuit_2 & \frac{1}{\sqrt{1-\eta_{2s}}\sqrt{\eta_{3s}}} \Psi_s \mathcal{E}_h (B_l \hat{D}_c E_j + D_l) & \bar{\psi}_s^2 - \frac{1}{\eta_{2s}} \Psi_s \mathcal{E}_h \mathcal{E}_h^T \Psi_s^T \end{bmatrix},$$

$$\spadesuit_1 = \frac{1}{\sqrt{1-\eta_{2s}}\sqrt{1-\eta_{3s}}} \Psi_s C_h (A_l + B_l \hat{D}_c C_j),$$

$$\spadesuit_2 = \frac{1}{\sqrt{1-\eta_{2s}}\sqrt{1-\eta_{3s}}} \Psi_s C_h (A_l Q_1 + B_l \hat{C}_c^j).$$

Algorithm 11.2 Main optimization problem (part 2)

At each time $k \geqslant 0$,

(a) for $k > 0$, apply (11.4) and (11.5) to obtain $\{A_c, B_c\}(k-1)$, then calculate $x_{c,k} = A_{c,k-1}x_{c,k-1} + B_{c,k-1}y_{k-1}$;
(b) for $k > 0$, calculate U_k and $M_{e,k}$ via later Algorithm 11.3;
(c) solve (11.6)–(11.11) to find $\{Q_1, M_1, \hat{A}_c^{lj}, \hat{B}_c^l, \hat{C}_c^j, \hat{D}_c\}_k^*$;
(d) take $\{Q_1, M_1\}_k = \{Q_1, M_1\}_k^*$, $Q_{2,k} = U_k^{-1}[Q_{1,k} - M_{1,k}^{-1}]$ and $M_{2,k} = -U_k^T M_{1,k}$;
(e) apply (11.4) and (11.5) to obtain $C_{c,k}$ and $D_{c,k}$, then implement $u_k = C_{c,k}x_{c,k} + D_{c,k}y_k$.

Algorithm 11.3 Main optimization problem (part 3)

Take $U_0 = I$. At each time $k > 0$,

(a) take

$$U_k = U_{k-1}, \tag{11.12}$$

$$\varrho_k = 1 - x_{c,k}^T \left[M_{3,k} - U_k^T M_{1,k-1}^* U_k\right] x_{c,k}, \tag{11.13}$$

$$M_{e,k} = \varrho_k^{-1} M_{1,k-1}^*, \tag{11.14}$$

where

$$M_{3,k} = M_{2,k}[M_{1,k-1}^* - Q_{1,k-1}^{*-1}]^{-1} M_{2,k}^T,$$

and $M_{2,k} = -U_k^T M_{1,k-1}^*$;
(b) obtain $\{M_e', U'\}_k$ satisfying

$$\{x_{k-1} - U_{k-1}x_{c,k-1} \in \varepsilon_{M_{e,k-1}}, \|w_{k-1}\| \leqslant 1\}, \Rightarrow x_k - U_k' x_{c,k} \in \varepsilon_{M_{e,k}'}, \tag{11.15}$$

$$1 - x_{c,k}^T \left[M_{3,k}' - (U_k')^T M_{1,k-1}^* U_k'\right] x_{c,k} \geqslant \varrho_k', \tag{11.16}$$

$$M_{e,k}' \geqslant (\varrho_k')^{-1} M_{1,k-1}^*, \tag{11.17}$$

$$M_{e,k}' \geqslant M_{e,k}, \tag{11.18}$$

where

$$M_{3,k}' = M_{2,k}'[M_{1,k-1}^* - Q_{1,k-1}^{*-1}]^{-1} M_{2,k}'^T,$$

and $M_{2,k}' = -(U_k')^T M_{1,k-1}^*$. If we obtain $\{M_e', U'\}_k$, then update $M_{e,k} = M_{e,k}'$ and $U_k = U_k'$.

Remark 11.1 The constraint (11.15) is introduced by updating the bound of x_k; the constraints (11.16) and (11.17) guarantee the recursive feasibility of (11.6)–(11.11); the constraint (11.18) means that $\varepsilon_{M_{e,k}'}$ is smaller than $\varepsilon_{M_{e,k}}$. Note that $x_k - U_k' x_{c,k} = x_k - U_k x_{c,k} - (U_k' - U_k)x_{c,k} \in \varepsilon_{M_{e,k}'}$ means that $x_k - U_k x_{c,k}$ lies in the ellipsoid defined by $M_{e,k}'$, the center of which is $(U_k' - U_k)x_{c,k}$. $x_k - U_k x_{c,k} \in \varepsilon_{M_{e,k}}$ means that $x_k - U_k x_{c,k}$ lies in the ellipsoid defined by $M_{e,k}$, the center of which is 0. $\{M_e', U'\}_k$ satisfying (11.15)–(11.18) is obtained by solving the optimization problem [Ding and Pan, 2017]

$$\min_{\phi_1, \phi_2, \{\varrho', Q_e, U'\}_k} \text{trace}(Q_{e,k}), \tag{11.19}$$

$$\text{s.t.} \begin{bmatrix} 1-\phi_1-\phi_2+\phi_1\|U_k x_{c,k-1}\|^2_{M_{e,k-1}} & \star & \star & \star \\ -\phi_1 M_{e,k-1} U_k x_{c,k-1} & \phi_1 M_{e,k-1} & \star & \star \\ 0 & 0 & \phi_2 I & \star \\ B_{k-1}C_{c,k-1}x_{c,k-1} - U'_k x_{c,k} & \heartsuit_{42} & \heartsuit_{43} & Q_{e,k} \end{bmatrix} \geqslant 0,$$
$$\heartsuit_{42} = A_{k-1} + B_{k-1}D_{c,k-1}C_{k-1},$$
$$\heartsuit_{43} = D_{k-1} + B_{k-1}D_{c,k-1}E_{k-1}, \tag{11.20}$$

$$\begin{bmatrix} 1-\varrho'_k & \star \\ U'_k x_{c,k} & Q^*_{1,k-1} - M^{*-1}_{1,k-1} \end{bmatrix} \geqslant 0, \tag{11.21}$$

$$Q_{e,k} \leqslant \varrho'_k M^{*-1}_{1,k-1}, \tag{11.22}$$

$$Q_{e,k} \leqslant M^{-1}_{e,k}, \tag{11.23}$$

$$\underline{\delta} I \leqslant (U'_k)^T + U'_k \leqslant \bar{\delta} I, \tag{11.24}$$

where $\underline{\delta}$ and $\bar{\delta}$ are the given positive scalars, and let $M'_{e,k} = Q^{-1}_{e,k}$.

In Ding and Pan [2017], take $\alpha_{lj} = \alpha$, $\mathcal{C}_k = \Psi_x$, $\mathcal{E}_k = \Psi_w$, $\Psi = I$, $\mathcal{F}_j = C_j$, $\mathcal{G}_j = E_j$. Moreover, in Ding and Pan [2017], assume that $\|w_k\|^2_{P_w} \leqslant 1$, where P_w is tunable, instead of $\|w_k\|^2 \leqslant 1$.

Example 11.1 See Ding and Pan [2017]. Handling CSTR model obtains the following discrete-time model:

$$A_1 = \begin{bmatrix} 0.8227 & -0.0017 \\ 6.1233 & 0.9367 \end{bmatrix}, A_2 = \begin{bmatrix} 0.9654 & -0.0018 \\ -0.6759 & 0.9433 \end{bmatrix},$$
$$A_3 = \begin{bmatrix} 0.8895 & -0.0029 \\ 2.9447 & 0.9968 \end{bmatrix}, A_4 = \begin{bmatrix} 0.8930 & -0.0006 \\ 2.7738 & 0.8864 \end{bmatrix},$$
$$B_1 = \begin{bmatrix} -0.0001 \\ 0.1014 \end{bmatrix}, \; B_2 = \begin{bmatrix} -0.0001 \\ 0.1016 \end{bmatrix},$$
$$B_3 = \begin{bmatrix} -0.0002 \\ 0.1045 \end{bmatrix}, \; B_4 = \begin{bmatrix} -0.000034 \\ 0.0986 \end{bmatrix},$$
$$C_1 = C_2 = C_3 = C_4 = \begin{bmatrix} 0 & 1 \end{bmatrix},$$
$$D_1 = D_2 = D_3 = D_4 = \begin{bmatrix} 0.0022 \\ 0.0564 \end{bmatrix},$$
$$E_1 = E_2 = E_3 = E_4 = 0, \; P_w = 100.$$

Besides, $\bar{u} = 10$, $\Psi_x = I$, $\Psi_w = 0$, $\bar{\psi} = \begin{bmatrix} 0.5 & 10 \end{bmatrix}^T$,

$$g_1(x_2) = 7.2 \times 10^{10} \exp\left(-\frac{8750}{x_2+350}\right),$$
$$g_2(x_2) = 3.6 \times 10^{10} \left[\exp\left(-\frac{8750}{x_2+350}\right) - \exp\left(-\frac{8750}{350}\right)\right]\frac{1}{x_2},$$

$$\omega_{1,k} = \frac{1}{2}\frac{g_1(x_2) - g_1(-10)}{g_1(10) - g_1(-10)}, \quad \omega_{2,k} = \frac{1}{2}\frac{g_1(10) - g_1(x_2)}{g_1(10) - g_1(-10)},$$
$$\omega_{3,k} = \frac{1}{2}\frac{g_2(x_2) - g_2(-10)}{g_2(10) - g_2(-10)}, \quad \omega_{4,k} = \frac{1}{2}\frac{g_2(10) - g_2(x_2)}{g_2(10) - g_2(-10)}.$$

Moreover, take $\mathcal{Q}_1 = 1$ and $\mathcal{R} = 0.25$. Take $M_e(0) = \text{diag}\{12.5, 0.125\}$ and $x_c(0) = \beta[0.05, 2.5]^T$ where $\beta \geqslant 0$ is the tunable term; $U(0) = I$ and $x(0) = x_c(0) + [0.2, 2]^T$ satisfy $x(0) - x^0(0) \in \varepsilon_{M_e(0)}$. Take $d = 0$, which means that $\sum_{l=1}^{r}\sum_{j=1}^{r}\omega_l\omega_j\Upsilon_{lj} \geqslant 0$ is substituted by $\Upsilon_{ll} \geqslant 0$, $l \in \{1, \dots, r\}$, and $\Upsilon_{lj} + \Upsilon_{jl} \geqslant 0$, $j > l$, $l, j \in \{1, \dots, r\}$. In solving (11.6)–(11.11), $\alpha = 0.001$ is given and w is produced randomly in the interval $[-0.1, 0.1]$. Utilize $\beta = 1.32$, and the disturbance sequence shown in Figure 11.1. The trajectories of $\tilde{x}$ and control input signal are shown in Figures 11.2 and 11.3, respectively. The evolutions of $\gamma(k)$ are shown in Figure 11.4, which can reflect the control performance since min–max optimization is utilized. In Figures 11.2–11.4, a is the algorithm in Theorem 11.1, b is an algorithm in Ding and Pan [2017], and c is an algorithm in Ding [2011c]. We add b and c for reference.

In (11.6)–(11.11), KBM formula is generalized, i.e., the constraint on x_k is generalized to (11.7) and (11.8), which restricts e_k and $x_{c,k}$; the stability/optimality constraint is generalized to (11.9), which contains the quadratic boundedness and the optimality conditions; the input constraint is generalized to (11.10); the state/input constraint is generalized to the constraint (11.11) on z.

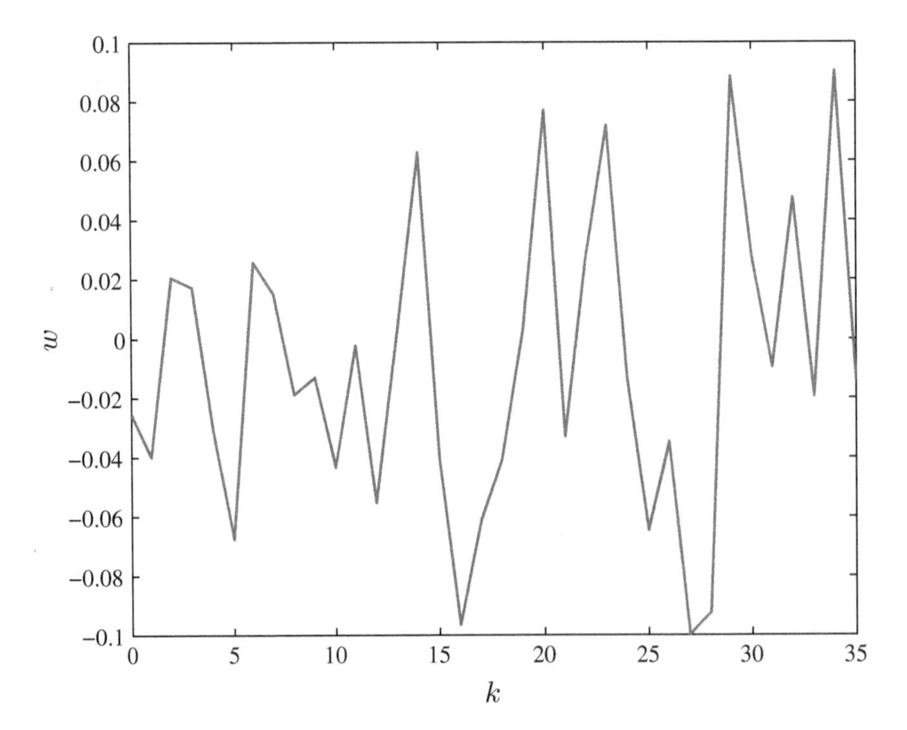

Figure 11.1 The disturbance utilized in the simulation.

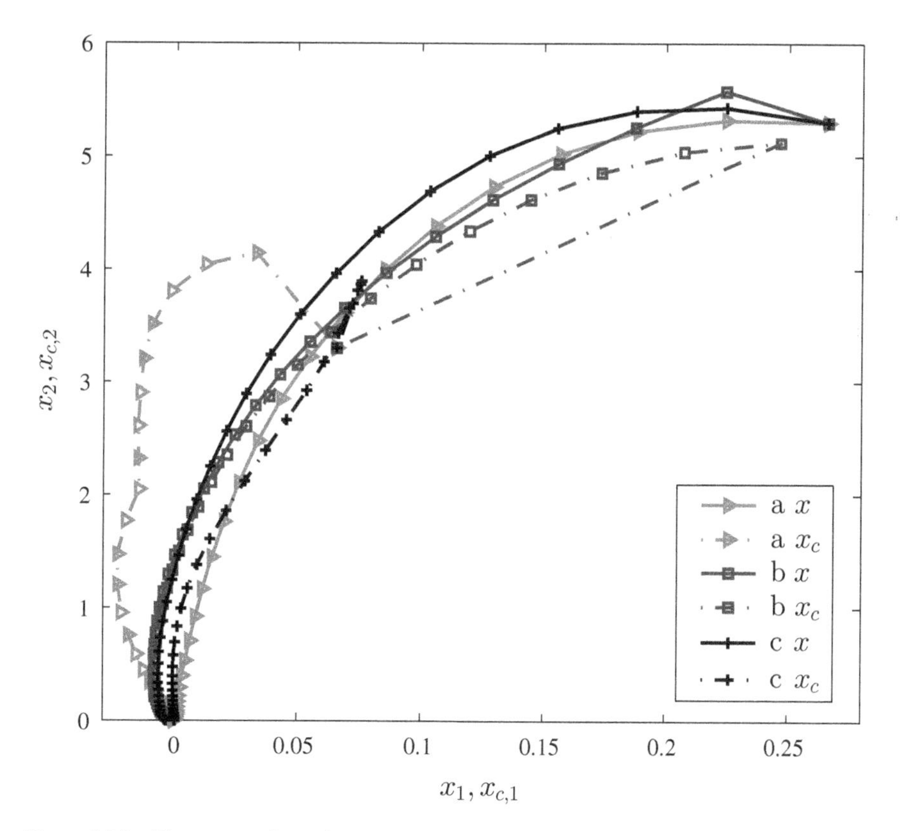

Figure 11.2 The state trajectories.

Next, let us take Theorem 11.1 as an example to illustrate how the above generalization happens.

11.1 Model and Controller Descriptions

The prediction equation, based on (11.1), is

$$\begin{aligned} x_{i+1|k} &= A_{i|k}x_{i|k} + B_{i|k}u_{i|k} + D_{i|k}w_{i|k}, \\ y_{i|k} &= C_{i|k}x_{i|k} + E_{i|k}w_{i|k}, i \geqslant 0. \end{aligned} \tag{11.25}$$

The prediction form of (11.2) is

$$|u_{i|k}| \leqslant \bar{u}, \quad |\Psi z_{i+1|k}| \leqslant \bar{\psi}, \quad i \geqslant 0, \tag{11.26}$$

where $z_{i|k} = \mathcal{C}_{i|k}x_{i|k} + \mathcal{E}_{i|k}w_{i|k}$. According to Assumption 9.2,

$$[A|B|C|D|E|\mathcal{C}|\mathcal{E}|\mathcal{F}|\mathcal{G}]_{i|k} = \sum_{l=1}^{L} \omega_{l,i|k}[A_l|B_l|C_l|D_l|E_l|\mathcal{C}_l|\mathcal{E}_l|\mathcal{F}_l|\mathcal{G}_l]. \tag{11.27}$$

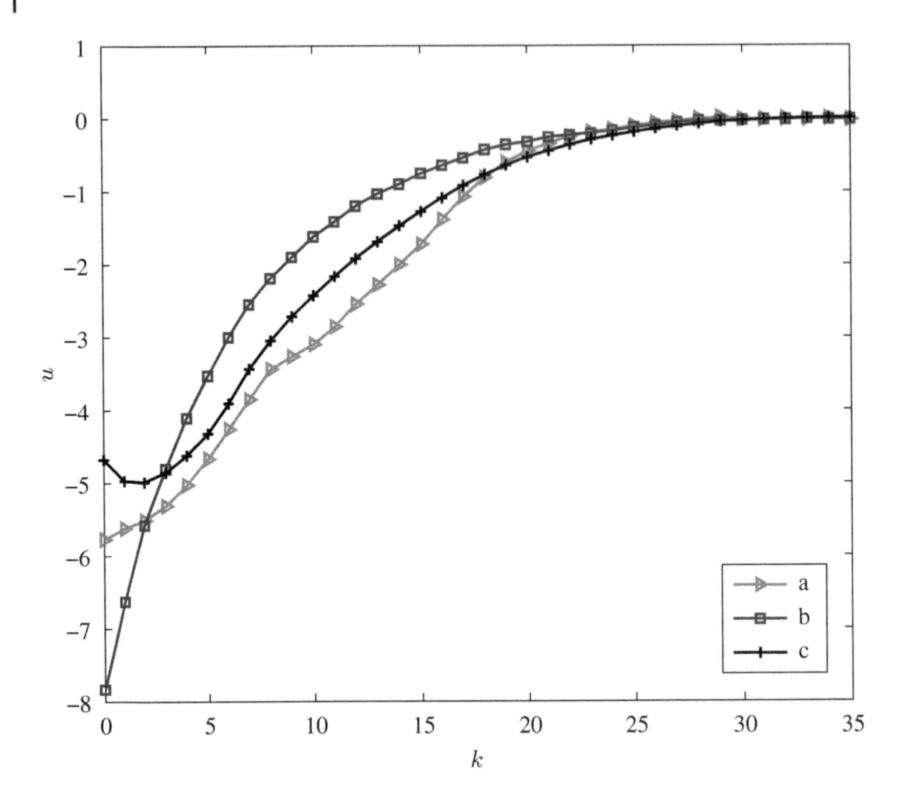

Figure 11.3 The control input signal.

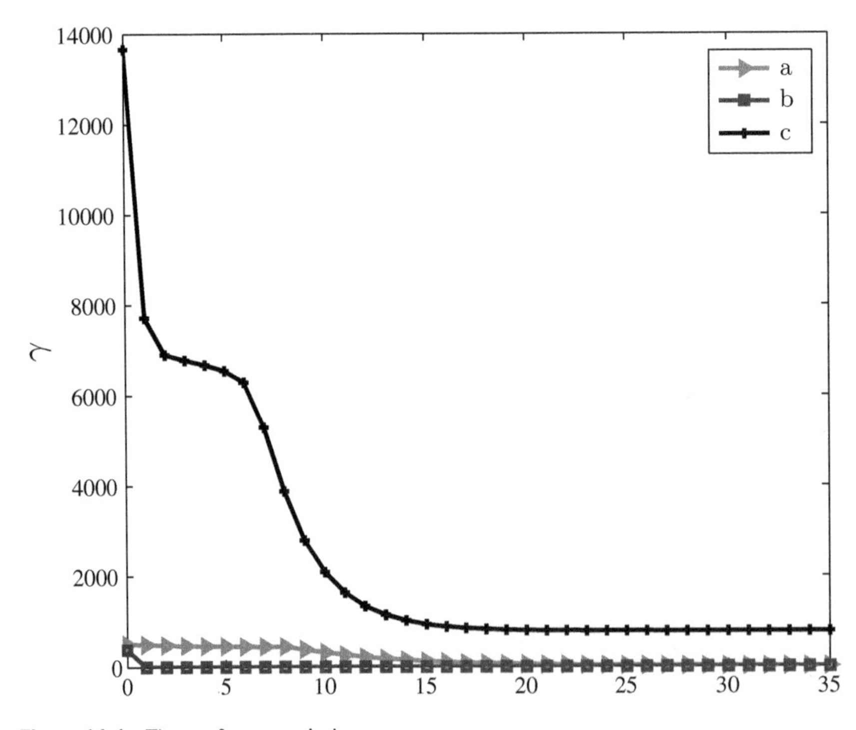

Figure 11.4 The performance index γ.

11.1.1 Controller for LPV Model

For LPV model (11.1), the dynamic output feedback controller is of the following form (see [Ding and Xie, 2009, Ding, 2013]):

$$\begin{cases} x_{c,k+1} = A_{c,k}x_{c,k} + L_{c,k}y_k \\ u_k = F_{x,k}x_{c,k} + F_{y,k}y_k \end{cases}, \tag{11.28}$$

where $\{A_c, L_c\}$ are the controller gain matrices; $\{F_x, F_y\}$ are the feedback gain matrices. It is unnecessary that $n_x = n_{x_c}$. The predictive equation of (11.28) is

$$\begin{cases} x_{c,i+1|k} = A_{c,k}x_{c,i|k} + L_{c,k}y_{i|k} \\ u_{i|k} = F_{x,k}x_{c,i|k} + F_{y,k}y_{i|k} \end{cases}. \tag{11.29}$$

Remark 11.2 There are four controller parameters $\{A_c, L_c, F_x, F_y\}$ in (11.28) and (11.29). In the literature, often there are only two controller parameters $\{L_c, F_x\}$ for the output feedback. We found that with only two controller parameters $\{L_c, F_x\}$, for the system (11.1), it is more difficult to find a feasible solution to the optimization problem of output feedback MPC. With 4 parameters $\{A_c, L_c, F_x, F_y\}$, the output feedback MPC can be applied to a much larger range of system models.

Define the augmented state $\tilde{x} = \begin{bmatrix} x \\ x_c \end{bmatrix}$. By applying (11.1) and (11.28), the augmented closed-loop system is

$$\tilde{x}_{k+1} = \Phi_k\tilde{x}_k + \Gamma_k w_k, \tag{11.30}$$

where

$$\Phi_k = \begin{bmatrix} A_k + B_kF_{y,k}C_k & B_kF_{x,k} \\ L_{c,k}C_k & A_{c,k} \end{bmatrix},$$

$$\Gamma_k = \begin{bmatrix} B_kF_{y,k}E_k + D_k \\ L_{c,k}E_k \end{bmatrix}.$$

The prediction based on (11.30) is

$$\tilde{x}_{i+1|k} = \Phi_{i,k}\tilde{x}_{i|k} + \Gamma_{i,k}w_{i|k}, \tag{11.31}$$

where

$$\Phi_{i,k} = \begin{bmatrix} A_{i|k} + B_{i|k}F_{y,k}C_{i|k} & B_{i|k}F_{x,k} \\ L_{c,k}C_{i|k} & A_{c,k} \end{bmatrix},$$

$$\Gamma_{i,k} = \begin{bmatrix} B_{i|k}F_{y,k}E_{i|k} + D_{i|k} \\ L_{c,k}E_{i|k} \end{bmatrix}.$$

By applying (11.27), it is shown that

$$\Phi_{i,k} = \sum_{l=1}^{L} \omega_{l,i|k} \sum_{j=1}^{L} \omega_{j,i|k} \Phi_{lj,k},$$

$$\Gamma_{i,k} = \sum_{l=1}^{L} \omega_{l,i|k} \sum_{j=1}^{L} \omega_{j,i|k} \Gamma_{lj,k},$$

$$\Phi_{lj,k} = \begin{bmatrix} A_l + B_l F_{y,k} C_j & B_l F_{x,k} \\ L_{c,k} C_j & A_{c,k} \end{bmatrix}, \ \Gamma_{lj,k} = \begin{bmatrix} B_l F_{y,k} E_j + D_l \\ L_{c,k} E_j \end{bmatrix}.$$

11.1.2 Controller for Quasi-LPV Model

For the quasi-LPV model (11.1), the dynamic output feedback controller is (11.3) and (11.4) (see [Ding and Xie, 2008, Ding, 2010a]), where $n_x = n_{x_c}$. The prediction equation, based on (11.3), is

$$\begin{cases} x_{c,i+1|k} = A_{c,i|k} x_{c,i|k} + B_{c,i|k} y_{i|k} \\ u_{i|k} = C_{c,i|k} x_{c,i|k} + D_{c,i|k} y_{i|k} \end{cases}, \tag{11.32}$$

where

$$\begin{cases} A_{c,i|k} = \sum_{l=1}^{L} \sum_{j=1}^{L} \omega_{l,i|k} \omega_{j,i|k} \bar{A}^{lj}_{c,k} \\ B_{c,i|k} = \sum_{l=1}^{L} \omega_{l,i|k} \bar{B}^{l}_{c,k} \\ C_{c,i|k} = \sum_{j=1}^{L} \omega_{j,i|k} \bar{C}^{j}_{c,k} \\ D_{c,i|k} = \bar{D}_{c,k}. \end{cases} \tag{11.33}$$

Remark 11.3 For the quasi-LPV, since $\omega_{l,k}$ are known, we can utilize $\{\bar{A}^{lj}_c, \bar{B}^l_c, \bar{C}^j_c, \bar{D}_c\}_k$ to calculate the parameter-dependent $\{A_c, B_c, C_c\}_k$. Such $\{A_c, B_c, C_c\}_k$ allows to find the convex optimization problem to simultaneously give $\{\bar{A}^{lj}_c, \bar{B}^l_c, \bar{C}^j_c, \bar{D}_c\}_k$. Hence, the parameter-dependent $\{A_c, B_c, C_c\}_k$ is considerably better than the non-parameter-dependent $\{A_c, L_c, F_x\}_k$.

Define the augmented state $\tilde{x} = \begin{bmatrix} x \\ x_c \end{bmatrix}$. By applying (11.1) and (11.3), the augmented closed-loop system is

$$\tilde{x}_{k+1} = \Phi_k \tilde{x}_k + \Gamma_k w_k, \tag{11.34}$$

where

$$\Phi_k = \begin{bmatrix} A_k + B_k D_{c,k} C_k & B_k C_{c,k} \\ B_{c,k} C_k & A_{c,k} \end{bmatrix},$$

$$\Gamma_k = \begin{bmatrix} B_k D_{c,k} E_k + D_k \\ B_{c,k} E_k \end{bmatrix}.$$

The prediction equation, based on (11.34), is

$$\tilde{x}_{i+1|k} = \Phi_{i,k}\tilde{x}_{i|k} + \Gamma_{i,k}w_{i|k}, \tag{11.35}$$

where

$$\Phi_{i,k} = \begin{bmatrix} A_{i|k} + B_{i|k}D_{c,i|k}C_{i|k} & B_{i|k}C_{c,i|k} \\ B_{c,i|k}C_{i|k} & A_{c,i|k} \end{bmatrix},$$

$$\Gamma_{i,k} = \begin{bmatrix} B_{i|k}D_{c,i|k}E_{i|k} + D_{i|k} \\ B_{c,i|k}E_{i|k} \end{bmatrix}.$$

By applying (11.33), it is shown that

$$\Phi_{i,k} = \sum_{l=1}^{L}\omega_{l,i|k}\sum_{j=1}^{L}\omega_{j,i|k}\Phi_{lj,k},$$

$$\Gamma_{i,k} = \sum_{l=1}^{L}\omega_{l,i|k}\sum_{j=1}^{L}\omega_{j,i|k}\Gamma_{lj,k},$$

$$\Phi_{lj,k} = \begin{bmatrix} A_l + B_l\bar{D}_{c,k}C_j & B_l\bar{C}^j_{c,k} \\ \bar{B}^l_{c,k}C_j & \bar{A}^{lj}_{c,k} \end{bmatrix}, \quad \Gamma_{lj,k} = \begin{bmatrix} B_l\bar{D}_{c,k}E_j + D_l \\ \bar{B}^l_{c,k}E_j \end{bmatrix}.$$

For the sake of unification, in the sequel the following four sets of parameters will be often abused:

(1) $A_{c,k}$, $A_{c,i|k}$, $\bar{A}^{lj}_{c,k}$ and $\hat{A}^{lj}_{c,k}$;
(2) $L_{c,k}$, $B_{c,i|k}$, $\bar{B}^{l}_{c,k}$ and $\hat{B}^{l}_{c,k}$;
(3) $F_{x,k}$, $C_{c,i|k}$, $\bar{C}^{j}_{c,k}$ and $\hat{C}^{j}_{c,k}$;
(4) $F_{y,k}$, $D_{c,i|k}$, $\bar{D}_{c,k}$ and $\hat{D}_{c,k}$.

The readers can transplant between LPV and quasi-LPV regarding the adjustment of notations.

11.2 Characterization of Stability and Optimality

Consider the closed-loop systems (11.31) and (11.35). They have the same form. They are composed of double convex combinations (i.e., the convex combinations by coefficients $\omega_{l,i|k}$ and $\omega_{j,i|k}$). We will borrow the notion of QB (Quadratic Boundedness) from Alessandri et al. [2004, 2006] to characterize the closed-loop stability of (11.31) and (11.35).

11.2.1 Review of Quadratic Boundedness

In Alessandri et al. [2004], the following model with nominal parametric matrices is considered:

$$x_{k+1} = Ax_k + Dv_k, \tag{11.36}$$

where A and D are the time-invariant (fixed) matrix, $v \in \mathbb{R}^{n_v}$ is the noise. In Alessandri et al. [2004], it is first assumed that $v \in \mathbb{V}$ where $\mathbb{V}$ is a compact (bounded and closed) set, and $\mathbb{V} \subset \mathbb{R}^{n_v}$.

Definition 11.1 See Alessandri et al. [2004]. The system (11.36) is said to be quadratically bounded with Lyapunov matrix $P > 0$ if

$$x^T Px \geqslant 1 \Rightarrow (Ax + Dv)^T P(Ax + Dv) \leqslant x^T Px, \quad \forall v \in \mathbb{V}.$$

The system (11.36) is said to be strictly quadratically bounded with Lyapunov matrix $P > 0$ if

$$x^T Px > 1 \Rightarrow (Ax + Dv)^T P(Ax + Dv) < x^T Px, \quad \forall v \in \mathbb{V}.$$

Lemma 11.1 See Alessandri et al. [2004]. Suppose there exists a $\xi \in \mathbb{V}$ such that $D\xi \neq 0$. If (11.36) is quadratically bounded with Lyapunov matrix $P > 0$, then it is strictly quadratically bounded with the same Lyapunov matrix.

Definition 11.2 The set $\mathbb{S}$ is a robust positively invariant set for (11.36), if

$$x \in \mathbb{S} \Rightarrow (Ax + Dv) \in \mathbb{S}, \quad \forall v \in \mathbb{V}.$$

Theorem 11.2 See Alessandri et al. [2004]. Suppose $v \in \varepsilon_{P_v}$ with $P_v > 0$. The following facts are equivalent.

(1) Equation (11.36) is quadratically bounded with Lyapunov matrix $P > 0$;
(2) Equation (11.36) is strictly quadratically bounded with Lyapunov matrix $P > 0$;
(3) The ellipsoid ε_P is a robust positively invariant set for (11.36);
(4) $x^T Px \geqslant v^T P_v v \Rightarrow (Ax + Dv)^T P(Ax + Dv) \leqslant x^T Px$;
(5) There exists $\alpha > 0$ such that

$$\begin{bmatrix} (1-\alpha)P - A^T PA & \star \\ -D^T PA & \alpha P_v - D^T PD \end{bmatrix} \geqslant 0; \tag{11.37}$$

(6) A is exponentially stable (i.e., there exists $\bar{P} > 0$ such that $\bar{P} - A^T \bar{P} A > 0$).

In Alessandri et al. [2006], the following model with uncertain parametric matrices is considered:

$$x_{k+1} = A_k x_k + D_k v_k, \tag{11.38}$$

where $[A_k|D_k]$ belongs to a known bounded set, i.e., $[A_k|D_k] \in \mathscr{P}$ for all $k \geqslant 0$, and $D \neq 0$ for at least one $[A|D] \in \mathscr{P}$.

Definition 11.3 See Alessandri et al. [2006]. Suppose $v_k \in \varepsilon_{P_v}$ for all $k \geqslant 0$, in (11.38). The system (11.38) is said to be strictly quadratically bounded with a common Lyapunov matrix $P > 0$, if

$$x^T Px > 1 \Rightarrow (Ax + Dv)^T P(Ax + Dv) < x^T Px, \ \forall v \in \varepsilon_{P_v}, \ \forall [A|D] \in \mathscr{P}.$$

Since $D \neq 0$ for at least one $[A|D] \in \mathscr{P}$, and $v \in \varepsilon_{P_v}$, there exists a $Dv \neq 0$. Similarly to Lemma 11.1, if (11.38) is quadratically bounded with Lyapunov matrix $P > 0$, then it is strictly quadratically bounded with the same Lyapunov matrix. The definition of quadratic boundedness is similar to Definition 11.1.

Definition 11.4 Suppose $v_k \in \varepsilon_{P_v}$ for all $k \geqslant 0$, in (11.38). The set $\mathbb{S}$ is a positively invariant set for (11.38), if

$$x \in \mathbb{S} \Rightarrow (Ax + Dv) \in \mathbb{S}, \ \forall v \in \varepsilon_{P_v}, \ \forall [A|D] \in \mathscr{P}.$$

Theorem 11.3 See Alessandri et al. [2006]. Suppose $v_k \in \varepsilon_{P_v}$ with $k \geqslant 0$. The following facts are equivalent.

(1) Equation (11.38) is strictly quadratically bounded with a common Lyapunov matrix $P > 0$;
(2) The ellipsoid ε_P is a positively invariant set for (11.38);
(3) There exists $\alpha_k \in (0, 1)$ such that

$$\begin{bmatrix} (1-\alpha_k)P - A_k^T P A_k & \star \\ -D_k^T P A_k & \alpha_k P_v - D_k^T P D_k \end{bmatrix} \geqslant 0. \tag{11.39}$$

Note that in the above theorem, it is necessary to use a time-varying α_k.

11.2.2 Stability Condition

In the output feedback MPC of this chapter, QB is equivalent to strict QB (see [Ding, 2009]). For the closed-loop systems (11.31) and (11.35), by generalizing the results in Section 11.2.1, we obtain the following results.

Definition 11.5 See first Ding and Xie [2008] and Ding [2010a] for quasi-LPV, and Ding and Xie [2009] and Ding et al. [2011] for LPV. Suppose (referring to Assumptions 9.1 and 9.2) at time k and for all $i > 0$,

(1) $\|w_{i|k}\| \leqslant 1$;
(2) there exist nonnegative coefficients $\omega_{l,i|k}$, $l = 1, \dots, L$ such that

$$\sum_{l=1}^{L} \omega_{l,i|k} = 1,$$

and $[A|B|C|D|E]_k = \sum_{l=1}^{L} \omega_{l,i|k} [A_l|B_l|C_l|D_l|E_l]$.

The system (11.31) or (11.35) is said to be quadratically bounded with a common Lyapunov matrix $M_k > 0$, if

$$\|\tilde{x}_{i|k}\|_{M_k}^2 \geqslant 1 \Rightarrow \|\tilde{x}_{i+1|k}\|_{M_k}^2 \leqslant \|\tilde{x}_{i|k}\|_{M_k}^2, \ \forall i \geqslant 0. \tag{11.40}$$

Definition 11.6 With the assumptions in Definition 11.5 satisfied, the set $\mathbb{S}$ is a positively invariant set for (11.31) or (11.35), if

$$\tilde{x}_{i|k} \in \mathbb{S} \Rightarrow \tilde{x}_{i+1|k} \in \mathbb{S}, \ \forall i \geqslant 0.$$

Theorem 11.4 See first Ding and Xie [2009] and Ding [2013] for LPV, and Ding et al. [2009] and Ding [2010a] for quasi-LPV. With the assumptions in Definition 11.5 satisfied, the following facts are equivalent.

(1) Equations (11.31) or (11.35) is quadratically bounded with a common Lyapunov matrix $M_k > 0$;
(2) The ellipsoid ε_{M_k} is a positively invariant set for (11.31) or (11.35);
(3) There exists $\alpha_{i,k} \in (0,1)$ such that

$$\begin{bmatrix} (1-\alpha_{i,k})M_k - \Phi_{i,k}^T M_k \Phi_{i,k} & \star \\ -\Gamma_{i,k}^T M_k \Phi_{i,k} & \alpha_{i,k} I - \Gamma_{i,k}^T M_k \Gamma_{i,k} \end{bmatrix} \geqslant 0, i \geqslant 0; \tag{11.41}$$

(4) $\Phi_{i,k}$ is exponentially stable for all $i > 0$, i.e., there exists $\bar{M}_k > 0$ such that $\bar{M}_k - \Phi_{i,k}^T \bar{M}_k \Phi_{i,k} > 0$.

In Ding and Ping [2012], the single-valued α is first replaced by

$$\alpha_{i,k} = \sum_{l=1}^{L}\sum_{j=1}^{L} \omega_{l,i|k}\omega_{j,i|k}\alpha_{lj}.$$

11.2.3 Optimality Condition

The disturbance-free form of (11.31) or (11.35) is

$$\tilde{x}_{u,i+1|k} = \Phi_{i,k}\tilde{x}_{u,i|k}, \quad \forall i \geqslant 0, \quad \tilde{x}_{u,0|k} = \tilde{x}_k. \tag{11.42}$$

Correspondingly,

$$\begin{aligned} u_{u,i|k} &= F_{x,k}x_{c,u,i|k} + F_{y,k}y_{u,i|k}, \\ y_{u,i|k} &= C_{i|k}x_{u,i|k}, \\ z_{u,i|k} &= C_{i|k}x_{u,i|k}, \ z'_{u,i|k} = \mathcal{F}_{i|k}x_{u,i|k}. \end{aligned}$$

Let us introduce the quadratic cost

$$\begin{aligned} J_k &= \sum_{i=0}^{\infty} J_{i,k}, \\ J_{i,k} &= \|z'_{u,i|k}\|^2_{\mathcal{Q}_1} + \|x_{c,u,i|k}\|^2_{\mathcal{Q}_2} + \|u_{u,i|k}\|^2_{\mathcal{R}}, \end{aligned}$$

where $\mathcal{Q}_1$, $\mathcal{Q}_2$, and $\mathcal{R}$ are positive-definite weight matrices and consider the condition

$$\|\tilde{x}_{u,i+1|k}\|^2_{M_k} - \|\tilde{x}_{u,i|k}\|^2_{M_k} \leqslant -\frac{1}{\gamma_k}J_{i,k}, \ \forall i \geqslant 0. \tag{11.43}$$

(In Theorem 11.1, it has taken $\mathcal{Q}_2 = 0$.) For exponentially stable $\Phi_{i,k}$, it will result in $\lim_{i\to\infty} z'_{u,i|k} = 0$, $\lim_{i\to\infty} x_{c,u,i|k} = 0$ and $\lim_{i\to\infty} u_{u,i|k} = 0$. Hence, summing (11.43) from $i = 0$ to $i = \infty$ yields

$$J_k \leqslant \gamma_k \|\tilde{x}_{u,0|k}\|^2_{M_k} = \gamma_k \|\tilde{x}_k\|^2_{M_k}. \tag{11.44}$$

Furthermore, let

$$\tilde{x}_k \in \varepsilon_{M_k}. \tag{11.45}$$

Then, applying (11.44) to (11.45) yields

$$J_k \leqslant \gamma_k, \tag{11.46}$$

i.e., γ_k is an upper bound of J_k. We will take γ_k as the cost function of the optimization problems which finds the controller parametric matrices.

The condition (11.43) can be rewritten as

$$\tilde{x}^T_{u,i|k} \Pi_{i,k} \tilde{x}_{u,i|k} \geqslant 0,$$

where

$$\Pi_{i,k} = M_k - \Phi^T_{i,k} M_k \Phi_{i,k} - \frac{1}{\gamma_k} \mathrm{diag}\{\mathcal{F}^T_{i|k} \mathcal{Q}_1 \mathcal{F}_{i|k}, \mathcal{Q}_2\} - \frac{1}{\gamma_k} \left[F_{y,k} C_{i|k} \ F_{x,k}\right]^T \mathcal{R} \left[F_{y,k} C_{i|k} \ F_{x,k}\right].$$

Hence, (11.43) is guaranteed by $\Pi_{i,k} \geqslant 0$. By applying Schur complement, it is shown that $\Pi_{i,k} \geqslant 0$ can be transformed into

$$\begin{bmatrix} M_k - \Phi^T_{i,k} M_k \Phi_{i,k} & \star & \star \\ \mathcal{Q}^{1/2} \mathrm{diag}\{\mathcal{F}_{i|k}, I\} & \gamma_k I & \star \\ \mathcal{R}^{1/2} \left[F_{y,k} C_{i|k} \ F_{x,k}\right] & 0 & \gamma_k I \end{bmatrix} \geqslant 0, \ i \geqslant 0, \tag{11.47}$$

where $\mathcal{Q} = \mathrm{diag}\{\mathcal{Q}_1, \mathcal{Q}_2\}$.

The condition (11.43) or (11.47) is for optimality, not primarily for stability. However, if

$$\mathrm{diag}\{\mathcal{F}^T_{i|k} \mathcal{Q}_1 \mathcal{F}_{i|k}, \mathcal{Q}_2\} + \left[F_{y,k} C_{i|k} \ F_{x,k}\right]^T \mathcal{R} \left[F_{y,k} C_{i|k} \ F_{x,k}\right] > 0,$$

then (11.47) means that $\bar{M}_k - \Phi^T_{i,k} \bar{M}_k \Phi_{i,k} > 0$, i.e., that $\Phi_{i,k}$ is exponentially stable (referring to point iv) of Theorem 11.4). We can indeed combine the optimality and stability conditions by imposing (see first [Ding and Xie, 2009, Ding et al., 2011] for LPV, and [Ding and Xie, 2008, Ding, 2010a] for quasi-LPV)

$$\|\tilde{x}_{i|k}\|^2_{M_k} \geqslant 1 \Rightarrow \|\tilde{x}_{i+1|k}\|^2_{M_k} - \|\tilde{x}_{i|k}\|^2_{M_k} \leqslant -\frac{1}{\gamma_k} \left[\|z'_{i|k}\|^2_{\mathcal{Q}_1} + \|x_{c,i|k}\|^2_{\mathcal{Q}_2} + \|u_{i|k}\|^2_{\mathcal{R}}\right], \ \forall i \geqslant 0. \tag{11.48}$$

It is easy to show that (11.48) is equivalent to (in the sense for any $\tilde{x}_{i|k}$ and $w_{i|k}$)

$$\begin{bmatrix} (1-\alpha_{i,k}) M_k - \Phi^T_{i,k} M_k \Phi_{i,k} & \star & \star & \star \\ -\Gamma^T_{i,k} M_k \Phi_{i,k} & \alpha_{i,k} I - \Gamma^T_{i,k} M_k \Gamma_{i,k} & \star & \star \\ \mathcal{Q}^{1/2} \mathrm{diag}\{\mathcal{F}_{i|k}, I\} & \mathcal{Q}^{1/2} \begin{bmatrix} \mathcal{G}_{i|k} \\ 0 \end{bmatrix} & \gamma_k I & \star \\ \mathcal{R}^{1/2} \left[F_{y,k} C_{i|k} \ F_{x,k}\right] & \mathcal{R}^{1/2} F_{y,k} E_{i|k} & 0 & \gamma_k I \end{bmatrix} \geqslant 0, i \geqslant 0. \tag{11.49}$$

Remark 11.4 It is apparent that feasibility of (11.49) guarantees both (11.41) and (11.47). With γ_k free (i.e., as a decision variable), feasibility of (11.41) guarantees both (11.47) and (11.49). Therefore, on the feasibility aspect, (11.49) and (11.41) are equivalent.

11.2.4 A Paradox for State Convergence

Consider the condition group $\{(11.45), (11.49)\}$ or $\{(11.45), (11.41)\}$. The condition (11.49) or (11.41) means that, if the augmented state $\tilde{x}_k$ lies outside of the ellipsoid ε_{M_k}, then $\tilde{x}_{i|k}$ will converge to ε_{M_k} with the increase of $i \geqslant 0$. However, the condition (11.45) requires that the initial augmented state lies within the ellipsoid ε_{M_k}. With the satisfaction of (11.45), the condition (11.49) or (11.41) cannot guarantee the convergence of $\tilde{x}_{i|k}$; the condition (11.49) or (11.41) only guarantees the invariance of $\tilde{x}_{i|k}$ within ε_{M_k}.

Although there is no guarantee that $\tilde{x}_{i|k}$ will converge, the convergence of $\tilde{x}_{i|k}$ will happen when $\|\tilde{x}_k\|$ is not small (see first [Ding and Ping, 2012] for LPV, and [Ding, 2011c] for quasi-LPV). The main reason lies in that (11.49) or (11.41) is a robust condition.

Let us impose that, if the augmented state $\tilde{x}_k$ lies outside of the ellipsoid $\varepsilon_{\beta_k^{-1}M_k}$, then $\tilde{x}_{i|k}$ converges to $\varepsilon_{\beta_k^{-1}M_k}$ with the increase of $i \geqslant 0$. Here, $\varepsilon_{\beta_k^{-1}M_k}$ is an ellipsoid not larger than ε_{M_k} since $0 < \beta_k \leqslant 1$ (see first [Ding et al., 2011] for LPV, and [Ding and Xie, 2008, Ding, 2011c] for quasi-LPV). By applying such β_k, we can change (11.40) as

$$\|\tilde{x}_{i|k}\|_{M_k}^2 \geqslant \beta_k \Rightarrow \|\tilde{x}_{i+1|k}\|_{M_k}^2 \leqslant \|\tilde{x}_{i|k}\|_{M_k}^2, \ \forall i \geqslant 0, \tag{11.50}$$

which is equivalent to (in the sense for any $\tilde{x}_{i|k}$ and $w_{i|k}$)

$$\begin{bmatrix} (1-\alpha_{i,k})M_k - \Phi_{i,k}^T M_k \Phi_{i,k} & \star \\ -\Gamma_{i,k}^T M_k \Phi_{i,k} & \alpha_{i,k}\beta_k I - \Gamma_{i,k}^T M_k \Gamma_{i,k} \end{bmatrix} \geqslant 0, \ i \geqslant 0. \tag{11.51}$$

We can also change (11.48) as

$$\begin{aligned} \|\tilde{x}_{i|k}\|_{M_k}^2 \geqslant \beta_k \Rightarrow & \ \|\tilde{x}_{i+1|k}\|_{M_k}^2 - \|\tilde{x}_{i|k}\|_{M_k}^2 \\ & \leqslant -\frac{1}{\gamma_k}\left[\|z'_{i|k}\|_{\mathcal{Q}_1}^2 + \|x_{c,i|k}\|_{\mathcal{Q}_2}^2 + \|u_{i|k}\|_{\mathcal{R}}^2\right], \quad \forall i \geqslant 0, \end{aligned} \tag{11.52}$$

which is equivalent to (in the sense for any $\tilde{x}_{i|k}$ and $w_{i|k}$)

$$\begin{bmatrix} (1-\alpha_{i,k})M_k - \Phi_{i,k}^T M_k \Phi_{i,k} & \star & \star & \star \\ -\Gamma_{i,k}^T M_k \Phi_{i,k} & \alpha_{i,k}\beta_k I - \Gamma_{i,k}^T M_k \Gamma_{i,k} & \star & \star \\ \mathcal{Q}^{1/2}\mathrm{diag}\{\mathcal{F}_{i|k}, I\} & \mathcal{Q}^{1/2}\begin{bmatrix} \mathcal{G}_{i|k} \\ 0 \end{bmatrix} & \gamma_k I & \star \\ \mathcal{R}^{1/2}\left[F_{y,k}Ci|k \ F_{x,k}\right] & \mathcal{R}^{1/2}F_{y,k}E_{i|k} & 0 & \gamma_k I \end{bmatrix} \geqslant 0, i \geqslant 0. \tag{11.53}$$

Adding $\beta_k \in (0, 1]$ as a free variable, due to the special position of β_k in either (11.51) or (11.53), does not affect the minimization of γ_k and feasibility. It is suggested to minimize β_k after the minimization of γ_k (see first [Ding and Ping, 2012] for LPV, and [Ding, 2011c] for quasi-LPV). If the controller parametric matrices are not re-optimized in minimizing β_k, it is easy to know that we do not need β_k, i.e., we can simply remove it.

11.3 General Optimization Problem

The dynamic output feedback robust MPC (OFRMPC) aims at solving, at each time k,

$$\min_{\{\gamma,M,A_c,L_c,F_x,F_y\}_k} \left\{ \max_{[A|B|C|D|E|C|\mathcal{E}|\mathcal{F}|\mathcal{G}]_{i|k}\in\Omega,\|w_{i|k}\|\leqslant 1} \gamma_k \right\}$$
$$\text{s.t. } (11.26),\ (11.45),\ (11.40),\ (11.43). \tag{11.54}$$

Lemma 11.2 See first Ding and Ping [2012] for LPV, and Ding et al. [2009] and Ding [2011c] for quasi-LPV. This is the result for recursive feasibility. Assume that the state x is measurable. At each time $k \geqslant 0$, solve (11.54) and implement u_k. The problem (11.54) is feasible for any time $k > 0$ if and only if it is feasible at time $k = 0$.

Theorem 11.5 See first Ding [2011c] for quasi-LPV, and Ding and Ping [2012] for LPV. This is the result for stability. Assume that the state x is measurable. At each time $k \geqslant 0$, solve (11.54) and implement u_k. If (11.54) is feasible at time $k = 0$, then with the evolution of k, $\{\gamma, z', x_c, u\}$ will converge to a neighborhood of the origin and stay in this neighborhood thereafter, and the constraints in (11.26) are satisfied for all $k \geqslant 0$.

According to Section 11.2, (11.54) is transformed into (equivalently in the sense for any $\tilde{x}_{i|k}$ and $w_{i|k}$)

$$\min_{\{\gamma,\alpha_{ij},M,A_c,L_c,F_x,F_y\}_k} \left\{ \max_{[A|B|C|D|E|C|\mathcal{E}|\mathcal{F}|\mathcal{G}]_{i|k}\in\Omega} \gamma_k \right\},$$
$$\text{s.t. } (11.26),\ (11.45),\ (11.41),\ (11.47), \tag{11.55}$$

with recursive feasibility and stability properties retained.

11.3.1 Handling Physical Constraints

In Ding and Huang [2007] and Ding et al. [2008], the following lemma is utilized to handle the physical constraints (e.g., the magnitude constraints on x, y and u).

Lemma 11.3 Suppose a and b are vectors with appropriate dimensions. Then for any scalar $\eta \in (0, 1)$, $\|a + b\|^2 \leqslant (1 - \eta)\|a\|^2 + \frac{1}{\eta}\|b\|^2$.

In Ding and Pan [2016a,b] and Ding et al. [2012a,b], it is found that applying the above lemma, although simple, can greatly reduce the conservativeness for physical constraint handling. In essence, the physical constraints are handled based on the invariance of $\tilde{x}_{i|k}$ within ε_{M_k}.

Theorem 11.6 See first Ding et al. [2012a] and Ding and Pan [2016a] for LPV, and Ding and Pan [2017] for quasi-LPV. Suppose at time k, there exist scalars $\alpha_{i,k} \in (0, 1)$ and η_{rs}, and matrix $M_k > 0$, such that (11.45) and (11.41) hold, and

$$\begin{bmatrix} M_k & \star & \star \\ 0 & I & \star \\ \frac{1}{\sqrt{1-\eta_{1s}}}\xi_s \left[F_{y,k}C_{i|k} \ F_{x,k}\right] & \frac{1}{\sqrt{\eta_{1s}}}\xi_s F_{y,k}E_{i|k} & \bar{u}_s^2 \end{bmatrix} \geqslant 0, s = 1, \dots, n_u, \ i \geqslant 0, \tag{11.56}$$

$$\begin{bmatrix} M_k & \star & \star \\ 0 & I & \star \\ \frac{1}{\sqrt{(1-\eta_{2s})(1-\eta_{3s})}}\Psi_s \mathcal{C}_{i+1|k}\Phi^1_{i,k} & \frac{1}{\sqrt{(1-\eta_{2s})\eta_{3s}}}\Psi_s \mathcal{C}_{i+1|k}\Gamma^1_{i,k} & \clubsuit \end{bmatrix} \geqslant 0,$$

$$\clubsuit = \bar{\psi}_s^2 - \frac{1}{\eta_{2s}}\Psi_s \mathcal{E}_{i+1|k}\mathcal{E}^T_{i+1|k}\Psi^T_s, \ s = 1, \dots, q, \ i \geqslant 0, \tag{11.57}$$

where $\Phi^1_{i,k}$ ($\Gamma^1_{i,k}$) is the first of the two rows of $\Phi_{i,k}$ ($\Gamma_{i,k}$). Take care of the special cases as follows.

(1) If $\mathcal{E}_{i+1|k} = 0$, then take $\frac{1}{\eta_{2s}}\Psi_s \mathcal{E}_{i+1|k}\mathcal{E}^T_{i+1|k}\Psi^T_s = 0$ and $\eta_{2s} = 0$;
(2) If $E_{i|k} = 0$, then take $\frac{1}{\sqrt{\eta_{1s}}}\xi_s F_{y,k}E_{i|k} = 0$ and $\eta_{1s} = 0$;
(3) If $D_{i|k} = 0$ and $E_{i|k} = 0$, then take $\frac{1}{\sqrt{\eta_{3s}}}\Psi_s \mathcal{C}_{i+1|k}\Gamma^1_{i,k} = 0$ and $\eta_{3s} = 0$.

Then, (11.26) is satisfied.

In the above theorem, one may want to choose η_{rs} to be time-varying. However, we have not found good method to online optimize η_{rs}, so we take η_{rs} as time-invariant.

According to Theorem 11.6, the problem (11.55) is approximated as (by no means equivalent to)

$$\min_{\{\gamma,\alpha_y,M,A_c,L_c,F_x,F_y\}_k} \left\{ \max_{[A|B|C|D|E|\mathcal{C}|\mathcal{E}|\mathcal{F}|\mathcal{G}]_{i|k}\in\Omega} \gamma_k \right\},$$
$$\text{s.t. } (11.45), \ (11.41), \ (11.47), \ (11.56), \ (11.57), \tag{11.58}$$

with recursive feasibility and stability properties retained. In (11.58), η_{rs} is prespecified (see first [Ding et al., 2012a, Ding and Pan, 2016a] for LPV, and [Ding and Pan, 2017] for quasi-LPV).

11.3.2 Current Augmented State

The condition (11.45) (i.e.,$\|\tilde{x}_k\|^2_{M_k} \leqslant 1$ or $\tilde{x}_k \in \varepsilon_{M_k}$) is the current condition on the augmented state. At time k, in $\tilde{x}_k = [x_k^T, x_{c,k}^T]^T$, x_k can be unmeasurable while $x_{c,k}$ is always known. When x_k is unmeasurable, we need to remove it from (11.45) for the sake of solving (11.58).

Let us define an error signal

$$e_k = x_k - x_{0,k},$$

where

$$x_{0,k} = U_k x_{c,k},$$

with U_k being a known transformation matrix. When $U_k = I$, defining e_k is usual; when $U_k = E_0^T$ is fixed, see first Ding [2011b] and Ding and Pan [2016b]; when U_k is on-line refreshed, see first Ding and Pan [2015, 2016a] for LPV and Ding and Pan [2017] for quasi-LPV. When x_k is unmeasurable, e_k is unknown (nondeterministic). If we can obtain the outer bound set of e_k, say $\mathscr{D}_{e,k}$, then we can utilize $x_{0,k} \oplus \mathscr{D}_{e,k}$ to replace x_k. Since $\mathscr{D}_{e,k}$ is known (deterministic), by replacing x_k with $x_{0,k} \oplus \mathscr{D}_{e,k}$, (11.45) becomes deterministic.

Using $x = e + Ux_c$, we obtain

$$\begin{aligned}
&\tilde{x}^T M \tilde{x} \\
&= (e + Ux_c)^T M_1 (e + Ux_c) + 2(e + Ux_c)^T M_2^T x_c + x_c^T M_3 x_c \\
&= e^T M_1 e + 2e^T (M_1 U + M_2^T) x_c + x_c^T (U^T M_1 U + 2U^T M_2^T + M_3) x_c. \qquad (11.59)
\end{aligned}$$

If we can remove the cross item $2e^T(M_1 U + M_2^T)x_c$, then the treatment of (11.45) will become easier, and the treatment of recursive feasibility of the resultant optimization will become simpler.

Lemma 11.4 In order to remove the cross item $2e^T(M_1 U + M_2^T)x_c$ in $\tilde{x}^T M \tilde{x}$, we need to take $U = -M_1^{-1} M_2^T$.

For quasi-LPV, Ding et al. [2009] and Ding [2010a] first impose $M_2 = -M_1$, and Ding and Pan [2017] first imposes $U = -M_1^{-1} M_2^T$, both removing the cross item. For LPV, Ding and Ping [2012] first imposes $M_2 = -M_1$, Ding [2011b] and Ding and Pan [2016b] first impose $M_2 = -E_0 M_1$, and Ding and Pan [2015, 2016a] first impose $U = -M_1^{-1} M_2^T$, all removing the cross item.

By substituting $U = -M_1^{-1} M_2^T$ into (11.59), we obtain

$$\tilde{x}^T M \tilde{x} = e^T M_1 e + x_c^T (M_3 - U^T M_1 U) x_c. \qquad (11.60)$$

By introducing a scalar ϱ_k and imposing

$$e_k^T M_{1,k} e_k \leqslant \varrho_k, \qquad (11.61)$$

$$x_{c,k}^T [M_{3,k} - U_k^T M_{1,k} U_k] x_{c,k} \leqslant 1 - \varrho_k, \qquad (11.62)$$

it is apparent that (11.45) is guaranteed. The condition (11.61) is guaranteed by (11.7) if we can first guarantee that $e_k \in \varepsilon_{M_{e,k}}$. The condition $e_k \in \varepsilon_{M_{e,k}}$ can be guaranteed by appropriately refreshing $M_{e,k}$ at each time $k > 0$. However, for the initial time $k = 0$, $e_k \in \varepsilon_{M_{e,k}}$ has to be assumed.

Assumption 11.1 $e_0 = x_0 - x_{0,0} \in \varepsilon_{M_{e,0}}$.

Based on Assumption 11.1 and the fact $e_k \in \varepsilon_{M_{e,k}}$, the problem (11.58) is approximated as (by no means equivalent to)

$$\begin{aligned}
&\min_{\{\gamma, \alpha_{lj}, \varrho, M, A_c, L_c, F_x, F_y\}_k} \left\{ \max_{[A|B|C|D|E|\mathcal{C}|\mathcal{E}|\mathcal{F}|\mathcal{G}]_{il|k} \in \Omega} \gamma_k \right\}, \\
&\text{s.t. } (11.62),\ (11.7),\ (11.41),\ (11.47),\ (11.56),\ (11.57), \qquad (11.63)
\end{aligned}$$

with recursive feasibility and stability properties retained in case $M_{e,k}$ is appropriately refreshed.

Lemma 11.5 For quasi-LPV, see Ding [2011c] first with $M_2 = -M_1$, and Ding and Pan [2017] first with $U = -M_1^{-1}M_2^T$; for LPV, see Ding and Ping [2012] first with $M_2 = -M_1$, Ding [2011b], Ding and Pan [2016b] first with $M_2 = -E_0M_1$, and Ding and Pan [2015, 2016a] first with $U = -M_1^{-1}M_2^T$. At each time $k > 0$,

(1) if we choose (11.12)–(11.14), then (11.7) and (11.62) are guaranteed by

$$M_{1,k} = \varrho_k M_{e,k},$$
$$x_{c,k}^T\left[M_{3,k} - U_k^T M_{1,k} U_k\right] x_{c,k} = 1 - \varrho_k;$$

(2) if we choose (11.15)–(11.18), then (11.7) and (11.62) are guaranteed by

$$M_{1,k} = \varrho_k' M_{e,k}',$$
$$x_{c,k}^T\left[M_{3,k}' - (U_k')^T M_{1,k} U_k'\right] x_{c,k} = 1 - \varrho_k'.$$

11.3.3 Some Usual Transformations

In order to solve (11.63), we need to transform (11.41) and (11.47) into familiar forms (comparing with e.g. [Kothare et al., 1996]). Define $Q = M^{-1}$ and

$$Q = \begin{bmatrix} Q_1 & Q_2^T \\ Q_2 & Q_3 \end{bmatrix}.$$

By applying Schur complement, (11.41) is transformed into

$$\begin{bmatrix} (1-\alpha_{i,k})M_k & \star & \star \\ 0 & \alpha_{i,k} I & \star \\ \Phi_{i,k} & \Gamma_{i,k} & Q_k \end{bmatrix} \geqslant 0, \quad i \geqslant 0. \tag{11.64}$$

By applying Schur complement, (11.47) is transformed into

$$\begin{bmatrix} M_k & \star & \star & \star \\ \Phi_{i,k} & Q_k & \star & \star \\ \mathcal{Q}^{1/2}\mathrm{diag}\{\mathcal{F}_{i|k}, I\} & 0 & \gamma_k I & \star \\ \mathcal{R}^{1/2}\left[F_{y,k}C_{i|k}\ F_{x,k}\right] & 0 & 0 & \gamma_k I \end{bmatrix} \geqslant 0, \quad i \geqslant 0. \tag{11.65}$$

Then, we need to remove or handle the convex combinations in {(11.64), (11.65), (11.56), (11.57)}. By invoking double convex combinations (DbCCs), (11.64) and (11.65) are equivalent to, respectively,

$$\sum_{l=1}^{L}\sum_{j=1}^{L} \omega_{l,i|k}\omega_{j,i|k}\Upsilon_{lj,k}^{\mathrm{QB}} \geqslant 0, \quad i \geqslant 0, \tag{11.66}$$

and

$$\sum_{l=1}^{L}\sum_{j=1}^{L} \omega_{l,i|k}\omega_{j,i|k}\Upsilon_{lj,k}^{\mathrm{opt}} \geqslant 0, \quad i \geqslant 0, \tag{11.67}$$

where

$$\Upsilon_{lj,k}^{\mathrm{QB}} = \begin{bmatrix} (1-\alpha_{lj})M_k & \star & \star \\ 0 & \alpha_{lj} I & \star \\ \Phi_{lj,k} & \Gamma_{lj,k} & Q_k \end{bmatrix},$$

$$\Upsilon^{\text{opt}}_{lj,k} = \begin{bmatrix} M_k & \star & \star & \star \\ \Phi_{lj,k} & Q_k & \star & \star \\ \mathcal{Q}^{1/2}\text{diag}\{\mathcal{F}_j, I\} & 0 & \gamma_k I & \star \\ \mathcal{R}^{1/2}\left[F_{y,k}C_j \ F_{x,k}\right] & 0 & 0 & \gamma_k I \end{bmatrix}.$$

By removing the single convex combination, (11.56) is guaranteed by

$$\Upsilon^u_{j,k} \geqslant 0, \ j = 1, \dots, L, \ s = 1, \dots, n_u, \ i \geqslant 0, \tag{11.68}$$

where

$$\Upsilon^u_{j,k} = \begin{bmatrix} M_k & \star & \star \\ 0 & I & \star \\ \frac{1}{\sqrt{1-\eta_{1s}}}\xi_s\left[F_{y,k}C_j \ F_{x,k}\right] & \frac{1}{\sqrt{\eta_{1s}}}\xi_s F_{y,k}E_j & \bar{u}_s^2 \end{bmatrix}.$$

By removing the single convex combination, and invoking DbCC, (11.57) is guaranteed by

$$\sum_{l=1}^{L}\sum_{j=1}^{L}\omega_{l,i|k}\omega_{j,i|k}\Upsilon^z_{hlj,k} \geqslant 0, \ h = 1, \dots, L, \ s = 1, \dots, q, \ i \geqslant 0, \tag{11.69}$$

where

$$\Upsilon^z_{hlj,k} = \begin{bmatrix} M_k & \star & \star \\ 0 & I & \star \\ \frac{1}{\sqrt{(1-\eta_{2s})(1-\eta_{3s})}}\Psi_s C_h \Phi^1_{lj,k} & \frac{1}{\sqrt{(1-\eta_{2s})\eta_{3s}}}\Psi_s C_h \Gamma^1_{lj,k} & \bar{\psi}_s^2 - \frac{1}{\eta_{2s}}\Psi_s \mathcal{E}_h \mathcal{E}_h^T \Psi_s^T \end{bmatrix}.$$

In summary, the problem (11.63) is approximated as (not strictly equivalent to)

$$\min_{\{\gamma,\alpha_{lj},\varrho,M,Q,A_c,L_c,F_x,F_y\}_k} \left\{ \max_{[A|B|C|D|E|\mathcal{C}|\mathcal{E}|\mathcal{F}|\mathcal{G}]_{i|k}\in\Omega} \gamma_k \right\},$$
$$\text{s.t. } (11.62), \ (11.7), \ (11.66) - (11.69) \text{ and } Q = M^{-1}, \tag{11.70}$$

with recursive feasibility and stability properties retained in case $M_{e,k}$ is appropriately refreshed.

11.3.4 Handling Double Convex Combinations

In the literatures of fuzzy control based on Takagi-Sugeno model and robust feedback control, DbCCs as in {(11.66), (11.67), (11.69)} have been extensively studied. Some well-known examples include [Kim and Lee, 2000] (being invoked by MPC in [Ding et al., 2009, Ding and Xie, 2009]), Montagner et al. [2007] and Oliveira and Peres [2005] (being invoked by MPC first in Ding [2011b, 2013]), Sala and Ariño [2007] (being invoked by MPC firstly in [Ding, 2010a]).

By analogy to "Theorem 11.1" in Montagner et al. [2007], the following result can be obtained.

Lemma 11.6 See Ding [2011b, 2013]. The conditions

$$\sum_{l=1}^{L}\sum_{j=1}^{L}\omega_{l,i|k}\omega_{j,i|k}\Upsilon_{lj,k} \geqslant 0, \ i \geqslant 0, \tag{11.71}$$

hold if and only if there exists a sufficiently large $d \geqslant 0$ such that

$$\sum_{l=1}^{L} C_l^{\ell}(d,2)\Upsilon_{ll,k} + \sum_{l=1}^{L-1}\sum_{j=l+1}^{L} C_{lj}^{\ell}(d,1,1)\left[\Upsilon_{lj,k} + \Upsilon_{jl,k}\right] \geqslant 0, \ell \in \{1,\dots,|\mathcal{K}(d+2)|\}. \tag{11.72}$$

Moreover, if (11.72) holds for a $d = \hat{d}$, then they hold for any $d > \hat{d}$.

This lemma has been utilized in Theorem 11.1. In this lemma, $\Upsilon_{lj,k} \in \{\Upsilon_{lj,k}^{\mathrm{QB}}, \Upsilon_{lj,k}^{\mathrm{opt}}, \Upsilon_{hlj,k}^{z}\}$. Equivalently, the techniques for the positivity of DbCC, as in Sala and Ariño [2007], can be exactly utilized to obtain finite dimensional sufficient conditions for the nonnegativity of DbCC in (11.71). For example, (11.71) is guaranteed by any one set of the following sets of conditions (see "Proposition 2" of Sala and Ariño [2007]):

Set 1: $(d = 0)$ (i) $\Upsilon_{ll,k} \geqslant 0, l \in \{1,\dots,L\}$, (ii) $\Upsilon_{lj,k} + \Upsilon_{jl,k} \geqslant 0, j > l, l,j \in \{1,\dots,L\}$;

Set 2: $(d = 1)$ (i) $\Upsilon_{ll,k} \geqslant 0, l \in \{1,\dots,L\}$, (ii) $\Upsilon_{ll,k} + \Upsilon_{lj,k} + \Upsilon_{jl,k} \geqslant 0, j \neq l, l,j \in \{1,\dots,L\}$, (iii) $\Upsilon_{lj,k} + \Upsilon_{jl,k} + \Upsilon_{jt,k} + \Upsilon_{tj,k} + \Upsilon_{tl,k} + \Upsilon_{lt,k} \geqslant 0, t > j > l, l,j,t \in \{1,\dots,L\}$.

In Sets 1 and 2, d is the complexity parameter of Sala and Ariño [2007]. With a larger d, the conditions are less conservative but the computational burden is heavier. There exists a finite d such that necessary and sufficient conditions for satisfaction of (11.71) can be obtained for a concrete model.

11.4 Solutions to Output Feedback MPC

For solving (11.70), LPV is much more difficult than quasi-LPV. For quasi-LPV, by setting

$$Q = \begin{bmatrix} Q_1 & -(Q_1 - M_1^{-1})M_1 M_2^{-1} \\ -M_2^{-T}M_1(Q_1 - M_1^{-1}) & M_2^{-T}M_1(Q_1 - M_1^{-1})M_1 M_2^{-1} \end{bmatrix}, \tag{11.73}$$

$$M = \begin{bmatrix} M_1 & M_2^T \\ M_2 & M_2(M_1 - Q_1^{-1})^{-1}M_2^T \end{bmatrix}, \tag{11.74}$$

which naturally satisfies $M = Q^{-1}$, and using the transformation (equivalent to (11.5), with "(k)" being omitted for brevity)

$$\begin{cases} \hat{D}_c = \bar{D}_c \\ \hat{C}_c^j = \bar{D}_c C_j Q_1 + \bar{C}_c^j Q_2 \\ \hat{B}_c^l = M_1 B_l \bar{D}_c + M_2^T \bar{B}_c^l \\ \hat{A}_c^{lj} = M_1 A_l Q_1 + M_1 B_l \bar{D}_c C_j Q_1 + M_2^T \bar{B}_c^l C_j Q_1 + M_1 B_l \bar{C}_c^j Q_2 + M_2^T \bar{A}_c^{lj} Q_2 \end{cases}, \tag{11.75}$$

a solution to (11.70) can be obtained through a single optimization problem (11.6)–(11.11). For the prespecified $\{\alpha, \eta_{1s}, \eta_{2s}, \eta_{3s}\}$, (11.6)–(11.11) is an LMI optimization problem. Before Ding and Pan [2017], for the quasi-LPV, some special solutions to (11.70) can be found in Ding [2011c] and Ding et al. [2013]. For LPV, even with the prespecified $\{\alpha_{lj}, \eta_{1s}, \eta_{2s}, \eta_{3s}\}$, one cannot find all the parameters $\{A_c, L_c, F_x, F_y\}_k$ in a single LMI optimization problem. In the following, we give two solutions to (11.70) for LPV.

11.4.1 Full Online Method for LPV

By applying the block-matrix inversion on Q, it is easy to show that

$$M = \begin{bmatrix} M_1 & -M_1 Q_2^T Q_3^{-1} \\ -Q_3^{-1} Q_2 M_1 & Q_3^{-1} + Q_3^{-1} Q_2 M_1 Q_2^T Q_3^{-1} \end{bmatrix}.$$

Take $U = -M_1^{-1} M_2^T$. Then, it is easy to show that $U = -Q_2^T Q_3^{-1}$ and

$$\tilde{x}_k^T M_k \tilde{x}_k = [x_k - x_k^0]^T M_{1,k} [x_k - x_k^0] + x_{c,k}^T Q_{3,k}^{-1} x_{c,k}.$$

Lemma 11.7 Let Assumption 11.1 hold and, at each time $k > 0$, find $\{x^0, M_e\}_k$ such that $x_k - x_k^0 \in \varepsilon_{M_{e,k}}$. Choose $\{U, x_c\}_0$ such that $U_0 x_{c,0} = x_0^0$ and, at each time $k > 0$, U_k such that $U_k x_{c,k} = x_k^0$. Then, the condition (11.62) holds if

$$\begin{bmatrix} 1 - \varrho_k & \star \\ x_{c,k} & Q_{3,k} \end{bmatrix} \geqslant 0. \tag{11.76}$$

Further define $N_1 = M_1^{-1}$ and $P_3 = Q_3^{-1}$. Then,

$$Q = \begin{bmatrix} N_1 + U Q_3 U^T & U Q_3 \\ Q_3 U^T & Q_3 \end{bmatrix}, \quad M = \begin{bmatrix} M_1 & -M_1 U \\ -U^T M_1 & P_3 + U^T M_1 U \end{bmatrix}, \tag{11.77}$$

which naturally satisfies $M = Q^{-1}$. By applying (11.77), the problem (11.70) becomes (equivalently)

$$\min_{\{\gamma, \alpha_{lj}, \varrho, N_1, M_1, P_3, Q_3, A_c, L_c, F_x, F_y\}_k} \left\{ \max_{[A|B|C|D|E|\mathcal{C}|\mathcal{E}|\mathcal{F}|\mathcal{G}]_{i|k} \in \Omega} \gamma_k \right\},$$
$$\text{s.t. (11.76), (11.7), (11.66) − (11.69), (11.77), } N_{1,k} = M_{1,k}^{-1}, \; P_{3,k} = Q_{3,k}^{-1}. \tag{11.78}$$

This approach is proposed by Ding et al. [2014] and Ding and Pan [2016b] where $U_k = E_0^T$ and, hence,

$$Q = \begin{bmatrix} Q_1 & E_0^T Q_3 \\ Q_3 E_0 & Q_3 \end{bmatrix}, \quad M = \begin{bmatrix} M_1 & -M_1 E_0^T \\ -E_0 M_1 & M_3 \end{bmatrix}.$$

In solving (11.78), usually $\alpha_{lj,k} = \alpha_k$ can be prespecified. One can line search α_k over the interval $(0, 1)$. Indeed, we found that the improvement on control performance is negligible by online optimizing α_k. The problem (11.78) has been solved by ICCA (the Iterative Cone-Complementary Approach) (see first in Ding [2011b, 2013]). ICCA has two major loops. The inner loop is CCA (the Cone-Complementary Approach) which minimizes trace$\{M_{1,k}N_{1,k} + N_{1,k}M_{1,k} + Q_{3,k}P_{3,k} + P_{3,k}Q_{3,k}\}$ in order to achieve $N_{1,k} = M_{1,k}^{-1}$ and $P_{3,k} = Q_{3,k}^{-1}$. The outer loop gradually reduces γ_k. Note that, even with α_k being prespecified, (11.78) cannot be transformed into LMI optimization problem.

Algorithm 11.4 Full dynamic OFRMPC

At each time $k > 0$,

(a) for $k = 0$, take $U_0 = I$;
(b) for $k > 0$, calculate $x_{c,k} = A_{c,k-1}x_{c,k-1} + L_{c,k-1}y_{k-1}$, and refresh $\{M_e, U, x^0\}_k$ as in (11.12)–(11.14);
(c) for $k > 0$, find $M'_{e,k}$ satisfying (11.15)–(11.18); and, if (11.15)– (11.18) are feasible, then $M_{e,k} = M'_{e,k}$, $U_k = U'_k$, $x^0_k = U'_k x_{c,k}$;
(d) solve (11.78) to find $\{A_c, L_c, F_x, F_y, M_1, N_1, Q_3, P_3\}^*_k$;
(e) implement $u_k = F_{x,k}x_{c,k} + F_{y,k}y_k$.

Remark 11.5 Finding $M'_{e,k}$ satisfying (11.15)–(11.18) can be replaced by solving (11.19)–(11.24), where for LPV model, (11.20) is replaced by

$$\sum_{l=1}^{L}\sum_{j=1}^{L}\omega_{l,k-1}\omega_{j,k-1}\begin{bmatrix} \heartsuit_1 & \star & \star & \star \\ \heartsuit_{21} & \phi_1 M_{x,k-1} & \star & \star \\ 0 & 0 & \phi_2 I & \star \\ \heartsuit_{41} & A_l + B_l F_{y,k-1} C_j & \heartsuit_{43} & Q_{e,k} \end{bmatrix} \geqslant 0,$$

$$\begin{aligned} \heartsuit_1 &= 1 - \phi_1 - \phi_2 + \phi_1 \|x^0_{k-1}\|^2_{M_{e,k-1}}, \\ \heartsuit_{21} &= -\phi_1 M_{e,k-1} x^0_{k-1}, \\ \heartsuit_{41} &= B_l F_{x,k-1} x_{c,k-1} - U'_k x_{c,k}, \\ \heartsuit_{43} &= D_l + B_l F_{y,k-1} E_j. \end{aligned} \tag{11.79}$$

Imposing (11.24) guarantees the nonsingular and boundedness of U'_k, which replaces the condition $\underline{\delta}I \leqslant (U'_k)^T U'_k \leqslant \bar{\delta}I$ that is more direct but cannot be handled using LMI. Eq. (11.24) requires $(U'_k)^T + U'_k$ be positive-definite. We can remove the requirement of positivity through replacing (11.24) by

$$\underline{\delta}I \leqslant (U'_k)^T U_k + U_k^T U'_k \leqslant \bar{\delta}I. \tag{11.80}$$

Moreover, (11.23) can be relaxed as

$$\text{trace}(Q_{e,k}) \leqslant \text{trace}(M_{e,k}^{-1}). \tag{11.81}$$

Theorem 11.7 See Ding and Pan [2015, 2016a]. Adopt Algorithm 11.4. Suppose Assumption 11.1 holds, and (11.78) is feasible at time $k = 0$. Then,

(1) (11.78) is feasible at any time $k > 0$;
(2) $\{\gamma, z', x_c, u\}$ will converge to neighborhood of 0, and the constraints in (11.2) are satisfied for all $k \geqslant 0$.

11.4.2 Partial Online Method for LPV

In order to alleviate the computational burden, we can prespecify $\{L_c, F_y\}$ in (11.78). In this way, $\{M_1, P_3\}$ are no longer the decision variables. Therefore, $\{(11.7), (11.66)–(11.69)\}$ will be modified accordingly.

By applying Schur complement, (11.7) is equivalent to

$$\begin{bmatrix} \varrho_k M_{e,k} & I \\ I & N_{1,k} \end{bmatrix} \geqslant 0. \tag{11.82}$$

Taking congruence transformations via diag$\{Q_k, I\}$ on (11.66) and (11.67), yields

$$\sum_{l=1}^{L} \omega_{l,i|k} \sum_{j=1}^{L} \omega_{j,i|k} \begin{bmatrix} (1-\alpha_{lj,k})Q_k & \star & \star \\ 0 & \alpha_{lj,k} I & \star \\ \check{\Phi}_{lj,k} & \Gamma_{lj,k} & Q_k \end{bmatrix} \geqslant 0, \ i \geqslant 0, \tag{11.83}$$

$$\sum_{l=1}^{L} \omega_{l,i|k} \sum_{j=1}^{L} \omega_{j,i|k} \begin{bmatrix} Q_k & \star & \star \\ \check{\Phi}_{lj,k} & Q_k & \star \\ \spadesuit & 0 & \gamma_k I \end{bmatrix} \geqslant 0, \ i \geqslant 0,$$
$$\spadesuit = \begin{bmatrix} \mathcal{Q}_1^{1/2} \mathcal{F}_j \heartsuit & \mathcal{Q}_1^{1/2} \mathcal{F}_j U_k Q_{3,k} \\ \mathcal{Q}_2^{1/2} Q_{3,k} U_k^T & \mathcal{Q}_2^{1/2} Q_{3,k} \\ \mathcal{R}^{1/2}[F_y C_j \heartsuit + \check{F}_{x,k} U_k^T] & \mathcal{R}^{1/2}[F_y C_j U_k Q_{3,k} + \check{F}_{x,k}] \end{bmatrix},$$
$$\heartsuit = [N_{1,k} + U_k Q_{3,k} U_k^T]. \tag{11.84}$$

Taking congruence transformations on (11.68)–(11.69), via diag$\{Q_k, I\}$, and applying the Schur complement, yields

$$\begin{bmatrix} Q_k & \star \\ \frac{1}{\sqrt{1-\eta_{1s}}} \xi_s \left[\spadesuit \ \ F_y C_j U_k Q_{3,k} + \check{F}_{x,k}\right] & \bar{u}_s^2 - \frac{1}{\eta_{1s}} \xi_s F_y E_j E_j^T F_y^T \xi_s^T \end{bmatrix} \geqslant 0,$$
$$\spadesuit = F_y C_j [N_{1,k} + U_k Q_{3,k} U_k^T] + \check{F}_{x,k} U_k^T,$$
$$j = 1, \dots, L, \ \ s = 1, \dots, n_u, \tag{11.85}$$

$$\sum_{l=1}^{L} \sum_{j=1}^{L} \omega_{l,i|k} \omega_{j,i|k}, \begin{bmatrix} Q_k & \star & \star \\ 0 & I & \star \\ \spadesuit_1 & \spadesuit_2 & \spadesuit_3 \end{bmatrix} \geqslant 0, \ h = 1, \dots, L, \ \ s = 1, \dots, q, \ \ i \geqslant 0,$$
$$\spadesuit_1 = \frac{1}{\sqrt{(1-\eta_{2s})(1-\eta_{3s})}} \Psi_s C_h \check{\Phi}_{lj,k}^1, \ \ \spadesuit_2 = \frac{1}{\sqrt{(1-\eta_{2s})\eta_{3s}}} \Psi_s C_h \Gamma_{lj,k}^1,$$
$$\spadesuit_3 = \bar{\psi}_s^2 - \frac{1}{\eta_{2s}} \Psi_s \mathcal{E}_h \mathcal{E}_h^T \Psi_s^T. \tag{11.86}$$

In (11.83)–(11.86),

$$\check{\Phi}_{lj,k} = \begin{bmatrix} (A_l + B_l F_y C_j) Q_{1,k} + B_l \check{F}_{x,k} U_k^T & (A_l + B_l F_y C_j) Q_{2,k}^T + B_l \check{F}_{x,k} \\ L_c C_j Q_{1,k} + \check{A}_{c,k} U_k^T & L_c C_j Q_{2,k}^T + \check{A}_{c,k} \end{bmatrix},$$
$$\check{\Phi}_{lj,k}^1 = \left[(A_l + B_l F_y C_j) Q_{1,k} + B_l \check{F}_{x,k} U_k^T \ \ (A_l + B_l F_y C_j) Q_{2,k}^T + B_l \check{F}_{x,k} \right],$$
$$\check{A}_{c,k} = A_{c,k} Q_{3,k}, \ \ \check{F}_{x,k} = F_{x,k} Q_{3,k}.$$

In summary, the problem (11.78) is simplified as

$$\min_{\{\gamma,\alpha_{lj},\varrho,N_1,Q_3,\breve{A}_c,\breve{F}_x\}_k} \left\{ \max_{[A|B|C|D|E|C|\mathcal{E}|\mathcal{F}|\mathcal{G}]_{i|k}\in\Omega} \gamma_k \right\},$$
$$\text{s.t. (11.76), (11.82) and (11.83)} - \text{(11.86)}, \tag{11.87}$$

with $\{A_{c,k}, F_{x,k}\}$ being calculated by

$$A_{c,k} = \breve{A}_{c,k} Q_{3,k}^{-1}, \quad F_{x,k} = \breve{F}_{x,k} Q_{3,k}^{-1}. \tag{11.88}$$

The solution to (11.87) can be obtained by LMI toolbox. Since CCA is not involved, it is computationally less expensive than (11.78).

Algorithm 11.5 Partial dynamic OFRMPC

At each time $k \geqslant 0$,

(a) see steps (a)-(c) in Algorithm 11.4;
(b) solve (11.87) to find $\{\breve{A}_c, \breve{F}_x, N_1, Q_3\}_k^*$;
(c) calculate $\{A_c, F_x\}_k^*$ via (11.88), and implement $u_k = F_{x,k} x_{c,k} + F_y y_k$.

Theorem 11.8 See Ding and Pan [2015, 2016a]. Adopt Algorithm 11.5. Suppose Assumption 11.1 holds, and (11.87) is feasible at time $k = 0$. Then,

(1) (11.87) will be feasible at each time $k > 0$;
(2) $\{\gamma, z', x_c, u\}$ will converge to a neighborhood of 0, and the constraints in (11.2) are satisfied for all $k \geqslant 0$.

11.4.3 Relaxed Variables in Optimization Problem

The scalars η_{rs} appear non-affine and nonlinear in (11.78) and (11.87). Although it is suggested that η_{rs} can be line searched over the interval $(0, 1)$ for online optimizations, in this way the computational burden will be considerably increased. An alternative is to offline optimize η_{rs}. In Ding and Pan [2015, 2016a], we offline calculated η_{rs} by applying the norm-bounding technique.

The condition (11.85) is satisfied if

$$\begin{bmatrix} Q_k & \star \\ \xi_s \left[F_y C_j [N_{1,k} + U_k Q_{3,k} U_k^T] + \breve{F}_{x,k} U_k^T \;\; F_y C_j U_k Q_{3,k} + \breve{F}_{x,k}\right] & \tilde{u}_s^2 \end{bmatrix} \geqslant 0,$$
$$j = 1, \dots, L, \quad s = 1, \dots, n_u, \tag{11.89}$$
$$\frac{1}{1-\eta_{1s}} \tilde{u}_s^2 + \frac{1}{\eta_{1s}} (\zeta_s^u)^2 \leqslant \bar{u}_s^2, \tag{11.90}$$

where $\zeta_s^u = \max\{(\xi_s F_y E_j E_j^T F_y^T \xi_s^T)^{1/2} | j = 1, \dots, L\}$. The maximum $\tilde{u}_s$ satisfying (11.90) is calculated by

$$\tilde{u}_s = \bar{u}_s - \zeta_s^u, \tag{11.91}$$

by taking $\eta_{1s} = \frac{\zeta_s^u}{\bar{u}_s}$.

The condition (11.86) is satisfied if

$$\sum_{l=1}^{L}\omega_{l,i|k}\sum_{j=1}^{L}\omega_{j,i|k}\begin{bmatrix} Q_k & \star \\ \Psi_s C_h \breve{\Phi}^1_{lj,k} & \tilde{\psi}_s^2 \end{bmatrix} \geqslant 0, \ h=1,\ldots,L, \ s=1,\ldots,q, \ i\geqslant 0, \tag{11.92}$$

$$\frac{1}{(1-\eta_{2s})(1-\eta_{3s})}\tilde{\psi}_s^2 + \frac{1}{(1-\eta_{2s})\eta_{3s}}(\bar{\zeta}_s^z)^2 + \frac{1}{\eta_{2s}}(\zeta_s^z)^2 \leqslant \bar{\psi}_s^2, \tag{11.93}$$

where $\zeta_s^z = \max\{(\Psi_s \mathcal{E}_h \mathcal{E}_h^T \Psi_s^T)^{1/2} | h=1,\ldots,L\}$, and

$$\bar{\zeta}_s^z = \min_{\bar{\zeta}_s^z} \bar{\zeta}_s^z, \ \text{s.t.} \ \sum_{l=1}^{L}\omega_{l,i|k}\sum_{j=1}^{L}\omega_{j,i|k}\begin{bmatrix} \bar{\zeta}_s^z I & \star \\ \Psi_s C_h \Gamma^1_{lj,k} & \bar{\zeta}_s^z \end{bmatrix} \geqslant 0,$$

$$h=1,\ldots,L, \ i\geqslant 0.$$

The maximum $\tilde{\psi}_s$ satisfying (11.93) is calculated by

$$\tilde{\psi}_s = \bar{\psi}_s - \zeta_s^z - \bar{\zeta}_s^z, \tag{11.94}$$

by taking $\eta_{2s} = \frac{\zeta_s^z}{\bar{\psi}_s}$ and $\eta_{3s} = \frac{\bar{\zeta}_s^z}{\bar{\psi}_s - \zeta_s^z}$.

In the above, since $\{\zeta_s^u, \zeta_s^z, \bar{\zeta}_s^z\}$ are the norms of the disturbance-related items, the method for optimizing $\{\eta_{1s}, \eta_{2s}, \eta_{3s}\}$ has been called the norm-bounding technique. In this way, we obtain the second best values of η_{rs} (though may not be the best).

The constraints (11.89) and (11.92) will not be utilized in the optimization problems (though they could be utilized), since they are more conservative than (11.85)–(11.86). In (11.85), the item $\bar{u}_s^2 - \frac{1}{\eta_{1s}}\xi_s F_y E_j E_j^T F_y^T \xi_s^T$ applies for each j, while in (11.89), the item $\tilde{u}_s^2$ imposes for all $j=1,\ldots,L$. Similarly, in (11.86), the item $\left[\frac{1}{\sqrt{(1-\eta_{2s})\eta_{3s}}}\Psi_s C_h \Gamma^1_{lj,k}, \bar{\psi}_s^2 - \frac{1}{\eta_{2s}}\Psi_s \mathcal{E}_h \mathcal{E}_h^T \Psi_s^T\right]$ applies for each pair of $\{l,j\}$, while in (11.92), the item $\tilde{\psi}_s^2$ imposes for all $l,j=1,\ldots,L$.

11.4.4 Alternative Forms Based on Congruence Transformation

Take $n_x = n_{x_c}$ and $\mathcal{Q}_2 = 0$ in this subsection. Based on (11.73) and (11.74), let us define

$$N_1 := M_1^{-1}, \ P_1 := Q_1^{-1}, \ U := -M_1^{-1}M_2^T, \ e := x - Ux_c,$$

$$\hat{A}_c := -UA_cQ_2, \ \hat{L}_c := -UL_c, \ \hat{F}_x = F_xQ_2,$$

$$\bar{A}_c := -UA_cM_2^{-T}(M_1 - P_1), \ \bar{F}_x := F_xM_2^{-T}(M_1 - P_1),$$

$$T_0 := \begin{bmatrix} I & 0 \\ 0 & M_2^{-T}(M_1 - P_1) \end{bmatrix}, \ T_1 := \begin{bmatrix} Q_1 & N_1 \\ Q_2 & 0 \end{bmatrix}, \ T_2 := \begin{bmatrix} I & 0 \\ 0 & -U^T \end{bmatrix},$$

$$\mathcal{M}_P := \begin{bmatrix} M_1 & \star \\ M_1 - P_1 & M_1 - P_1 \end{bmatrix}, \ \mathcal{Q}_N := \begin{bmatrix} Q_1 & \star \\ N_1 - Q_1 & Q_1 - N_1 \end{bmatrix},$$

$$\mathcal{N}_Q := \begin{bmatrix} Q_1 & \star \\ N_1 & N_1 \end{bmatrix}, \ \bar{\Phi}_{lj} := \begin{bmatrix} A_l + B_lF_yC_j & B_l\bar{F}_x \\ \hat{L}_cC_j & \bar{A}_c \end{bmatrix},$$

$$\hat{\Phi}_{lj} := \begin{bmatrix} (A_l + B_lF_yC_j)Q_1 + B_l\hat{F}_x & (A_l + B_lF_yC_j)N_1 \\ \hat{L}_cC_jQ_1 + \hat{A}_c & \hat{L}_cC_jN_1 \end{bmatrix},$$

$$\hat{\Gamma}_{lj} := \begin{bmatrix} D_l + B_l F_y E_j \\ \hat{L}_c E_j \end{bmatrix},$$
$$\bar{\Phi}^1_{lj} := \left[A_l + B_l F_y C_j \ \ B_l \bar{F}_x\right], \ \hat{\Gamma}^1_{lj} := D_l + B_l F_y E_j,$$
$$\hat{\Phi}^1_{lj} = \left[(A_l + B_l F_y C_j) Q_1 + B_l \hat{F}_x \ \ (A_l + B_l F_y C_j) N_1\right].$$

Based on these notations, we have

$$A_c = -U^{-1}\bar{A}_c(M_1 - P_1)^{-1}M_2^T, \ L_c = -U^{-1}\hat{L}_c, \ F_x = \bar{F}_x(M_1 - P_1)^{-1}M_2^T,$$
$$M_2 = -U^T M_1, \tag{11.95}$$

$$A_c = -U^{-1}\hat{A}_c Q_2^{-1}, \ L_c = -U^{-1}\hat{L}_c, \ F_x = \hat{F}_x Q_2^{-1}, \ Q_2 = U^{-1}(Q_1 - N_1). \tag{11.96}$$

According to (11.73), we have $Q_3 = U^{-1}(Q_1 - N_1)U^{-T}$. Applying a congruence transformation on (11.76), via $\text{diag}\{I, U_k^T\}$, yields

$$\begin{bmatrix} 1 - \varrho_k & \star \\ U_k x_{c,k} & Q_{1,k} - N_{1,k} \end{bmatrix} \geqslant 0. \tag{11.97}$$

Based on (11.73) and (11.74), applying the congruence transformations on (11.66) and (11.67), via $\text{diag}\{T_{0,k}, I, T_{2,k}\}$ and $\text{diag}\{T_{0,k}, T_{2,k}, I, I\}$, respectively, yields

$$\sum_{l=1}^{L} \omega_{l,i|k} \sum_{j=1}^{L} \omega_{j,i|k} \Upsilon^{\text{qb}}_{lj,k} \geqslant 0, \quad i \geqslant 0, \tag{11.98}$$

$$\sum_{l=1}^{L} \omega_{l,i|k} \sum_{j=1}^{L} \omega_{j,i|k} \Upsilon^{\text{opt}}_{lj,k} \geqslant 0, \quad i \geqslant 0, \tag{11.99}$$

where $\Upsilon^{\text{qb}}_{lj,k} := \begin{bmatrix} (1-\alpha_{lj,k})\mathcal{M}_{P,k} & \star & \star \\ 0 & \alpha_{lj,k} I & \star \\ \bar{\Phi}_{lj,k} & \hat{\Gamma}_{lj,k} & \mathcal{Q}_{N,k} \end{bmatrix}$, $\Upsilon^{\text{opt}}_{lj,k} := \begin{bmatrix} \mathcal{M}_{P,k} & \star & \star & \star \\ \bar{\Phi}_{lj,k} & \mathcal{Q}_{N,k} & \star & \star \\ \left[\mathcal{Q}_1^{1/2} F_j \ \ 0\right] & 0 & \gamma_k I & \star \\ \mathcal{R}^{1/2}[F_{y,k} C_j \ \ \bar{F}_{x,k}] & 0 & 0 & \gamma_k I \end{bmatrix}$.

Applying the congruence transformations on (11.68) and (11.69), via $\text{diag}\{T_{0,k}, I\}$, yields

$$\begin{bmatrix} \mathcal{M}_{P,k} & \star & \star \\ 0 & I & \star \\ \frac{1}{\sqrt{1-\eta_{1s}}}\xi_s[F_{y,k} C_j \ \ \bar{F}_{x,k}] & \frac{1}{\sqrt{\eta_{1s}}}\xi_s F_{y,k} E_j & \bar{u}_s^2 \end{bmatrix} \geqslant 0, j = 1, \dots, L, \ s = 1, \dots, n_u, \tag{11.100}$$

$$\sum_{l=1}^{L}\sum_{j=1}^{L} \omega_{l,i|k}\omega_{j,i|k} \begin{bmatrix} \mathcal{M}_{P,k} & \star & \star \\ 0 & I & \star \\ \frac{1}{\sqrt{(1-\eta_{2s})(1-\eta_{3s})}}\Psi_s C_h \bar{\Phi}^1_{lj,k} & \frac{1}{\sqrt{(1-\eta_{2s})\eta_{3s}}}\Psi_s C_h \hat{\Gamma}^1_{lj,k} & \bar{\psi}_s^2 - \frac{1}{\eta_{2s}}\Psi_s \mathcal{E}_h \mathcal{E}_h^T \Psi_s^T \end{bmatrix} \geqslant 0,$$
$$h = 1, \dots, L, \ s = 1, \dots, q, \ i \geqslant 0. \tag{11.101}$$

Summarizing the above, an equivalent transformation of (11.78) is (see [Ding and Pan, 2016a])

$$\min_{\{\gamma,\alpha_{lj},\varrho,M_1,N_1,Q_1,P_1,\bar{A}_c,\hat{L}_c,\bar{F}_x,F_y\}_k} \left\{ \max_{[A|B|C|D|E|\mathcal{C}|\mathcal{E}|\mathcal{F}|\mathcal{G}]_{i|k}\in\Omega} \gamma_k \right\},$$
$$\text{s.t. } (11.97),\ (11.7),\ (11.98)-(11.101),\ M_{1,k}=N_{1,k}^{-1}, Q_{1,k}=P_{1,k}^{-1}. \quad (11.102)$$

$\{A_c, L_c, F_x\}_k$ are calculated by (11.95). The optimization problem (11.102) is non-convex, but its near-optimal solution arbitrarily close to the theoretically optimal one, can be found by applying CCA, iterative optimization, and LMI toolbox.

Based on (11.73) and (11.74), applying the congruence transformations on (11.66) and (11.67), via $\text{diag}\{T_{1,k}, I, T_{2,k}\}$ and $\text{diag}\{T_{1,k}, T_{2,k}, I, I\}$, respectively, yields

$$\sum_{l=1}^{L}\omega_{l,i|k}\sum_{j=1}^{L}\omega_{j,i|k}\begin{bmatrix}(1-\alpha_{lj,k})\mathcal{N}_{Q,k} & \star & \star\\ 0 & \alpha_{lj,k}I & \star\\ \hat{\Phi}_{lj,k} & \hat{\Gamma}_{lj,k} & \mathcal{Q}_{N,k}\end{bmatrix}\geqslant 0,\ i\geqslant 0, \quad (11.103)$$

$$\sum_{l=1}^{L}\omega_{l,i|k}\sum_{j=1}^{L}\omega_{j,i|k}\begin{bmatrix}\mathcal{N}_{Q,k} & \star & \star\\ \hat{\Phi}_{lj,k} & \mathcal{Q}_{N,k} & \star\\ \begin{bmatrix}\mathscr{Q}_1^{1/2}\mathcal{F}_jQ_{1,k} & \mathscr{Q}_1^{1/2}\mathcal{F}_jN_{1,k}\\ \mathscr{R}^{1/2}(F_yC_jQ_{1,k}+\hat{F}_{x,k}) & \mathscr{R}^{1/2}F_yC_jN_{1,k}\end{bmatrix} & 0 & \gamma_kI\end{bmatrix}\geqslant 0, i\geqslant 0.$$
(11.104)

Applying the congruence transformations on (11.68)–(11.69), via $\text{diag}\{T_{1,k}, I\}$, yields

$$\begin{bmatrix}\mathcal{N}_{Q,k} & \star\\ \frac{1}{\sqrt{1-\eta_{1s}}}\xi_s\left[F_yC_jQ_{1,k}+\hat{F}_{x,k}\ \ F_yC_jN_{1,k}\right] & \bar{u}_s^2-\frac{1}{\eta_{1s}}\xi_sF_yE_jE_j^TF_y^T\xi_s^T\end{bmatrix}\geqslant 0, \quad (11.105)$$
$$j=1,\dots,L,\ s=1,\dots,n_u,$$

$$\sum_{l=1}^{L}\sum_{j=1}^{L}\omega_{l,i|k}\omega_{j,i|k}\begin{bmatrix}\mathcal{N}_{Q,k} & \star & \star\\ 0 & I & \star\\ \frac{1}{\sqrt{(1-\eta_{2s})(1-\eta_{3s})}}\Psi_sC_h\hat{\Phi}^1_{lj,k} & \frac{1}{\sqrt{(1-\eta_{2s})\eta_{3s}}}\Psi_sC_h\hat{\Gamma}^1_{lj,k} & \bar{\psi}_s^2-\frac{1}{\eta_{2s}}\Psi_s\mathcal{E}_h\mathcal{E}_h^T\Psi_s^T\end{bmatrix}\geqslant 0,$$
$$h=1,\dots,L,\ s=1,\dots,q,\ i\geqslant 0. \quad (11.106)$$

In summary, an equivalent transformation of (11.87) is (see [Ding and Pan, 2016a])

$$\min_{\{\gamma,\alpha_{lj},\varrho,N_1,Q_1,\hat{A}_c,\hat{F}_x\}_k} \left\{ \max_{[A|B|C|D|E|\mathcal{C}|\mathcal{E}|\mathcal{F}|\mathcal{G}]_{i|k}\in\Omega} \gamma_k \right\},$$

$$\text{s.t. } (11.97),\ (11.82),\ (11.103)-(1.106). \quad (11.107)$$

$\{A_c, L_c, F_x\}_k$ are calculated by (11.96) and $\{\hat{L}_c, F_y\}$ are prespecified. The solution to (11.107) can be obtained by LMI toolbox.

Example 11.2 See Ding and Pan [2016a]. Consider the nonlinear model of a continuous stirred tank reactor (CSTR) for an exothermic, irreversible reaction. We add the disturbance

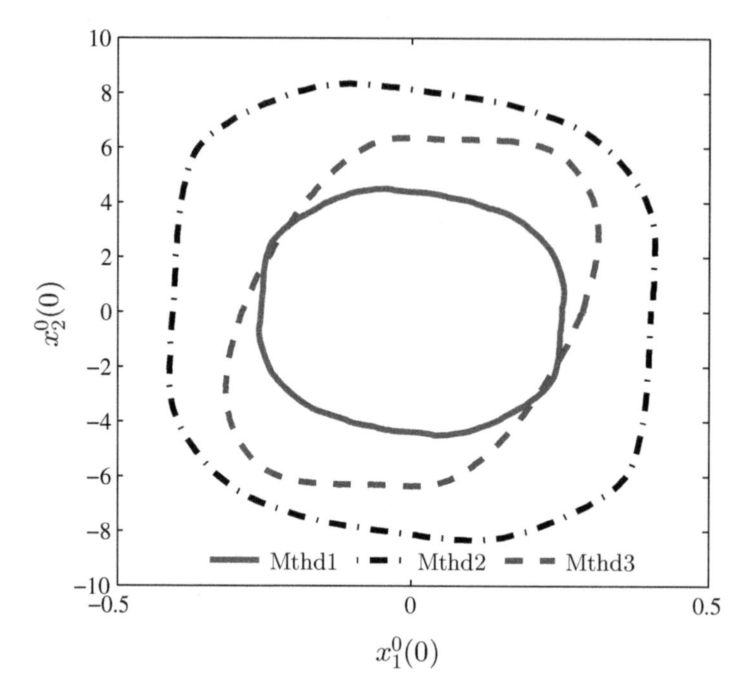

Figure 11.5 The regions of attraction, set (i).

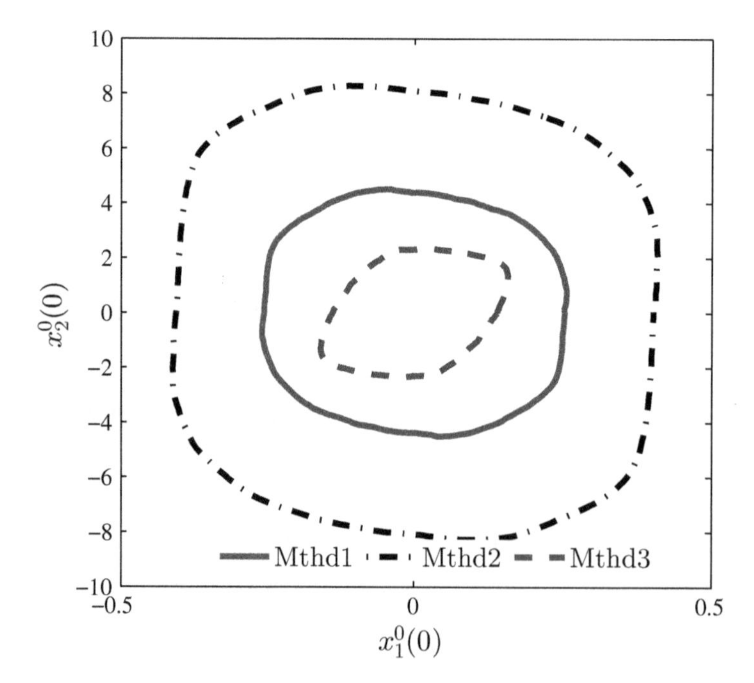

Figure 11.6 The regions of attraction, set (ii).

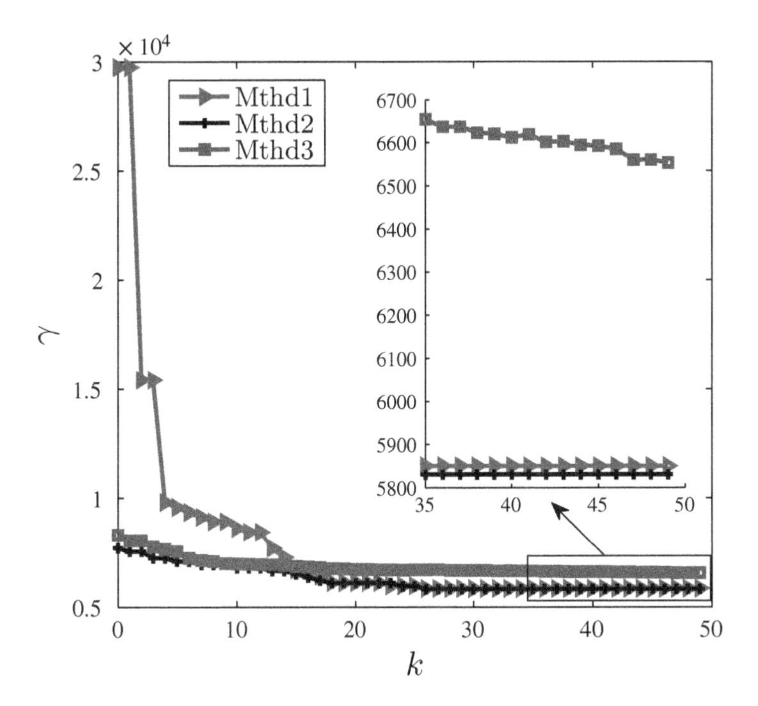

Figure 11.7 The evolutions of γ.

w in the original model. The sampling period is 0.05min. The parameters of the model are

$$L = 4,$$

$$A_1 = \begin{bmatrix} 0.8227 & -0.00168 \\ 6.1233 & 0.9367 \end{bmatrix}, \; A_2 = \begin{bmatrix} 0.9654 & -0.00182 \\ -0.6759 & 0.9433 \end{bmatrix},$$

$$A_3 = \begin{bmatrix} 0.8895 & -0.00294 \\ 2.9447 & 0.9968 \end{bmatrix}, \; A_4 = \begin{bmatrix} 0.8930 & -0.00062 \\ 2.7738 & 0.8864 \end{bmatrix},$$

$$B_1 = \begin{bmatrix} -0.000092 \\ 0.1014 \end{bmatrix}, \; B_2 = \begin{bmatrix} -0.000097 \\ 0.1016 \end{bmatrix}, B_3 = \begin{bmatrix} -0.000157 \\ 0.1045 \end{bmatrix},$$

$$B_4 = \begin{bmatrix} -0.000034 \\ 0.0986 \end{bmatrix}, \; C_l = \begin{bmatrix} 0 & 1 \end{bmatrix}, \; D_l = \begin{bmatrix} 0.00223 & 0 \\ 0 & 0.0564 \end{bmatrix},$$

$$E_l = \begin{bmatrix} 0.5 & 0.5 \end{bmatrix}, \; l = 1, \ldots, 4.$$

The physical constraints are

$$|u_k| \leq 10, \;\; |x_{1,k+1}| \leqslant 0.5, \;\; |y_{k+1}| \leqslant 10, \;\; k \geqslant 0.$$

For generating the true state, let $\omega_1 = \frac{1}{2}\frac{\varphi_1(y)-\varphi_1(-\bar{\psi}^y)}{\varphi_1(\bar{\psi}^y)-\varphi_1(-\bar{\psi}^y)}$, $\omega_2 = \frac{1}{2}\frac{\varphi_1(\bar{\psi}^y)-\varphi_1(y)}{\varphi_1(\bar{\psi}^y)-\varphi_1(-\bar{\psi}^y)}$, $\omega_3 = \frac{1}{2}\frac{\varphi_2(y)-\varphi_2(-\bar{\psi}^y)}{\varphi_2(\bar{\psi}^y)-\varphi_2(-\bar{\psi}^y)}$ and $\omega_4 = \frac{1}{2}\frac{\varphi_2(\bar{\psi}^y)-\varphi_2(y)}{\varphi_2(\bar{\psi}^y)-\varphi_2(-\bar{\psi}^y)}$, where $\varphi_1(y) = 7.2 \times 10^{10} e^{-\frac{8750}{y+350}}$ and $\varphi_2(y) = 3.6 \times 10^{10}(e^{-\frac{8750}{y+350}} - e^{-\frac{8750}{350}})/y$, and $\bar{\psi}^y = 10$.

We utilize LMI toolbox of Matlab to simulate three methods: Mthd2 corresponding to (11.102); Mthd3 corresponding to (11.107); Mthd1, which is the same as Mthd2 except

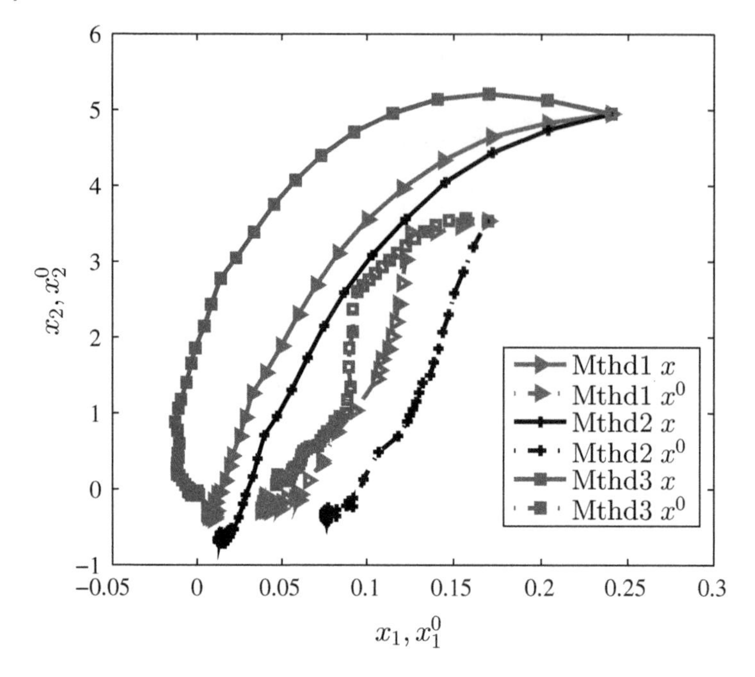

Figure 11.8 The state trajectories of closed-loop system.

that η_{rs} take fixed values (for $\mathcal{E}_k \neq 0$, $\{\eta_{1s}, \eta_{2s}, \eta_{3s}\} = \{\frac{1}{2}, \frac{1}{3}, \frac{1}{2}\}$ as in Ding et al. [2011] and Ding [2011b]; for $\mathcal{E}_k = 0$, $\{\eta_{1s}, \eta_{2s}, \eta_{3s}\} = \{\frac{1}{2}, 0, \frac{1}{2}\}$). Therefore, Mthd1 and Mthd2 differ in whether or not η_{rs} are optimized, while Mthd2 and Mthd3 in whether or not $\{\hat{L}_c, F_y\}$ are online decision variables. Let us first show the difference in RoAs. Here, RoA refers to a region, say $\mathcal{X}^0$, such that whenever $x_0^0 \in \mathcal{X}^0$, the optimization problem is feasible at time $k = 0$. Choose $M_{e,0} = \text{diag}\{100, 0.25\}$ and $\{t^0, n, \kappa_1, \kappa_2, \mathcal{Q}, \mathcal{R}\} = \{100, 2, 0.5, 0.99, 16, 9\}$ (these are parameters for ICCA, see e.g. Ding [2011b]). Consider two sets of parameters: (i) $\hat{L}_c = [-0.0001, 0.0025]^T$ and $F_y = -1.1557$; (ii) $\hat{L}_c = [0, 0.0001]^T$ and $F_y = -2.0975$. For sets (i) and (ii), RoAs by applying Mthd1–Mthd3 are depicted in Figures 11.5 and 11.6, respectively.

Let us then compare the control performance. The parameters, if not re-chosen, will be the same as above. Choose $\{\underline{\delta}, \bar{\delta}\} = \{1, 3\}$. $w(k)$ is randomly generated, with $w(0) = 0$. Furthermore, choose $x_{c,0} = \frac{1}{\sqrt{2}}[-0.49, 8]^T$ and $x_0 = \frac{1}{\sqrt{2}}[-0.59, 10]^T$. The evolutions of $\gamma(k)$, as in Figure 11.7, represent the control performance since we consider the min-max optimization.

By observing, we obtain the following conclusions.

(1) Offline optimizing η_{rs} can enlarge RoA and improve the control performance;
(2) Online optimizing $\{\hat{L}_c, F_y\}$ can enlarge RoA and improve the control performance.

For Mthd1–Mthd2, since $\kappa_2 = 0.99$, i.e., 2 percent minimization error can be observed, the comparisons should be made by allowing this inaccuracy. While online optimizing $\{\hat{L}_c, F_y\}$

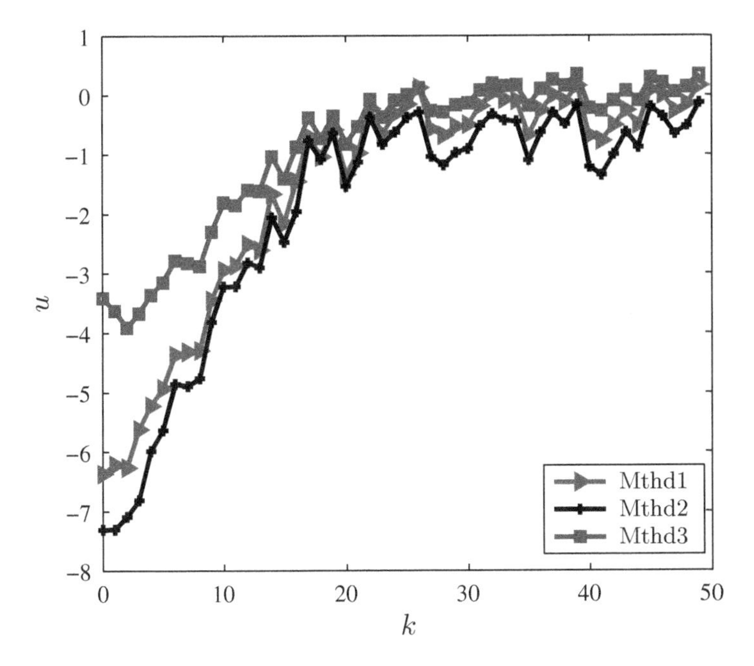

Figure 11.9 The control input signals.

is advantageous for the control performance and RoA, it is disadvantageous for computation. For the state trajectories and the control input signals, see Figures 11.8 and 11.9, which verify the closed-loop QB and constraint satisfaction.

11.4.5 Description of Bound on True State

In the above, the ellipsoid outer bound of e or x is adopted. We can also use the polyhedron outer bound of x, such as

(1) the polyhedron represented by plane, i.e., (see first [Ding et al., 2016])

$$x_k \in \mathscr{P}_{x,k} := \{x| - G_k\bar{e} \leqslant Hx - \check{x}_k \leqslant G_k\bar{e}\}, \tag{11.108}$$

where $\check{x}$ is a bias term; $H = \begin{bmatrix} H_a \\ H_b \end{bmatrix}$ is the given transformation matrix and H_a is nonsingular, G_k is the diagonal matrix, $\bar{e} = [\bar{e}_1, \bar{e}_2, \ldots, \bar{e}_p]^T$ where $p > n_x$, and for all $j = 1, \ldots, p$, $\bar{e}_j > 0$ is prespecified;

(2) the polyhedron represented by vertex, i.e.,

$$x_k \in \overline{\mathscr{P}}_{x,k} = \text{Co}\{\vartheta_{j,k}|j = 1, 2, \ldots, n_{\vartheta,k}\},$$

which is the general form of convex polyhedron (see first [Ding, 2011c] for quasi-LPV, and [Ding and Ping, 2012] for LPV).

The details of using polyhedron bounds are not discussed. The following points, however, are meaningful:

(1) For the output feedback MPC in this chapter, $\mathscr{P}_{x,k}$ in (11.108) is the general form of convex polyhedron, and is equivalent to $\mathscr{P}_{x,k} = \{\xi | \mathcal{H}\xi \leqslant G_k\vec{1}\}$ ($\mathcal{H} \in \mathbb{R}^{p \times n_x}$ is prespecified and $\vec{1} = [1, 1, \dots, 1]^T$) in Ding [2013]. Other polyhedron sets (e.g., [Ding and Xie, 2008, 2009, Ding et al., 2008, 2009, Ding, 2010a]) are special cases.

(2) Before Ding et al. [2013], only ellipsoid set or polyhedron set is used in the optimization problems. The recursive feasibility is guaranteed by refreshing the ellipsoid set simply, while using polyhedron set may loose the recursive feasibility. In Ding et al. [2013], either ellipsoid bound or polyhedron bound is adopted at each time, of which the polyhedron bound is contained by the ellipsoid bound. Moreover, Ding et al. [2013] provide some sufficient conditions guaranteeing the recursive feasibility for using the polyhedron set. Ding et al. [2019] further discuss the potential of using ellipsoid bound and polyhedron bound at same time.

References

A. Alessandri, M. Baglietto, and G. Battistelli. On estimation error bounds for receding-horizon filters using quadratic boundedness. *IEEE Transactions on Automatic Control*, 49:1350–1355, 2004.

A. Alessandri, M. Baglietto, and G. Battistelli. Design of state estimators for uncertain linear systems using quadratic boundedness. *Automatica*, 42:497–502, 2006.

H. H. J. Bloemen, T. J. J. Van Den Boom, and H. B. Verbruggen. Model-based predictive control for Hammerstein–Wiener systems. *International Journal of Control*, 74(5):482–495, 2001.

H. H. J. Bloemen, T. J. J. van de Boom, and H. B. Verbruggen. Optimizing the end-point state-weighting matrix in model-based predictive control. *Automatica*, 38:1061–1068, 2002.

H. Chen and F. Allgöwer. A quasi-infinite horizon nonlinear model predictive control scheme with guaranteed stability. *Automatica*, 34:1205–1217, 1998.

D. W. Clarke and R. Scattolini. Constrained receding-horizon predictive control. *IEE Proceedings, Part D (Control Theory and Applications)*, 138:347–354, 1991.

D. W. Clarke, C. Mohtadi, and P.S. Tuffs. Generalized predictive control—part I. The basic algorithm. *Automatica*, 23(2):137–148, 1987.

F. A. Cuzzola, J. C. Geromel, and M. Morari. An improved approach for constrained robust model predictive control. *Automatica*, 38:1183–1189, 2002.

M. T. Cychowski and T. O'Mahony. Feedback min-max model predictive control using robust one-step sets. *International Journal of Systems Science*, 41(7):813–823, 2010.

B. S. Dayal and J. F. MacGregor. Identification of finite impulse response models: methods and robustness issues. *Industrial & Engineering Chemistry Research*, 35(11):4078–4090, 1996.

B. C. Ding. Quadratic boundedness via dynamic output feedback for constrained nonlinear systems in Takagi–Sugeno's form. *Automatica*, 45(9):2093–2098, 2009.

B. C. Ding. Constrained robust model predictive control via parameter-dependent dynamic output feedback. *Automatica*, 46:1517–1523, 2010a.

B. C. Ding. Properties of parameter-dependent open-loop MPC for uncertain systems with polytopic description. *Asian Journal of Control*, 12(1):58–70, 2010b.

B. C. Ding. Homogeneous polynomially nonquadratic stabilization of discrete-time Takagi–Sugeno systems via nonparallel distributed compensation law. *IEEE Transactions on Fuzzy Systems*, 18(5):994–1000, 2010c.

B. C. Ding. Stabilization of linear systems over networks with bounded packet loss and its use in model predictive control. *Automatica*, 47(11):2526–2533, 2011a.

Model Predictive Control, First Edition. Baocang Ding and Yuanqing Yang.

B. C. Ding. Dynamic output feedback MPC for LPV systems via near-optimal solutions. In *Proceedings of the 30th Chinese Control Conference*, pages 3340–3345, Yantai, China, 2011b.

B. C. Ding. Dynamic output feedback predictive control for nonlinear systems represented by a Takagi–Sugeno model. *IEEE Transactions on Fuzzy Systems*, 19:831–843, 2011c.

B. C. Ding. Comments on "Constrained infinite-horizon model predictive control for fuzzy-discrete-time systems". *IEEE Transactions on Fuzzy Systems*, 19(3):598–600, 2011d.

B. C. Ding. New formulation of dynamic output feedback robust model predictive control with guaranteed quadratic boundedness. *Asian Journal of Control*, 15(1):302–309, 2013.

B. C. Ding and B. Huang. Output feedback model predictive control for nonlinear systems represented by Hammerstein–Wiener model. *IET Control Theory and Applications*, 1:1302–1310, 2007.

B. C. Ding and H. G. Pan. Synthesis approaches of dynamic output feedback robust MPC for LPV system with unmeasurable polytopic model parametric uncertainty - Part I. Norm-bounded disturbance. In *Proceedings of the 27th Chinese Control and Decision Conference*, pages 73–78, Qingdao, China, 2015.

B. C. Ding and H. G. Pan. Output feedback robust MPC for LPV system with polytopic model parametric uncertainty and bounded disturbance. *International Journal of Control*, 89:1554–1571, 2016a.

B. C. Ding and H. G. Pan. Output feedback robust model predictive control with unmeasurable model parameters and bounded disturbance. *Chinese Journal of Chemical Engineering*, 24:1431–1441, 2016b.

B. C. Ding and H. G. Pan. Dynamic output feedback predictive control of Takagi–Sugeno model with bounded disturbance. *IEEE Transactions on Fuzzy Systems*, 25:653–667, 2017.

B. C. Ding and X. B. Ping. Dynamic output feedback model predictive control for nonlinear systems represented by Hammerstein–Wiener model. *Journal of Process Control*, 22:1773–1784, 2012.

B. C. Ding and Y. G. Xi. Design and analysis of the domain of attraction for generalized predictive control with input nonlinearity. *Acta Automatica Sinica*, 30(6):954–960, 2004a.

B. C. Ding and Y. G. Xi. Stability analysis of generalized predictive control based on Kleinman's controllers. *Science in China Series F-Information Science*, 47(4):458–474, 2004b.

B. C. Ding and Y. G. Xi. A two-step predictive control design for input saturated Hammerstein systems. *International Journal of Robust and Nonlinear Control*, 16(7):353–367, 2006.

B. C. Ding and L. H. Xie. Robust model predictive control via dynamic output feedback. In *Proceedings of the 7th World Congress on Intelligent Control and Automation*, pages 3388–3393, Chongqing, China, 2008.

B. C. Ding and L. H. Xie. Dynamic output feedback robust model predictive control with guaranteed quadratic boundedness. In *Proceedings of the Joint 48th IEEE Conference on Decision and Control & 28th Chinese Control Conference*, pages 8034–8039, Shanghai, China, 2009.

B. C. Ding, S. Y. Li, and Y. G. Xi. Stability analysis of generalized predictive control with input nonlinearity based on Popov's theorem. *Acta Automatica Sinica*, 29(4):582–588, 2003a.

B. C. Ding, Y. G. Xi, and S. Y. Li. Stability analysis on predictive control of discrete-time systems with input nonlinearity. *Acta Automatica Sinica*, 29(6):827–834, 2003b.

B. C. Ding, Y. G. Xi, and S. Y. Li. On the stability of output feedback predictive control for systems with input nonlinearity. *Asian Journal of Control*, 6(3):388–397, 2004.

B. C. Ding, Y. G. Xi, M. T. Cychowski, and T. O'Mahony. A synthesis approach for output feedback robust constrained model predictive control. *Automatica*, 44:258–264, 2008.

B. C. Ding, L. H. Xie, and F. Z. Xue. Improving robust model predictive control via dynamic output feedback. In *Proceedings of Chinese Control and Decision Conference*, pages 2116–2121, Guilin, China, 2009.

B. C. Ding, B. Huang, and F. W. Xu. Dynamic output feedback robust model predictive control. *International Journal of Systems Science*, 42:1669–1682, 2011.

B. C. Ding, X. B. Ping, and Y. G. Xi. A general reformulation of output feedback MPC for constrained LPV systems. In *Proceedings of the 31st Chinese Control Conference*, pages 4195–4200, Hefei, China, 2012a.

B. C. Ding, Y. G. Xi, and X. B. Ping. A comparative study on output feedback MPC for constrained LPV systems. In *Proceedings of the 31st Chinese Control Conference*, pages 4189–4194, Hefei, China, 2012b.

B. C. Ding, X. B. Ping, and H. G. Pan. On dynamic output feedback robust MPC for constrained quasi-LPV systems. *International Journal of Control*, 86(12):2215–2227, 2013.

B. C. Ding, Y. G. Xi, X. B. Ping, and T. Zou. Dynamic output feedback robust MPC with relaxed constraint handling for LPV system with bounded disturbance. In *Proceedings of the 7th World Congress on Intelligent Control and Automation*, pages 2624–2629, Shenyang, China, 2014.

B. C. Ding, Y. G. Xi, and H. G. Pan. Synthesis approaches of dynamic output feedback robust MPC for LPV system with unmeasurable polytopic model parametric uncertainty - Part II. Polytopic disturbance. In *Proceedings of the 27th Chinese Control and Decision Conference*, pages 95–100, Qingdao, China, 2015.

B. C. Ding, C. B. Gao, and X. B. Ping. Dynamic output feedback robust MPC using general polyhedral state bounds for the polytopic uncertain system with bounded disturbance. *Asian Journal of Control*, 18:699–708, 2016.

B. C. Ding, J. Dong, and J. C. Hu. Output feedback robust MPC using general polyhedral and ellipsoidal true state bounds for LPV model with bounded disturbance. *International Journal of Systems Science*, 50(3):625–637, 2019.

K. P. Fruzzetti, A. Palazoğlu, and K. A. McDonald. Nolinear model predictive control using Hammerstein models. *Journal of Process Control*, 7(1):31–41, 1997.

E. Garone and A. Casavola. Receding horizon control strategies for constrained LPV systems based on a class of nonlinearly parameterized Lyapunov functions. *IEEE Transactions on Automatic Control*, 57(9):2354–2360, 2012.

E. G. Gilbert and K. T. Tan. Linear systems with state and control constraints: the theory and application of maximal output admissible sets. *IEEE Transactions on Automatic Control*, 36(9):1008–1020, 1991.

A. H. González, E. J. Adam, and J. L. Marchetti. Conditions for offset elimination in state space receding horizon controllers: a tutorial analysis. *Chemical Engineering and Processing*, 47:2184–2194, 2008.

T. Hu, D. E. Miller, and L. Qiu. Controllable regions of LTI discrete-time systems with input saturation. In *Proceedings of the 37th IEEE Conference on Decision and Control*, volume 1, pages 371–376. IEEE, 1998.

T. Hu, Z. L. Lin, and Y. Shamash. Semi-global stabilization with guaranteed regional performance of linear systems subject to actuator saturation. *Systems & Control Letters*, 43(3):203–210, 2001.

D. E. Kassmann, T. A. Badgwell, and R. B. Hawkins. Robust steady-state target calculation for model predictive control. *AIChE Journal*, 46(5):1007–1024, 2000.

E. C. Kerrigan and J. M. Maciejowski. Designing model predictive controllers with prioritised constraints and objectives. In *Proceedings of the IEEE International Symposium on Computer Aided Control System Design*, pages 33–38, 2002.

E. Kim and H. Lee. New approaches to relaxed quadratic stability condition of fuzzy control systems. *IEEE Transactions on Fuzzy Systems*, 8:523–533, 2000.

M. V. Kothare, V. Balakrishnan, and M. Morari. Robust constrained model predictive control using linear matrix inequalities. *Automatica*, 32:1361–1379, 1996.

W. H. Kwon and D. G. Byun. Receding horizon tracking control as a predictive control and its stability properties. *International Journal of Control*, 50(5):1807–1824, 1989.

W. H. Kwon, H. Choi, D. G. Byun, and S. Noh. Recursive solution of generalized predictive control and its equivalence to receding horizon tracking control. *Automatica*, 28(6):1235–1238, 1992.

Y. I. Lee and B. Kouvaritakis. Constrained robust model predictive control based on periodic invariance. *Automatica*, 42:2175–2181, 2006.

J. H. Lee and J. Xiao. Use of two-stage optimization in model predictive control of stable and integrating systems. *Computers and Chemical Engineering*, 24:1591–1596, 2000.

J. H. Lee, M. Morari, and C. E. Garcia. State-space interpretation of model predictive control. *Automatica*, 30(4):707–717, 1994.

D. H. Lee, J. B. Park, and Y. H. Joo. Improvement on nonquadratic stabilization of discrete-time Takagi–Sugeno fuzzy systems: multiple-parameterization approach. *IEEE Transactions on Fuzzy Systems*, 18(2):425–429, 2010.

D. W. Li, Y. G. Xi, and P. Y. Zheng. Constrained robust feedback model predictive control for uncertain systems with polytopic description. *International Journal of Control*, 82(7):1267–1274, 2009.

Z. L. Lin and A. Saberi. Semi-global exponential stabilization of linear systems subject to input saturation via linear feedbacks. *Systems & Control Letters*, 21(3):225–239, 1993.

Z. L. Lin, A. Saberi, and A. A. Stoorvogel. Semiglobal stabilization of linear discrete-time systems subject to input saturation via linear feedback-an ARE-based approach. *IEEE Transactions on Automatic Control*, 41(8):1203–1207, 1996.

Y. Lu and Y. Arkun. Quasi-min-max MPC algorithms for LPV systems. *Automatica*, 36:527–540, 2000.

P. Lundström, J. H. Lee, M. Morari, and S. Skogestad. Limitations of dynamic matrix control. *Computers & Chemical Engineering*, 19(4):409–421, 1995.

W. J. Mao. Robust stabilization of uncertain time-varying discrete systems and comments on "an improved approach for constrained robust model predictive control". *Automatica*, 39:1109–1112, 2003.

D. Q. Mayne. Model predictive control: recent developments and future promise. *Automatica*, 50:2967–2986, 2014.

D. Q. Mayne, J. B. Rawlings, C. V. Rao, and P. O. M. Scokaert. Constrained model predictive control: stability and optimality. *Automatica*, 36:789–814, 2000.

V. F. Montagner, R. C. L. F. Oliveira, and P. L. D. Peres. Necessary and sufficient LMI conditions to compute quadratically stabilizing state feedback controllers for Takagi–Sugeno systems. In *Proceedings of the 2007 American Control Conference*, pages 4059–4064, New York City, USA, 2007.

E. Mosca and J. Zhang. Stable redesign of predictive control. *Automatica*, 28(6):1229–1233, 1992.

K. R. Muske and T. A. Badgwell. Disturbance modeling for offset-free linear model predictive control. *Journal of Process Control*, 12:617–632, 2002.

A. Nikandrov and C. L. E. Swartz. Sensitivity analysis of LP-MPC cascade control systems. *Journal of Process Control*, 19(1):16–24, 2009.

R. C. L. F. Oliveira and P. L. D. Peres. Stability of polytopes of matrices via affine parameter-dependent Lyapunov functions: asymptotically exact LMI conditions. *Linear Algebra and its Applications*, 405:209–228, 2005.

R. C. L. F. Oliveira and P. L. D. Peres. Parameter-dependent LMIs in robust analysis: characterization of homogeneous polynomially parameter-dependent solutions via LMI relaxations. *IEEE Transactions on Automatic Control*, 52(7):1334–1340, 2007.

J. M. Ortega and W. C. Rheinboldt. *Iterative Solution of Nonlinear Equations in Several Variables*. SIAM, 2000.

G. Pannocchia and J. B. Rawlings. Disturbance models for offset-free model-predictive control. *AIChE Journal*, 49(2):426–437, 2003.

R. K. Pearson and M. Pottmann. Gray-box identification of block-oriented nonlinear models. *Journal of Process Control*, 10(4):301–315, 2000.

B. Pluymers, J.A.K. Suykens, and B. de Moor. Min-max feedback MPC using a time-varying terminal constraint set and comments on "Efficient robust constrained model predictive control with a time-varying terminal constraint set". *Systems and Control Letters*, 54:1143–1148, 2005.

M. A. Poubelle, R. R. Bitmead, and M. R. Gevers. Fake algebraic Riccati techniques and stability. *IEEE Transactions on Automatic Control*, 33(4):379–381, 1988.

S. J. Qin and T. A. Badgwell. A survey of industrial model predictive control technology. *Control Engineering Practice*, 11(7):733–764, 2003.

C. V. Rao and J. B. Rawlings. Steady states and constraints in model predictive control. *AIChE Journal*, 45(6):1266–1278, 1999.

A. Sala and C. Ariño. Asymptotically necessary and sufficient conditions for stability and performance in fuzzy control: application of Polya's theorem. *Fuzzy Sets and Systems*, 158:2671–2686, 2007.

J. Schuurmans and J. A. Rossiter. Robust predictive control using tight sets of predicted states. *IEE Control Theory and Applications*, 147(1):13–18, 2000.

E. D. Sontag. *Mathematical Control Theory*. Springer, New York, 1990.

Y. J. Wang and J. B. Rawlings. A new robust model predictive control method I: theory and computation. *Journal of Process Control*, 14:231–247, 2004.

C. P. Yu, C. S. Zhang, and L. H. Xie. A new deterministic identification approach to Hammerstein systems. *IEEE Transactions on Signal Processing*, 62(1):131–140, 2013.

Q. M. Zhu, K. Warwick, and J. L. Douce. Adaptive general predictive controller for nonlinear systems. *IEE Proceedings, Part D (Control Theory and Applications)*, 138:33–40, 1991.

Index

Model Predictive Control, First Edition. Baocang Ding and Yuanqing Yang.